The Future of War

THE Future OF War

POWER,
TECHNOLOGY,
AND
AMERICAN
WORLD
DOMINANCE
IN THE
21ST CENTURY

George & Meredith Friedman

ST. MARTIN'S GRIFFIN ⚞ NEW YORK

Library of Congress Cataloging-in-Publication Data

Friedman, George.
    The future of war : power, technology & American
world dominance in the 21st century / by George and
Meredith Friedman.
        p.   cm.
    Includes index.
    ISBN 0-312-18100-0
    1. United States—Armed Forces—Weapons
systems.   2. Precision guided munitions—United
States.   3. International relations.   4. Twenty-first
century—Forecasts.   I. Friedman, Meredith.
II. Title.
UF503.F75   1998
355'.033073—dc21   97-47097
                                                  CIP

First published in the United States by Crown
Publishers, Inc.

First St. Martin's Griffin Edition: March 1998

10   9   8   7   6   5   4   3   2   1

Dedicated to the memory of

NORMAN M. FINE

Whose life was bound up with the revolution in warfare

and to

1ST LT. MEREDITH M. LEBARD   &   CADET DAVID A. FRIEDMAN
United States Army   United States Air Force ROTC

Who have chosen to serve their country and participate in

that revolution's future

Contents

Preface

We began this book in response to the euphoria of 1991–1992. In the wake of the collapse of the Soviet Union, several truths were commonplace: the American era was coming to an end because of the failures of the American economy; military power was less important than economic power; history had certainly taken a dramatic turn. There are, of course, many who continue to believe these things—but fewer than before. The American economy has proven far more robust than its critics believed, war remains ubiquitous, and nations continue to struggle with one another over moral and material issues.

We take a very traditional view of the international system. Wars will continue to be waged, military power will matter a great deal, the American economic miracle that defined the twentieth century will continue to define the twenty-first. We are also making a radical argument. While warfare will continue to dominate and define the international system, the manner in which wars are waged is undergoing a dramatic transformation, which will greatly enhance American power. Indeed, the twenty-first century will be defined by the overwhelming and persistent power of the United States. We are arguing that the rise of American power is not merely another moment in a global system spanning five hundred years but is actually the opening of an entirely new global system. We are in a profoundly new epoch in which the world that revolved around Europe is being replaced by a world revolving around North America.

Part of the reason for this historical transformation is rooted in geopolitics. The United States is the only major power native to both the Atlantic and Pacific Oceans—and the Pacific has come to rival the economic importance of the Atlantic. But part of our reasoning is strictly military in nature. Something extraordinary happened during Operation Desert Storm. The sheer one-sidedness of the victory, the devastation of the Iraqi Army compared to a handful of casualties on the American side, points to a qualitative shift in military power. On the surface, Iraq had a formidable force. By the standards of 1970 or even 1980, Iraq should have been able to inflict substantial casualties on allied forces. We have seen lopsided victories in the past—Poland in 1939 or Egypt in 1967—but never such low casualties on one side. In fact, the Iraqi Army, a reasonable force for the postwar world, had met an enemy from the post–Cold War world. A generational shift had taken place.

Like others, we focus on a new class of munitions that played a critical part in Desert Storm—precision-guided munitions (PGM). Precision-guided munitions differ from other munitions in one fundamental respect. Traditional munitions, once fired, were under the control of the laws of gravity and ballistics. In contrast, precision-guided munitions could correct their course after they were fired. Whether guided by their own sensors and computers or by human control, precision-guided munitions transformed the statistical foundations of war and with it the calculus of both military and political power. Where previously it would have been necessary to fire thousands of rounds to destroy a target, now it only took a handful.

Many noted this event. In our mind, however, this innovation ranks with the introduction of firearms, the phalanx, and the chariot as a defining moment in human history. The phalanx catapulted Greece to glory; the chariot gave us the Persian Empire. Europe conquered the world with the gun. Each weapon defined not only the manner in which wars were waged but the very texture of human history.

For one thing, the new class of weapons hold open the possibility of an end to the age of total war. Total war was built on two characteristics of gun technology: inaccuracy and massed explosive power designed to compensate for it. Masses of weapons had to be produced and fired or dropped together in order to hit elusive targets. This required the total mobilization of society to make war and made society as a whole the target of warriors. The result has been an unprecedented and unbearable slaughter. The accuracy of PGM promises to give us a very different age; perhaps even a more humane one. It is odd to speak favorably about the moral character of a weapon, but the image of a Tomahawk missile slamming precisely into its target when contrasted with the strategic bombardments of World War II does in fact contain a deep moral message and meaning. War may well be a ubiquitous and terrible part of the human condition, but war's permanence does not necessarily mean that the slaughters of the twentieth century are also permanent. It is possible that our situation is not hopeless.

We believe that we are entering a new epoch of war. But let us be emphatic that while the manner in which wars are waged is about to change dramatically, the eternal foundation of war remains unchanged. What was true for the Greek hoplite was true for the American GI. It will also be true in the epoch of precision-guided munitions. The warrior's trade will remain one of courage, dedication, and suffering. Precision-guided munitions will not render war antiseptic, any more than did the tank or crossbow or bronze armor. Technology changes how men fight and die, but it does not change the horror and glory of battle, nor does it change the reality of death.

This book, then, is about one part of the future of war. Another part, the eternal truth of warfare, is not discussed here. This is not because it is of no importance. Quite the contrary, it is probably more significant than the part we have selected to write about. We are content to address one part of the future of war—its technology. In a way, it is simpler to write about this than to explain why a soldier would choose to place himself between home and war's desolation. That is a matter that touches the depths of the human condition. We will content ourselves with a simpler theme—the manner in which the coming centuries are being shaped by a new, still primitive, technology.

The new generations of weapons are controversial. No radical new technology is ever uncontroversial, and nowhere are they as controversial as in the military. Both the tank and the airplane were denigrated by serious military thinkers in the decade after their introduction. It is therefore no surprise that the General Accounting Office, long noted for missing the forest while busily counting the trees, issued a report in July 1996 claiming to show that. "Many of DoDs and manufacturers' postwar claims about weapon system performance—particularly the F-117,

TLAM Tomahawk and laser guided bombs—were overstated, misleading, inconsistent with the best available data or unverifiable." The GAO particularly took issue with claims in advertisement that munitions had achieved one bomb, one hit accuracy. Obviously, this had not been achieved. Nevertheless, the worst case evaluations were extraordinary in shifting the statistics of warfare. Put simply: in Desert Storm, we saw the Model-Ts of precision-guided munitions. If these were the Model-Ts, then what will fully mature systems achieve? The forest is far more important than an accountant's trees.

We must bear in mind that it took centuries for firearms to move from the periphery of the battlefield to the core. There is a deep chasm between the advent of technology and its full implementation in doctrine and strategy. The process that began with the application of Star Wars technology to the tactical battlefield first showed itself in Desert Storm in a limited and immature way. In due course, the older technologies of armor, manned aircraft, and aircraft carriers will pass away. The key is in the term *in due course*. This process is under way, but the new technologies can only supplement and not yet supplant existing technologies.

It is important to us that our readers understand that we are speaking of a generational and even millennial shift. Aircraft carriers and tanks will be replaced, but nothing is available today that can replace them. To dismantle the existing forces of the United States in anticipation of the new technology would be nothing short of criminal folly. We are here examining the future measured in decades. Nothing will damage our national security more than thinking that the future, in its fullness, is here today.

Therefore, we ask the reader to bear three things in mind as he reads this book:

- This book focuses on the United States not only because we are Americans but also because we see America as the still-emerging center of gravity of the global system. The military America shapes will be the archetype of military power.
- This is a book about part of war—its technology—not the whole of war. Our focus on technology should not be mistaken for a belief on our part that the soul of the warrior and the brutality of war have been abolished by an antiseptic technology. At the same time, we harbor genuine hopes that the terrors of the twentieth century might be behind us.
- This book is not intended and must not be taken as a prescription for next year's defense budget. The United States is not in a

position by any means to slash defense budgets and manpower in the hope that new technologies will do the job at lower costs. Quite the contrary. Prudent policies will require higher budgets as older forces are maintained and even modernized, while new technologies are prepared and matured.

This book, therefore, is about one part of the future of war. It is not about the whole of war, and it is not really about war today. It is not about the art of soldiering and it is not about the problems of budgets. It is about the end of a five-hundred-year epoch—the epoch of Europe and guns—and about the beginning of a new epoch—the epoch of America and precision-guided munitions.

We are, of course, indebted to the many people who, over the past years, helped us in writing this book and in thinking about war. We are grateful to those at Louisiana State University and elsewhere who helped us along, particularly Leonard Hochberg, Matt Baker, Tim Reynolds, and the many who helped with editing—Laurel, Becky, Amanda, Amy, and others. We are also grateful to Kitty Ross for her help in reshaping this manuscript. Above all, we are grateful to our editor, Michael Denneny, who stuck by us with patience above and beyond the call of duty, and to his assistant, John Clark.

Finally, we would like to thank two people, dear friends who were indispensable in writing this book but who passed away from the same terrible disease before this work could be finished. Bob Oskam, our first literary agent, started us on our way and helps us still, if only through memory. Paul Olsen, our dear friend of many years, helped shape many of the ideas in this book. We hope he would be pleased. We will always be indebted to him. To these and all the others who aided us, our thanks.

Finally, we would like to thank our children, Memi, David, Edward, and Jonathan, and pray that we will never again have to hear their sweet voices declaim, "Are you guys ever gonna finish that book?"

Introduction

THE DAWN OF THE AMERICAN EPOCH

The twenty-first century will be the American century. This may seem an odd thing to say, since it is commonly believed that the twentieth century was the American century and that, with its end, American preeminence is drawing to a close. But the period since American intervention determined the outcome of World War I to the present was merely a prologue. Only the rough outlines of American power have become visible in the last hundred years, not fully formed and always cloaked by transitory problems and trivial challenges—Sputnik, Vietnam, Iran, Japan. In retrospect, it will be clear that America's clumsiness and failures were little more than an adolescent's stumbling—of passing significance and little note. With the dismantling of the Soviet Union, the American Century truly begins, as the patterns of economic and military life make the United States the center of gravity of the international system. The year 2000 brings to a close the first global geopolitical epoch and opens the door to the second, an epoch that was developing for almost a century before it burst into the open on the streets of Moscow.

The great event for which the twentieth century will be remembered was the collapse of the European imperial system, the result of the Old World's disastrous civil wars, which began with the disputes between Spain and Portugal during the fifteenth century and ended with the two world conflicts. Nearly five hundred years of history—a history written by European warriors and merchants—came to a close. This collapse also created a short and intense competition between the great Eurasian power that had managed for the first time to unite eastern and central Europe under a single rule and the great Western Hemispheric power that ruled the world's oceans more completely than they had ever been ruled before.

Now the struggle is over. The Soviet Union, unable to break out of its encirclement and mired in economic inefficiency, broke apart, and the dream of a united Eurasia shattered. Only one great power has survived—the United States—suddenly and quite unexpectedly alone and victorious. It dominates North America completely. The American Navy goes where it wants and does what it wants, and American troops land in Eurasia and brush aside enemy armies and they intervene in Latin America at will, for reasons as trivial as arresting drug dealers. Other nations may temporarily produce and sell things more efficiently than America, but only with American sufferance.

The collapse of the Soviet Union and the end of the Cold War came almost exactly five hundred years after Columbus's great voyage of discovery. As unlikely as it may seem, these two events were intimately connected. Columbus's encounter with the Western Hemisphere, along with the voyages of men like Vasco da Gama and Magellan, opened an unprecedented era in human history. Under Europe's rule humanity became fully aware of itself for the first time. The fall of the Soviet Union, five hundred years after the founding of the first global system, represents the close of that epoch. The world built on Europe's domination has ended; a new global order founded on American power has begun.

A decade before the collapse of the Soviet Union, an extraordinary event took place, almost unnoticed. In 1980, for the first time in history, the value of transpacific trade equaled the value of transatlantic trade.[1] This meant that control of the Atlantic, by itself, was no longer the key to global wealth; domination of both oceans was now necessary. While a nation that is native to either ocean has a natural advantage over nations that are far away, a nation that is native to both the Atlantic and Pacific is, at least geographically, in a position of unparalleled superiority, assuming, of course, that it is politically and militarily capable of taking advantage of its geography.

Only one nation is both native to the Atlantic and Pacific *and* capable of exploiting that advantage—the United States of America. With the collapse of the Soviet Union, the only power capable of challenging American politico-military might, the focus of the international system has finally and definitively shifted from Europe to the United States.

Great geopolitical transformations are always accompanied by radical shifts in the way of making war. Geography dictates the nature of military technology and culture by determining the type of force required for armed combat. Landlocked nations conduct wars differently than do island nations, and countries whose enemies are nearby fight differently from those whose enemies are far away. The military needs of European powers were wildly different from those of the Mongol hordes.

Because the United States is separated from Europe and Asia, military interventions in Eurasia will always be at long distances, with U.S. forces necessarily smaller than indigenous forces. Force multipliers, weapons that multiply the effectiveness of a smaller number of troops— ours or our allies—have been the foundation of American grand strategy. Therefore, one of the keys to America's power has been a revolution in weaponry.

For the past five centuries, firearms have dominated warfare. All firearms are ballistic in nature—once fired, their projectiles cannot be controlled. And because they are inherently inaccurate, many weapons have to be fired at the same time to have effect. This has meant large armies, large industrial plants, large merchant marines. During the last decades, with the invention of highly accurate weapons whose projectiles' path can be directed *after* being fired, the United States has laid the groundwork for eliminating ballistic weapons. As we saw in Desert Storm, these new weapons made it possible for the United States and its allies to overwhelm the numerically superior Iraqi troops.

This book is about the new epoch in warfare that has been ushered in by the new geopolitical epoch. However, before we can deal with the way future wars will be fought, we must consider whether war has a future at all. This seems an odd way to begin, since the history of mankind has been the history of war. Yet some serious people today contend that war has become obsolete and others contend that the advent of nuclear weapons has made war unthinkable. Still others contend that the end of the Cold War has made political and military considerations far less important than economic ones. **We are truly living in the giddy springtime of the bourgeoisie, in which the radicalism underlying bourgeois ideology has come irresistibly to the fore. Money and personal ambition are seen as having corroded every other human institution—**

including the nation-state—thereby making war irrelevant. Indeed, in the endless, earnest discussions of economies without borders, of capital flowing without consideration of politics, sober businessmen have made pronouncements that sound as extreme as those of Marxists, who for a century had predicted the end of the nation-state and the creation of an era of universal peace based on economic development.

Robert Reich, currently a key adviser to President Clinton, put it powerfully: "We are living through a transformation that will rearrange the politics and economics of the coming century. There will be no national products or technologies, no national corporations, no national industries. All that will remain rooted within national borders are the people who comprise a nation."[2]

This belief in the declining importance of nations is not confined to businessmen or economists, or even to Americans, but is shared by leaders as diverse as former secretary of state George Shultz, former French president Valéry Giscard d'Estaing, and former Israeli prime minister Shimon Peres.[3]

The most compelling argument against war's having a future goes roughly like this: The great powers—the United States, Europe, and Japan—are economically interconnected, and war would disrupt vitally important relationships among them. All of these nations enjoy a sufficiently great advantage from the international economic system that the risk of disruption far outweighs any possible advantages that war might provide. It follows, therefore, that there will be no systemic war—that is, no war between the great powers. There might be conflicts between and within lesser nations—and at times the great powers might join forces in limiting these conflicts—but the great conflicts of the past will not occur in the future.

This is not a preposterous argument. On the contrary, it is persuasive and reasonable. Its sole defect is that it is wrong. Economic cooperation breeds economic interdependence. Interdependence breeds friction. The search for economic advantage is a desperate game that causes nations to undertake desperate actions, a fact that can be demonstrated historically. Far greater levels of international economic interdependence existed earlier in this century, and, far from preventing war, helped cause it.

Also, bear in mind that this is not the first time the bourgeoisie has decided war is obsolete. They did the same thing in 1913, on the eve of World War I. In an enormously influential book, *The Great Illusion*, published in 1910, Norman Angell argued that, because of economic factors, war among the great powers was genuinely impossible.

Even if we could annihilate Germany we should annihilate such an important section of our debtors as to create hopeless panic in London. Such panic would so react on our own trade that it would be in no sort of condition to take the place which Germany had previously occupied in neutral markets, aside from the question that by such annihilation, a market equal to that of Canada and South Africa combined would be destroyed.[4]

Four years later, Angell was proven utterly wrong.

The new radical internationalists are asserting, as their pre–World War I predecessors had done, that the global integration of national and regional economies makes the cost of war so high that no rational nation would undertake it. So they envision the coming age as an extremely peaceful one, in which economic competition will be the dominant theme, political competition secondary, and war unthinkable. Accordingly, the great nations of the world will send salesmen forth with order books rather than soldiers with weapons. Businessmen are seen as more important than soldiers, economists as more insightful than political scientists, and corporations as more powerful than states.

But is the basic assertion true?—Is the current level of interdependence actually unprecedented? We can roughly calculate the degree to which nations have been economically interdependent. We can also determine whether or not conflict was possible at that level of interdependence, simply because we know whether or not wars broke out. For example, we know how much trade took place among the great European powers prior to 1914, and we also know that it was not sufficient to prevent the outbreak of a terrible war. Therefore, if the argument for the declining significance of military power and the declining risk of war is to be taken seriously, it is necessary to demonstrate that international trade and finance today are at much higher levels than they were in 1913, just prior to World War I.

The extraordinary fact is that, in terms of trade and investment, global economic activity has not surged ahead during the twentieth century but has stayed at roughly the same level as had been attained by the beginning of World War I. Moreover, the minimal increase in international economic activity has been less global than regional, pointing to stronger rather than weaker barriers to international economic integration.

For example, in 1913, Britain derived one-third of her national income from exports; in 1993, exports yielded less than one-fifth of British Gross Domestic Product (GDP). Taken as a whole, total world

MERCHANDISE EXPORTS AS A PERCENTAGE OF
GROSS DOMESTIC PRODUCT (GDP)[5]

	UK	FRA	GER	BEL	ITA	JAP	US	AUS
1913	33.9%	13.8%	19.2%	57.1%	12.4%	14.2%	7.2%	42.2%
1929	23.3	17.3	16.9	47.8	10.2	15.9	6.7	36.4
1956	22.3	14.5	15.5	32.6	8.1	9.2	5.5	14.2
1975	26.0	19.0	20.7	45.2	16.4	11.1	6.9	12.3
1993	19.4	16.5	23.1	52.6	18.5	8.4	7.2	15.0

exports as a percentage of the world's GDP in 1993 were only 12.2 percent—hardly a record-shattering amount. Even if we were to include service exports, little would change. On a global basis, service exports have never accounted for more than 4.3 percent of global GDP (in 1992) and were as low as 1.8 percent in 1988. In 1993, Japan's exports of nonfactor services equaled only 2 percent of GDP.[6] Thus, even assuming that there were no service exports whatsoever in 1913—clearly untrue—global international trade has not broken out beyond levels achieved in 1913.

The only countries whose dependence on trade has surged are the Europeans, particularly those who belong to the European Community.[7] As the European Community unites, it resembles more and more a single nation, or at least a single economic entity. Counting trade between Belgium and France as international trade today is about as reasonable as regarding California-Illinois business as international trade. Thus, if we exclude trade between members of the European Community, we find that reliance on exports falls to only 8.6 percent, a bit higher than the United States and much lower than Japan.

In short, net exports, even including services, are a small percentage of total world economic activity; moreover, they have not been growing in the past generation but have fallen to below 1913 levels. Claims for the increasing importance of international investment are similarly off the mark.

In 1913, leading nations exported about 4 to 8 percent of their GNP to other countries. Today, no country invests more than 2.3 percent of its GNP overseas, and the average is well below 2 percent. It is therefore extraordinary to hear people talk about the high levels of interdependence in the world today and how this will prevent war. Increased interdependence and globalization are an illusion. There was certainly a surge in global activity in the immediate postwar period, but this was an aberration. As the devastation wrought by war was cleared, economic activity in

ANNUAL FLOW OF FOREIGN DIRECT INVESTMENT
AS A PERCENTAGE OF GNP[8]

	1913	1992
UK	8.2%	1.4%
France	3.7	2.3
Germany	4.3 (est)	0.8
US	net borrower	0.5
Japan	net borrower	0.4

general, and international activity in particular, regained its equilibrium and returned to **levels that had been reached prior to World War I**.

The argument that interdependence gives rise to peace is flawed in theory as well as in practice. Conflicts arise from friction, particularly friction involving the fundamental interests of different nations. The less interdependence there is, the fewer the areas of serious friction. The more interdependence there is, the greater the areas of friction, and, therefore, the greater the potential for conflict. Two widely separated nations that trade little with each other are unlikely to go to war—Brazil is unlikely to fight Madagascar precisely because they have so little to do with each other. France and Germany, on the other hand, which have engaged in extensive trade and transnational finance, have fought three wars with each other over about seventy years. Interdependence was the *root* of the conflicts, not the deterrent.

There are, of course, cases of interdependence in which one country effectively absorbs the other or in which their interests match so precisely that the two countries simply merge. In other cases, interdependence remains peaceful because the economic, military, and political power of one country is overwhelming and inevitable. In relations between advanced industrialized countries and third-world countries, for example, this sort of asymmetrical relationship can frequently be seen.

All such relationships have a quality of unease built into them, particularly when the level of interdependence is great. When one or both nations attempt, intentionally or unintentionally, to shift the balance of power, the result is often tremendous anxiety and, sometimes, real pain. Each side sees the other's actions as an attempt to gain advantage and becomes frightened. In the end, *precisely because the level of interdependence is so great,* the relationship can, and frequently does, spiral out of control.

Consider the seemingly miraculous ability of the United States and Soviet Union to be rivals and yet avoid open warfare. These two powers could forgo extreme measures because they were not interdependent. Neither relied on the other for its economic well-being, and therefore, its social stability. This provided considerable room for maneuvering. Because there were few economic linkages, neither nation felt irresistible pressure to bring the relationship under control; neither felt any time constraint. Had one country been dependent on the other for something as important as oil or long-term investment, there would have been enormous fear of being held hostage economically. Each would have sought to dominate the relationship, and the result would have been catastrophic.

In the years before World War I, as a result of European interdependence, control of key national issues fell into the hands of foreign governments. Thus, decisions made in Paris had tremendous impact on Austria, and decisions made in London determined growth rates in the Ruhr. Each government sought to take charge of its own destiny by shifting the pattern of interdependence in its favor. Where economic means proved insufficient, political and military strategies were tried.

The international system following the Cold War resembles the pre–World War I system in some fundamental ways. First, there is a general prosperity. That is to say, the international economic system appears to be functioning extremely well, in spite of the normal cyclical downturns of the early 1990s. Second, almost no fundamental ideological issues divide the major powers; one could say there is general agreement on matters of political principle. Third, there is a long-standing pattern of interdependence, measured in both trade and financial flows—capital has become transnational. Fourth, and perhaps most important, beneath the apparent prosperity and stability there is a sense within each great power of a real and growing vulnerability to the actions of others. Some nations fear that growing protectionism will shift the balance of the system against them, while others are convinced that maintaining the current system will be devastating to their interests.

Today, observers focus on the first three phenomena, as they did prior to World War I, and argue that there is no economic basis for political conflict. What they miss is that the subsurface sense of insecurity—experienced by Japan, the United States, and Europe—marks the beginning of such conflict.

Thus, the argument that war is obsolete because of growing interdependence is unsupportable. War may be obsolete, but, if it is, it is not because of interdependence. As we have seen, World War I broke out at a time when interdependence was substantially higher than it is today; indeed, in all likelihood war broke out because interdependence was so

high. Today, war remains not only possible but, as a simple statistical matter, highly likely.

War, rather than being an exceptional condition, has been a regular and frequent part of the international system. In the eight decades since 1917, the United States has been in a state of all-out war for seventeen years, or over 20 percent of the time. Since the founding of our nation, some 220 years ago, the United States has been at war for about thirty-four years, 15.5 percent of the time.[9] The United States has averaged a major war about every twenty-five years, and from the Civil War to the Spanish American War the United States has never gone more than thirty-three years without one, and that only by not counting constant warfare with the Indian nations in the West and other minor skirmishes. In fact, in the past century, the frequency and intensity of major wars has been increasing.

This regularity of war has by no means been confined to the United States. In 1884, Russian prince Peter Kropotkin wrote that "war is the usual condition of Europe. A thirty-year supply of the causes of war is always at hand."[10] Most major powers engage in war between 10 and 20 percent of the time, on a frequency of about every twenty years. In the past century, Britain has been at war for twelve years, not counting limited conflicts in such places as India, the Suez, and Malaya. Japan has been at war for eighteen years, counting the war in China but not counting World War I, in which it was a minor player. The Soviet Union has been at war for twenty-two years, including Afghanistan. The immediate causes of wars change. Economic issues may possibly supplant ideological issues. But wars will still be waged. They will involve great powers and not merely "peacekeeping" operations. And the outcome of wars will determine the fate of nations.

Wars will continue, but, as we have suggested, that does not mean that they will be fought in the same way as before. The European conquest of the world would not have been possible without the advent of the chemically powered, projectile-firing tube—the gun. The gun revolutionized global political relations just as it revolutionized the theory and practice of war. The end of the first global epoch marks the end of the gun as the foundation of military culture. It marks the end of the ballistic epoch and the rise of a new form of weaponry. It also marks the beginning of a new military culture, one uniquely suited to the new preeminent power—the United States.

Great powers can be understood by their technology. Indeed, their technology—the walled cities of the ancient eastern Mediterranean, the civil engineering of Rome—comes to symbolize their power. Each was unprecedented, each unique to its particular civilization. Consider the manner in which the steam engine defined British culture—the very texture of what it meant to be British. Consider also how it brought Britain power and wealth and propelled Britain to preeminence within European civilization.

American civilization, it is now clear, is also founding itself on an unprecedented technology—the computer. The computer, like the steam engine, defines, in essence, what it means to be American. Certainly, like the steam engine, the computer diffuses to other societies—to be both used and produced. But just as the steam engine was quintessentially British—even though all of Europe used it—and served as the foundation of empire, so too the computer is quintessentially American and will serve as the foundation of American power in the coming century.

What makes the computer an expression of the American essence is its deeply embedded pragmatism. The computer does not deal with thought as the contemplation of the good or as the pursuit of beauty. Rather, it is a vast narrowing of the sphere of thought to what can be expressed in programming languages. But within that sphere, there is a dramatic deepening and magnification of reason's power. The computer focuses reason's ability to think about solving practical and immediate problems, declaring all other issues—aesthetic, metaphysical, or even moral—of little interest. In this sense, the computer expresses the quintessential pragmatic, unphilosophic spirit of America, in all spheres of life.

The computer has permeated the office, the factory, the home. It is also permeating American warfare. The computer provides us with an altogether new species of weapon—the intelligent weapon, the weapon that can, in some sense, think. Called precision-guided munitions, smart bombs, and the like, these weapons can both sense the world around them and guide themselves to the target. This marriage between the sensor and the projectile, fused together by the microchip, devastated the Iraqi Army. The computer helped define the Gulf War, by allowing the dumb projectile—its course determined on firing—to be transcended by the infinitely more accurate precision-guided munitions. Just as we have not yet begun to see the effect of computers on our homes and offices, we have not yet begun to see the effect of precision-guided munitions on the battlefield.

Great technologies permeate great civilizations. The computer is undoubtedly going to be the foundation of economic power in the twenty-first century. Other civilizations will design them, others will produce them, but no other civilization will find them so easy and obvious a fit—nowhere else will computers serve so easily as instruments of production, weapons of war, or toys to amuse children.

The computer really represents the first genuinely American expression of civilization. More than the auto or television, the computer redefines both thought and daily life. It also will redefine war—and serve as the foundation of American economic and military power in the twenty-first century. This book, then, is about one part of that revolution. It is about war's encounter with the computer and the manner in which national power will be redefined.

PART ONE

WEAPONS AND STRATEGY

Introduction

THE
CULTURE
OF WAR

A few hours before the war began, a ship on station hundreds of miles offshore fired a brace of cruise missiles. Subsonic, air-breathing, low-flying, the missiles would take several hours to travel to their targets. Embedded in each missile's computer was a map of the terrain over which it was to travel, along with a picture of the target and the exact point on the target that it was to strike, a picture downloaded from a reconnaissance satellite. They sped along at treetop level in a land without trees, examining the terrain below and comparing it to the maps in their brains, adjusting their courses as necessary. As the missiles approached their targets in the capital city, a helicopter hundreds of miles away on the border began moving forward several feet above the ground, well under the radar screen. A few Special Forces men who had infiltrated the area hours before turned on their laser designators, illuminating tiny points on the enemy's radar station. In the helicopter, the weapons officer focused the warheads of his rockets on those tiny points of brilliant light and fired. The rockets impacted seconds later, destroying the radar installation and opening a gap in the enemy's radar defense system through which hundreds of attacking aircraft raced forward.

At the same instant, one of the cruise missiles homed in on the telecommunications center. Almost simultaneously, other cruise missiles smashed into the air defense center, the ministry of defense, and key points in the country's electrical grid. In a matter of seconds, the enemy commander's ability to communicate with his troops was critically compromised, and the country's ability to power its industry was destroyed. Electronic and signals intelligence satellites orbiting as far out as 22,300 miles monitored the chaos below, reporting success. The enemy would be helpless against the air onslaught that would follow. For days and weeks ahead, the allies would smash enemy aircraft and tanks with devastating results.

The destruction of the Iraqi air defense system in the opening seconds of Operation Desert Storm was not the result of luck. It was the result of a revolution in warfare that had begun a generation before in Vietnam and will continue for centuries to come. In the same way that the primitive guns of the conquistadores revolutionized warfare for five hundred years following their introduction, the still primitive precision-guided munitions—from giant cruise missiles to tiny antitank rounds—will revolutionize warfare in the coming epoch. Nothing will ever be the same again.

Columbus and his heirs conquered the world because of the tube-fired, chemical-explosive-powered projectile—the gun and bullet. Guns were effective because they permitted warriors to kill their sword-, spear- and bow-wielding enemies before those enemies could come close enough to do them harm. The gun's lethality was limited only by its inaccuracy. Thousands of rounds had to be fired to inflict a single hit. The European solution was masses of men firing masses of guns produced by masses of factories.

Following its emergence as a global power after the Second World War, the United States has desperately searched for a way to fight wars that did not require masses of men and for weapons that did not require the firing of thousands of rounds to kill a single enemy. Accuracy, and the technology necessary to produce that accuracy, had been the driving force of American military culture. Weapons that hit their target in one or two tries have permitted smaller armies to overwhelm larger ones. This, above all else, has been what the United States needed.

Just as art derives from a culture's encounter with its circumstances, and great art from an encounter with extreme circumstance, so the manner in which a nation wages war is an encounter between a nation's culture and extreme and deadly circumstance. Some nations collapse under the pressure. Other nations master the modes of warfare at hand. Great nations, in unprecedented circumstances, find themselves creating unprecedented ways of waging war. Like the great blossoming of art, these nations frequently have little idea of the radical transformation they are bringing into being. Struggling to cope with circumstance, they are aware only of the next step, the next problem. That they are creating a new and comprehensive mode of war that will last for centuries is not their concern. This is the American condition today.

Throughout our history, America has been repelled by the horrors of war and fascinated by war's seductive nobility. Americans simultaneously feared and praised the warrior, were disgusted by and drawn to men like Patton and Sherman. And though the United States has constantly been either at war or preparing for war during this century—indeed, throughout its history—a simultaneous and paradoxical thrust of U.S. foreign policy has been the avoidance of war. Nowhere is this ambivalence more apparent than with our attraction to and revulsion for the technology of war—weapons.

America is by its nature a technological nation. The American regime is a technical contrivance intended to achieve an unnatural end—peace and tranquillity. In the same way, technical solutions to the problems of war are as natural to America as bravery was to Spartans. There are many parts to war—the virtues of the soldier, the endurance of the populace, the wisdom and skill of the general, and so on. All of these matter to Americans, but the core of the American experience of war is technological. For Americans, weaponry is even more essential than courage or generalship.

This obsession with weaponry in war showed itself late. It was only with the emergence of the United States as a leading power after the Second World War, our first truly global military adventure, that we looked into ourselves to find the thing that made us powerful. Our understanding of our victory in that war focused on our technical ability—the power to develop, produce, and supply weapons. This understanding carried forward to the Cold War and sustained us there as well. For us, war and weaponry became fused into a single, elemental construct. When we thought of war, we thought of the technology designed, like our regime, to keep disorder far away from our lives.

Most nations' culture of war is of consequence only to themselves and their neighbors. The Swiss and Israeli reserve systems, the Vietnamese dependence on the hardy light infantryman, are important to them and to their enemies, but these are not models to emulate. The United States, however, is not merely another specimen of the nation-state. At the dawn of the twenty-first century it is the preeminent politico-military power in the world. The American understanding of war, therefore, is more than an idiosyncratic response to culture and circumstance. Simply because of America's overwhelming power and presence, its military culture defines global military culture. Just as British naval experience stamped itself on every navy in the world and German general-staff experience stamped itself on every army, so too the American experience of war, its weapons and underlying technology, is systematically transforming the way in which everyone wages war.

Since the end of the Cold War, the United States is now the only great military power—the only nation able to project its power to any point on the globe. It is also the great economic power. The decline of Japanese economic fortunes, the inability of Europe to evolve into much more than a vast free-trade zone, have meant that the United States is not only the world's largest economy but the only one that is both continental in scope and politically united.

Because of its size, productivity, and power, the United States has become the center of gravity of the international system. The collapse of the Soviet Union from a potential global power into a purely regional Russian power finalized the emergence of the United States as the preeminent global power. The obsession with preventing the unification of Eurasia under a single power, which had transfixed the international system since Napoleon, has been rendered moot. For the moment, no Eurasian power is in a position to threaten U.S. hegemony, and Eurasian unification would mean far less in the new dual oceanic system than before. Thus, the United States has ushered in the second global system. The first system, created by Portuguese and Spanish explorers, centered on the Atlantic and the relation of Atlantic Europe to Central Europe. The second system, built on the equality of the Atlantic and Pacific Oceans, necessarily focuses on the single nation that is native to both, has vast resources to draw on, possesses an overwhelming military power, and appears to have the will to use that power in serious and sometimes frivolous exercises.

Nations make war in very national ways: Roman legionnaires constructing complex roads and fortifications; Mongolian horsemen moving as a rapid mass through the countryside; Japanese samurai perfecting one-on-one combat; German panzers attacking after meticulous planning and preparation. War, like art, is drawn from the soul of a nation. Most nations make wars in unmemorable ways. Whether they do it badly or they do it well, they merely emulate more original nations. Some nations, finding themselves at the crossroads of history and geography, are compelled to innovate, to transform the ways in which wars are fought.

The United States has today, far and away, the most successful military in the world. It has both global capabilities and the ability to bring overwhelming force to bear. But it has problems as well. First, the success of the United States in dominating the Western Hemisphere and controlling the world's oceans means that it must fight its wars on territory far away from its home bases—it must project its power great distances. Second, because it is fighting at great distances and must, therefore, devote vast resources to transporting and supplying its widely dispersed forces, it is always at a numerical disadvantage. In addition, the U.S. population, although large, is dwarfed by the masses of Eurasia. Superbly engineered, highly lethal weapons traveling thousands of miles and capable of multiplying the power of relatively small numbers of men have rendered these problems obsolete.

As the European gun-based weapons culture became increasingly senile—functional but straining the resources of Europe—the Americans simplified the social structure of war by imbuing their weapons with an unprecedented quality: intelligence. Smart weapons, weapons capable of seeing the enemy and steering themselves to the target, showed themselves first in Vietnam, then in several lesser conflicts, and finally appeared full blown during the Persian Gulf War. There, smart weapons destroyed radar sites, smashed telecommunications complexes, destroyed tanks, and overwhelmed a large force, well armed with guns.

To understand this revolution, it must be placed into context, for it is far from the first revolution, and certainly not the last. The rise and fall of weapons, intimately tied to the rise and fall of nations, is an ancient theme, and one fundamental to the human condition—as fundamental as war. Indeed, to fully understand this process, we ought to begin at the beginning, with the Bible.

1

David's Sling

ON THE RISE
AND FALL OF
WEAPONS

Consider David's defeat of Goliath with a sling. In principle it is not different from an air strike by an F-18 firing Maverick antitank missiles—both are primarily political instruments. Slings and antitank missiles are weapons originating in the politics of the moment and are designed to influence political life. But in addition to being political, both are also *technical* events. They would have been impossible without a prior technological revolution.

In the tenth century B.C., the Philistines had seized the coastal plain of the Levant, driving the Israelites into the poorer hill country of the interior. Because they controlled the coastal centers, the Philistines were able to engage in maritime commerce. Access to the eastern Mediterranean gave them a tremendous advantage in terms of military technology. By the time of Goliath, it had become the conventional wisdom that heavy bronze armor and iron spears were the central and immutable basis for military power. The emerging metallurgical industries—particularly those producing military-grade bronze and iron—were scattered through the eastern Mediterranean. Whoever controlled the coastal plain of the Levant could be part of this burgeoning trade in weapons-grade metals, purchasing the products outright or, alternatively, developing cutting-edge technologies of their own.

The occupying Philistines attempted to disarm the Hebrews by destroying their metallurgical industry. They accomplished this, according to 1 Samuel, by seizing and sending into exile all smiths. Not only did this make the production of weapons impossible, it also made the Hebrews dependent on the Philistines for the production and maintenance of their agricultural implements. So, the key to Philistine military power was industrial strength, which also increased the power of their own economy—a case of ancient industrial-military synergy.

The two principal offensive weapons of the period were the heavy iron spear, suitable for combat at close quarters, and the light throwing spear. Both weapons were moved by simple muscle power. The target-acquisition and fire-control system were the human eyes and hands, and the limits of the weapons were the limits of a human being. All armies had the same weapons, and all armies averaged the same range and accuracy. Given the technology, few offensive technical innovations were possible, except for increasing the spear's weight. Even new tactical moves were limited by the fixed nature of the weapon. The only way to be ahead of the game was on the defensive side, by improving the ability of the soldier to survive attack. Thus, as the dialectic of weapons progressed, the advantage went to the side with the greatest domestic economic and technological sophistication, the side able to fabricate armor that could block the spears.

When Goliath deployed for combat, he

> had a helmet of brass upon his head, and he was clad with a coat of mail; and the weight of the coat was five thousand shekels of brass. And he had greaves of brass upon his legs, and a javelin of brass between his shoulders. And the shaft of his spear was like a weaver's beam; and his spear's head weighed six hundred shekels of iron; and his shield bearer went before him. (1 Samuel 17:5–7)

Goliath's task—his offensive mission—was to hurl the javelin and cut with the spear. However, to live long enough to do this, he had to carry over 220 pounds of brass mail and a shield so large that it required a separate bearer to go into combat with him.[1] Save for the fact that it

21

allowed the offensive mission to proceed in the first place, the armor was purely parasitic. It did not increase his ability to carry out his mission.

Goliath had two weaknesses. First, the very weapons system that protected him from the enemy limited his mobility. The extraordinary weight of his armor meant that he was not going anywhere fast, especially if the terrain was hilly or extremely uneven. Now, this was not a problem if his enemy was similarly encumbered; both would be equally affected and limited in movement. But should either side make a breakthrough in strategic or tactical mobility, the entire equation would dramatically shift. Second, although Goliath was well armored, he was not perfectly armored. There were unprotected patches around his eyes, which, being his fire-control system, required a clear field of vision. This created a small, but real, area of vulnerability.

If Goliath—the archetype of the heavily armored infantryman—was to be defeated, then what was needed was a more mobile platform possessing a weapon of sufficient range to strike at Goliath from beyond his circle of lethality. This weapon ought to have sufficient accuracy to take advantage of weaknesses in Goliath's defensive system and sufficient power to disable him.

David was not a member of the military class and was therefore not encumbered by preconceptions about weaponry, nor had he already invested in expensive weapons systems. This meant that he was free to confront his political task without prejudice and free to craft technologies and tactics suitable to his mission.

> And Saul clad David with his apparel and he put a helmet of brass upon his head, and he clad him with a coat of mail. And David girded his sword upon his apparel and he essayed to go, but he could not, for he had not tried it. And David said unto Saul: I cannot go with these for I have not tried them. And David put them off him. And he took his staff in his hand and chose him five smooth stones out of the brook, and put them in the shepherd's bag which he had, even in his scrip; and his sling was in his hand and he drew near the Philistine. (1 Samuel 17:38–40)

David was small but mobile and could be destroyed by Goliath if he wandered into his kill zone. But Goliath, clad in armor, was slow moving and could not therefore move forward fast enough to trap David—he couldn't catch him. It was uneven combat. David was safe if he was disciplined and followed doctrine; only Goliath was at risk. David's lack of training in "state-of-the-art" military technologies saved his life as well as the freedom of Israel. Had he worn Saul's armor and tried to use his weapons, he would have been killed.

David himself was the weapons platform, providing mobility, target acquisition, and fire control. The weapon was a sling, the projectile was a pebble—a radically new system that was light but capable of long range and great accuracy. Its main advantage was that while it drew some of its energy from muscle power, it multiplied that energy by drawing on centrifugal force. David whirled the sling over his head, dramatically increasing the energy of the projectile and compensating for its small mass with dramatically increased velocity and his skillfulness.

The revolution brought about by the sling was made possible not only by technological innovation but also by social receptivity to change on the part of the military. This was not present among the Philistines. They were secure in the conventional wisdom. They had successfully consolidated their control over the coastal plain, had recovered from Saul's earlier offensive, and were now engaged in their own counteroffensive into the interior of the country. At the same time, the arms race, based on the premise of heavier and heavier weaponry, was attractive. Metallurgy was both cosmopolitan and high tech and reflected international trade links in the eastern Mediterranean. It also increased the power of military technicians and the officers who were allied with them. The Israelites were on the defensive and Saul's forces were reeling before the enemy. Why make a change now?

The great Philistine offensive had caused an internal crisis of confidence in the Hebrew chain of command. Under pressure, some military leaders become more rigid, clinging to old doctrines, increasing their dependence on outworn solutions. Saul was different. He accepted the introduction of a new technology and combat doctrine even though it undermined the intellectual and social foundation of his own military. The so-called revolution actually appeared to be a social regression—allowing low-cost combat, carried out by warriors from outside the appropriate social strata—as well as a radical and implausible simplification of combat doctrine. But by being able to accept such an upheaval, Saul was able to defeat the Philistines.

It is interesting to note that Saul wanted to have it both ways, accepting David's victory but attempting to break him afterward and revert to the old ways. The ensuing conflict between Saul's conventional force and David's highly mobile, lightly armed force proved the case for David's methods. Along with the new weaponry, David revised military doctrine, placing priority on mobility and stealth to compensate for the lack of technical sophistication and defensive protection. In the end, David's band broke the back of Saul's forces. A revolution in technology

and doctrine that had began when David defeated Goliath was affirmed
by Saul's inability to launch an effective offensive against David.

The story of David and his rise to power demonstrates eight points
about weapons development:

1. New technology frequently appears less sophisticated than old
   technology. For example, in the fourteenth century, guns
   appeared to be frail against fortifications. In the twentieth century, the battleship appeared to be the apex of technology while
   the aircraft that flew against it seemed flimsy and primitive.

2. Each weapon system (or general culture of weapons) has a life
   cycle. It begins with the simple purity of the offensive and culmi-
   nates in a weapon system overwhelmed by its own defensive
   measures. In the end, the effort required to defend the weapon
   dwarfs the weapon's offensive power.

3. The weapon system reaches its limit of usefulness when the
   defensive measures necessary for its survival destroy the
   weapon's cost effectiveness. That is to say, weapons reach their
   limits of usefulness when the cost of defending them is so great
   that it precludes purchasing other necessary weapons or crip-
   ples the civilian economy. This was the case with Goliath's armor
   and may be the case with stealth technology, today's armor.

4. The army least likely to recognize this point is the one that has
   been the most successful. Successful wars breed the illusion that
   particular technologies will always be effective. This illusion
   merges with the interests of the successful command structure,
   which uses the perception of technical infallibility to create a
   sense of its own superiority and even invincibility. Victory in
   World War I led the French to technical and command failure in
   World War II. Defeat short of annihilation is the best impetus to
   change—the U.S. Army after Vietnam showed the salutary effect
   of defeat, while the Israeli Army after 1967 showed the harm
   that victory can do. One wonders where the U.S. victory in the
   Persian Gulf will take it.

5. At its high point, just before disaster, the last generation's technology appears invincible. Fully armored knights, fortifications
   with enormous fixed-artillery emplacements, battleships, and
   intercontinental ballistic missiles—all appeared to be the final
   word in technology. They remained that way until events suddenly rendered them irrelevant or even burdensome.

6. The technologies that succeed in defeating the previous reigning weapons system share one characteristic: a simplification of

warfare, returning to the heart of warfare—the relentless offensive.

7. Parasitization is *always* under way—each weapon becomes senile. The length of the life cycle of a weapon is determined by the pace of the countermeasures and the ability to design defenses against the countermeasures.

8. A successful military is one that can constantly overthrow old weapons and doctrine and integrate new ideas and personnel without social upheaval. All successful military forces have been able to do this for a while. None has been able to do this permanently.

The lessons of the Hebrew-Philistine wars are relevant today. In all three spheres of contemporary warfare—strategic bombardment, sea-lane control, and armored warfare—the primary strategic function of the weapon has been obscured by the need to construct expensive defenses against threats to the weapons platform—aircraft, ships, tanks. The finest contemporary example is the B-2 Stealth bomber, which, in order to carry a destructive payload of about twenty-five tons, needs defensive measures costing over $1 billion each. The cost has made the basic mission possible only by a grotesque diversion of resources. Placing a billion-dollar plane at risk to drop twenty-five tons of explosives implies that the target *must* be destroyed and that *no* other means of achieving this end is possible—two unlikely propositions.

Of course, the tendency of the system to become overloaded with protective measures can also lead to another response—a radically new approach to the mission. The medieval knight, because of the weight and cost of his defensive apparatus, was increasingly unable to carry out his offensive function. He was freed from the problem by a revolution in warfare—the introduction of firearms. Even so, firearms did not abolish armor overnight! Armor continued to be used until its weight virtually immobilized the soldier. At this point, when the knight's offensive capability had been reduced to near zero, personal armor was abolished and a social revolution in military life took place.

The rise and fall of strategically significant weapons systems is the history of the rise and fall of nations and epochs. **The strategically significant weapon is the one that brings force to bear in such a way that it decisively erodes the war-making capability of the enemy.** For example, let us assume that the greatest threat presented by the enemy is the ability to move weapons platforms around quickly, and that by destroying his petrochemical production facilities we could impede his mobility. Strategically significant weapons would be those that would destroy petrochemical plants. So, in World War II, bombers with the range to reach these targets were such weapons. In Vietnam, the enemy's warmaking ability could not be decisively crippled by the same sort of long-range bombers. There, the strategic weapon was North Vietnamese infantry, able to move stealthily and impose a rate of attrition on American troops that was politically unacceptable to the United States.

The failure to recognize strategically decisive weapons is catastrophic. The Soviet Union's illusion that intercontinental ballistic missiles and swarms of tanks were strategically decisive led it to disaster. The American recognition that strategic power rested on superb command, control, communications, and intelligence, from the infantry squad to the office of the president, led it to victory.

Weapons have an inherent tendency to become more inefficient—either less effective or more expensive, or both. In the short run, the nation that can effectively bear the burden of defending increasingly inefficient weapons is the most effective militarily. In the long run, the nation that is able to redefine its strategic interests so that it can employ affordable weapons is in a stronger position.

The cost of designing and producing a weapon rises as the need to counter enemy countermeasures increases. The complexity of defending the weapon gives the illusion that the weapon is becoming more sophisticated. In fact, this is frequently an indication of its growing vulnerability. For example, the presence of the Aegis air defense system on carrier battle groups is less a sign of the growing sophistication of aircraft carriers than it is a sign that the aircraft carrier is now at enormous risk from its enemies. As a weapon moves toward senility, it continues to function—just as cavalry continued to be used until well into the twentieth century—but its effectiveness declines, and its cost soars until caring for it becomes an unbearable burden.

One of the signs of a senile weapon system is its divergence from the general economy. A weapon whose procurement detracts from economic growth is no longer on the cutting edge of technology—and therefore well past its peak efficiency. Building a modern armored fleet can be both extraordinarily costly and harmful to the economy. Creating an advanced military command system for space-based weapons may also be costly, but it can actually spur the civilian economy. Innovations at the cutting edge of military technology are usually at the cutting edge of civilian technology as well. But just as metallurgy became a trap for the Philistines, so too contemporary technology, seeking to make a tank stealthy or a carrier impregnable, can be our trap.

Consider this example: The purpose of an aircraft carrier is to control the sea-lanes by destroying enemy vessels. This is accomplished through the antiship munitions carried by the aircraft based on board the carrier. However, since the carrier must itself be defended against enemy action, a substantial part of the carrier's aircraft load is diverted from the strategic mission—sea-lane control—to a subsidiary mission, defending against enemy attacks. In addition, a fleet of other ships must deploy to protect the carrier from submarine, surface, and air attack. None of these vessels has a strategic purpose, none decisively threatens enemy power; instead, each is dedicated to protecting the aircraft carrier and facilitating the primary mission. In specific terms: Modern aircraft carriers have between twenty-four and thirty-six strike aircraft on board—more in some specialized cases. Some sixty other planes, a cruiser, one or two destroyers, and one or two nuclear attack submarines, along with supply vessels, shore facilities, and so forth, all exist so that the handful of aircraft can each drop eight to twelve tons of ordnance at a time. As threats against the aircraft carrier rise and the cost of keeping it operational soars, its offensive capability will tend toward zero. At that point, it will become senile.

When Goliath was outfitted in hundreds of pounds of armor so that he could throw a spear a few dozen feet; when cavalry was armored so that horses could charge firearms; when national treasures were spent on battleships so that six or nine guns could throw a few hundred pounds of explosive a few miles—these were moments of senility. The weapon continues to survive but with little purpose and at great expense and even danger to others. It must be remembered that weapons cling to life as do men. A senile weapon system can cling to life for a very long time, doing great harm to those who continue to champion it.

The Rise and Decline of the Great European Weapon System

The European age began with the great voyages of discovery. These voyages were made possible because of the weapons the explorers carried based on controlled, chemical explosives. After all, a handful of men landing in strange and dangerous places had to be prepared to fight for their lives. This was particularly true if, as was usually the case, their intention was to rob the natives.

Black powder, an explosive made from charcoal, sulfur, and saltpeter, creates a powerful explosion. This fact was known to several civilizations, but it was not until the fourteenth century in the European nations of the Mediterranean and Atlantic that the military applications of this chemical became apparent.² One use was as a simple explosive designed to destroy the stoutest fortifications. To survive, the walls had to be made even thicker and more complex. But simple, inexpensive improvements in explosives technology would eventually allow even the most apparently sophisticated construction to be breached. Black powder is an example of a simple new technology overwhelming an increasingly expensive defensive system.

The most important use of black powder as a military weapon occurred when it was exploded at one end of a metal tube, driving a projectile out an opening at the other end. The long tube controlled the direction of the projectile's movement, and the quantity of energy released in the explosion allowed it to travel for some distance according to a predictable trajectory and do damage to objects in its path. The three-part structure of the weapon—the tube, the propellant, the projectile—remained the essential technology of gunpowder warfare throughout the European age. Its use ranged from small weapons that could be held in the palm and fired by hand, to enormous metal tubes weighing tons, firing huge projectiles with some precision for twenty or more miles. The tubes underwent many refinements, from the loading of the projectile and propellant, to the aiming mechanism, to their metallurgical composition, and so on. And marginal improvements were made in the explosives. But the projectile itself remained essentially unrefined. It destroyed its target either by its own kinetic energy or by explosives contained within the projectile, but its basic structure remained unchanged over the past century, while its logic has remained intact for the past five hundred years.³

The principle of black powder's exploding at the base of a metal tube and propelling a projectile to a target became the foundation of European and Western military power. In every encounter with non-Western civilizations, the gun permitted inferior numbers to overwhelm much larger forces armed with nonexplosive weapons. From China to India to Africa and the Western Hemisphere, Europe slaughtered what it could not seduce.

Although the weapon was constant, the platform carrying it underwent a dramatic transformation during the second great wave of the European epoch. Just as chemical explosives had revolutionized the use of projectiles in war in the fourteenth century, so controlled chemical explosions revolutionized the manner in which weapons platforms subsequently moved. The internal combustion engine transformed weapons platforms driven by the wind (sailing ships) and muscle (the galley) into something extraordinary—self-moving weapons.

The coal-powered ship and railroad locomotive represented the first generation of hydrocarbon-driven weapons platforms. Petroleum-driven ships, automobiles, and finally aircraft represented the second great wave. This revolution took place during the late nineteenth century, at the height of Europe's second wave of global expansion. And it meant that the explosive-based technology could be taken virtually anywhere at relatively high speeds to find and destroy enemy forces. Indeed, the weapons platform was the foundation of Europe's conquest of the world.

Three principal weapons platforms emerged between the American Civil War and the end of World War I:

- *The battleship:* Powered first by coal and then by oil, it provided enormous tubes for firing huge projectiles. Primarily designed to destroy other ships, the battleship was also capable of shore bombardment. Its political purpose was to project power to the colonies and to protect those colonies from other European rivals.
- *The tank:* Powered by petroleum (or diesel or other derivatives), it emerged from the armored-train concept of the late nineteenth century as an anti-infantry weapon and a platform for transporting infantry.
- *The bomber:* Also powered by petroleum engines, the bomber delivered its explosive charge by dropping it rather than by firing it from a tube. To protect themselves and to combat enemy aircraft armed with machine guns and small cannon, bombs and reconnaissance aircraft also mounted guns and later evolved into fighters.

These three weapons platforms dominated the military world between 1914 and 1941 and, except for the substitution of the aircraft carrier for the battleship, continue to dominate the battlefield today.

The basic task of the weapons platform is to carry the projectile into range of the target, while allowing the target acquisition system (which could range from eyes to lasers) to spot the target and order the fire-control system to launch the projectile. Most of the improvements that have been made in the platform occurred not to make its weapons more deadly but simply to allow the platform to survive long enough to carry out its mission. Thus, the tank became more heavily gunned and armored to protect against enemy upgunning and armoring. The British Mark I tank, which was introduced in 1916, was armed in the female version with only machine guns and, in the male version, with both

machine guns and two six-pound cannon. It had maximum armor protection of 10 mm, enough to stop small-arms fire, light machine-gun bullets, and shrapnel, but not enough to stop direct cannon fire. As cannon
were mounted on other armored chassis, the tank grew in weight and
protectiveness, without becoming more effective at destroying enemy
infantry and enveloping enemy formations. The Pershing tank of World
War II carried 102 mm of armor and a 90-mm gun, both intended to
cope with the heavier armor carried by opposing tanks.[4] Yet its function
was no different from the Mark I's. The Pershing and other World War II
tanks spent most of their time trying to survive against enemy tanks and
antitank weapons. In due course, their primary mission was forgotten—
the tank became an antitank weapon.

The primary expense of land warfare did not involve the guns
themselves. Rather it became the design, construction, and constant
modification of the weapons platform to keep the gun crew safe and permit the gun to be maneuvered around the battlefield. In other words,
the cost of protecting the tank was greater than the cost of the actual
weaponry. It must be remembered that in addition to the cost of the platform, a tremendous logistical and industrial cost emerged as well.

Modern weapons platforms cannot be produced without a national
industrial and resource base. In the first place, a nation has to have available a wide array of minerals, particularly petrochemicals and metals,
and the ability to refine and distribute them. It needs a substantial manpower pool for mining and refining and an even larger pool for production. All of this rests on the foundations of a sophisticated civilian
industrial economy. The design and production of a weapons system cannot simply be grafted on top of an agrarian economy. The most important counterexample, the Soviet Union, has now shown that, in the long
run, such piggybacking is impossible to sustain. Dual-use economies—
that is, economies in which a substantial portion of the population is
devoted to a wide range of manufacture having little or nothing to do
with the direct waging of war and the same factories are used for both
civilian and military production—have a tremendous economic advantage over nations with a mass industry dedicated solely to military production.

One of the reasons the United States is the leading world power is
that it has both a huge mineral base and a vast industrial plant, which for
a very long time has integrated civilian and military-industrial production. General Motors produces both tanks and cars; McDonnel Douglas
produces both airliners and fighter planes; Litton Industries makes submarines and microwave ovens. Nations that are not able to produce the

broad spectrum of platforms and weapons that mobile warfare and their national interest requires must depend on the massive, continental nations for the means to defend themselves. Great power flows from that dependency.

Some smaller nations may have the intellectual capacity to design and even manufacture large numbers of components and, in some cases, entire weapons systems. But marshaling the brute strength necessary for the production of vast numbers of petrochemical engines for increasingly differentiated and complex weapons is beyond their ability. Thus Israel, which found itself with military demands far in excess of its industrial capacity, has had to have a patron, as did Vietnam, a nation with population and resources but without the cultural basis for organizing the vast enterprise of industrial warfare. So weapons platforms transfer—the transfer of aircraft, tanks, and warships—became a new system of dependency, imposed by the very complexity of the new weapons platforms themselves. In the end, continental powers able to impose industrial disciplines emerged as dominant powers, and economic power grew less and less distinct.

But underlying this power is the problem of modern warfare. The cost of fielding a warrior in a tank, plane, or ship is enormous. For each warrior who goes into battle, thousands upon thousands work to produce, maintain, and deliver his weapons, spare parts, munitions, and fuel. **Production and delivery have become the heart of modern warfare—ultimately more important even than combat. And the single most important element of modern warfare has been the production and delivery of oil.**

From World War II onward, the availability of petroleum has determined the outcome of battles, campaigns, and even wars. Tanks, planes, and ships (save a handful of nuclear-powered ships) are all moved by petroleum. Possession of oil fields has become the goal of strategy. Hitler's strategy in World War II was shaped by oil—that is why he gambled in the Caucasus in 1942, a gamble that led to Stalingrad and final defeat. Patton's drive into Germany in 1944 was halted by a lack of oil. And, of course, the threat to the world oil supply led the United States into the Persian Gulf War of 1990–91.

When petroleum-based engines replaced animal or wind power on land, sea, and in the air, the problem of warfare became quintessentially logistical: to provide the means for movement and combat, rather than to engage and destroy the enemy.[5] In fact, the problem of war became indirect—to destroy the enemy without engaging him by destroying his logistical system, by denying him the petroleum needed for movement.

Logistics has become, in effect, the tail wagging the dog. A Greek army could survive purely by forage if it stayed on the move. A division in the Franco-Prussian war of 1870 consumed about 50 tons a day of food and fodder. In 1916, this number had risen to 150 tons, due to the increased use and size of artillery and other projectiles. By early World War II, about 300 tons a day were consumed in North Africa, while Allied planners expected to spend 650 tons per day per division in France.[6] A contemporary American armored division requires in excess of 3,000 tons of supply per day.[7]

The ability to interfere with the economy of the modern battle-field—at any point from the mines and oil wells to the factory, to the pipeline that delivers supplies to the battlefield—is the ability to win the battle, the campaign, even the war. A revolution in warfare would occur at the point where an inexpensive, simple means could be found to consistently disrupt that supply line. The modern economy of warfare continues to function, but only with constant care, constant expenses, and the constant threat of complete collapse.

The epoch of the petrochemical/projectile economy of war is coming to an end because of the growing complexity and cost of the logistical system. At the same time, the three key weapons platforms—the tank, the bomber, and the aircraft carrier—are in grave danger of senility in their own right. All have been endangered by new, intelligent projectiles. This has made the cost of defending them soar, while the weapons' effectiveness has increased only slightly. At the same time, new weapons able to carry out their missions at far less cost are emerging. With the United States as the only global superpower, at the center of a two-ocean economy, the international system is undergoing its first radical transformation in five hundred years. The military system is undergoing a revolution as well.

A New Power, a New Way of War

Over the past century, the United States has moved to the center stage of history. As a global power, the United States suffers from a singular defect—it has a low population relative to its land mass and responsibilities. The population density of North America is only 68 people per square mile, while in Japan it's 844; France, 252; and Germany, 400. The manpower-intensive weapons of the European era put a substantial strain on the limited American population. So many men were needed to produce the weapons and to transport weapons and supplies to the battlefield that relatively few men were available for waging war. During

World War II, Korea, Vietnam, and the Persian Gulf, American combat forces were much smaller than their enemy's.

Throughout the past century, the United States has tried to create a political and a technical solution to this problem. Politically, the American strategy was to find allies who were either in difficult circumstances and, thus, were prepared to accept subjugation to American interests in return for immediate American help or were willing to undertake the American politico-military mission in return for long-term political or economic benefits. This was a brilliant stroke, but not one on which an empire could be based in the long run. It placed too much power in the hands of allies; in fact, the United States wound up the prisoner of its own relationships.

The technical solution was to develop weapons so efficient that they would compensate for the shortage of American fighting men. Beginning in the 1950s, American technical expertise took gun and engine technology to the limits of development, perfecting traditional weapons platforms, such as the aircraft carrier and submarine, the main battle tank, and the manned bomber, all invented by other nations. The excellence of the Nimitz-class carrier, the Los Angeles–class submarine, the Abrams tank, and the B-52 so towered over their competitors that the United States has become the unrivaled military power at the turn of the twenty-first century. It is important to note, however, that to increase the survivability of these platforms and deal with enemy threats, the United States also created a new class of weapons, precision-guided munitions. The goal was to enhance traditional weapons platforms—for example, the bomber could launch its projectile from a distance, without flying into enemy antiaircraft concentrations. The actual result was to introduce a new set of technologies that would, in due course, render the gun/engine platform combination obsolete.

One of the forces driving American weapons development was economic. During the First and Second World Wars, the United States compensated for inferior manpower by building vast quantities of somewhat inferior material. For example, the American Sherman tank of World War II fame was inferior to both the German Panther and the Soviet T-34; but America could mass-produce it in quantities large enough to overcome its limitations. After World War II, the United States lost the advantage of mass production of weapons, as other nations mastered the technology and developed the essential industries. The Soviets, for example, were able to produce greater quantities of armor and aircraft than the Americans. The American response was to raise the quality of the weapons. Now the United States had to compensate for both fewer men and fewer weapons.

The root problem was the relationship between the civilian economy and defense production. World War II ended the Great Depression by utilizing civilian industries to produce weapons and war supplies. The war also pushed American industry to innovate to accommodate the military's demands for cutting-edge equipment. As a result of the research and development that had taken place during the war, the postwar civilian economy boomed—a boom that continued through the mid-1960s. Somewhere about then, the key midcentury industries—steel, coal, and automobile manufacturing—became retrograde industries, no longer on the cutting edge of technology and no longer forcing civilian economic growth.

When the Soviets insisted on building some 5,000 tanks a year during the 1960s (down to 3,500 in 1989), they were forcing capital investment in steel, engines, coal—industries past their prime. The Soviets were diverting investment resources to support the previous generation's industries. When the United States pursued the same strategy during the 1970s, defense spending became a tremendous drag on the American economy. As steel production declined in America and began increasing in less developed nations, such as Korea and Brazil, defense analysts maintained that the decline threatened the industrial base of our defense system—how would we produce the vast quantity of tanks or planes necessary to cope with the Soviet threat? They argued—just as their Soviet counterparts had done—that defense needs must be met even if it meant crippling the economy.

The American decision was to move from the mass production of weapons to innovative technology. Although this involved an enormous commitment of money and training, it did not involve the vast commitment of manpower that weapons production had required in World War II. For example, American tank production concentrated on improved fire-control systems designed to increase the percentage of hits and kills on enemy tanks rather than concentrating on increasing the numbers of tanks. Superior avionics improved the survivability and lethality of fighter aircraft. As the result of such innovations, American defense research and development became a terrific goad to American industry. The 1990s has seen a surge of technology transfers from defense to the civilian sector, reminding us of the 1950s.

The transformation of fire-control systems, extending range and precision beyond what the unaided or minimally aided eye could achieve, was one indication of the technical revolution that was taking place. Another, far more dramatic and historically significant, was the introduction of a nonballistic projectile, a projectile whose trajectory was not determined at the moment of explosion in the tube but rather during the flight itself. During the Vietnam War, the United States introduced bombs, dropped from a plane, whose course was not determined solely by the laws of gravity; rather, they could be maneuvered to the target using television cameras, lasers, and radio signals. This innovation, along with rockets that offered continual-thrust propulsion instead of that from a single explosion—therefore allowing maneuverability by guidance systems—raised the possibility of dramatically increased hits and kills.

Attention was riveted on the new weapons when, in 1973, it was discovered that the Soviets had produced antitank missiles (the AT-3 Sagger) that were remarkably effective against Israeli tanks. A similar shock occurred in 1982 when it was discovered, during the Falklands War, that the Exocet antiship missile could destroy British warships. In the Persian Gulf, precision munitions appeared to have determined the outcome of the war.

As with many great breakthroughs, precision munitions initially appeared to be merely an extrapolation of existing technologies and not a fundamentally new approach to warfare. None of the basic weapons platforms was thought to be made redundant by the new munitions. Indeed, tanks, ships, and planes were the platforms from which the projectiles were fired. It is ironic that the new precision weapons were created to enhance the survivability of the traditional weapons system, when their final result will be to make the tank, surface vessel, and the airplane vulnerable to the point of annihilation. The quantitative increase of range and effectiveness becomes qualitative, as a new culture of warfare emerges.

Eventually, it became clear that there was something unprecedented in the new class of projectiles, something unheard of since firearms became preeminent in warfare some five hundred years ago. Whether rocket- or tube-based, the new projectile was not bound by the laws of ballistics. In theory, it could not miss the target, as it could be adjusted to compensate for the target's evasive maneuvers. In practice, it had a far greater probability of hitting the target than a conventional ballistic weapon. More important by far, there was no *theoretical* limit to the distance it could go. As shown by the Tomahawk cruise missile during the Persian Gulf campaign, a missile could be fired from hundreds of miles away, could guide itself to the target via terrain and satellite navigation, and could home in on the target through various terminal-guidance devices.

If the Tomahawk could perform this feat from hundreds of miles at subsonic speeds, then why couldn't future cruise missiles deliver explosives to the target from thousands of miles away, while traveling at supersonic and even hypersonic speeds? Combined with real-time intelligence from satellites, the truly guided cruise missile could become a tactical weapon, even at $1 million apiece.[8]

The fundamental assumption of warfare has been that the weapons platform must be on the battlefield, or close to it, to fire its projectile effectively. This was true for the Greeks at Marathon and for the Germans at the Bulge. This is no longer true. It was also assumed that to survive its encounter with the enemy, the weapons platform had to be heavily armored and defended, sometimes in extreme ways, as with the battleship or the Stealth bomber. But a platform dramatically removed from the battlefield does not need to be armored at all. A missile can be fired from a small house or an office building or from caves located ten thousand miles from the target. The entire concept of the battlefield changes and becomes unrecognizable.

By threatening the usefulness of traditional weapons, the new weapons threaten a revolution in warfare that will parallel the geopolitical revolution currently taking place. Indeed, they reinforce the geopolitical shift by putting power in certain hands and taking it away from others.

Assume, for example, that a man-portable missile were developed, inexpensive to produce and highly efficient in destroying armor. Assume further that while the missile's basic research and development would require a sophisticated industrial base, it could easily be mass-produced by less developed countries. This is, after all, the case with computers that are invented in the United States and then mass-produced in Malaysia. What would be the geostrategic consequences of such a weapon? The result would be that small nations with disciplined work-forces—nations that can manufacture a computer—can manufacture the means of war.

Thus, countries like Israel and Singapore could become quite powerful. The revolution in weapons derives from the geopolitical revolution that puts the United States at the center of the international system. In turn, these new weapons promise to transform the very meaning of geo-military power.

Conclusion: War and the Crisis of Vision

The modern system of warfare is undergoing a double crisis of vision. First is the problem of intellectual vision—the willingness (or unwillingness) of the military and defense analysts to accept the fact that we are facing a genuine technological revolution that will transform the American way of war as well as their own careers.

The crisis of vision is also literal: the ability of the warrior to see his enemy. The expansion of armies during the last few centuries has posed a fundamental command problem—how to command and control forces that are beyond one's sight. This management problem has been compounded by the introduction of the indirect-fire weapon—the howitzer—which could lob a shell beyond the visual range of the commander or the gunners. This has caused a crisis of intelligence that continues to afflict forces today and for which no sure solution has been found. Not only is the commander no longer able to see his enemies at one glance, he is no longer able to see his own forces. The fog of war, in the sharp phrase of the Prussian military theorist Karl von Clausewitz, has been multiplied exponentially in modern warfare.

The introduction of aircraft has made the problem of command, control, communications, and intelligence almost unbearably complex. In the attack against Iraq, B-52s took off from Louisiana along with cruise missiles launched from ships hundreds of miles away and helicopters a few hundred yards away. The size of the battlefield has expanded until it has become almost identical with the world—indeed, in nuclear warfare, the battlefield is the globe.

The management of a battle that is waged over thousands of miles poses basic epistemological questions. How can the commander be sure that what he learns is true; how can he be sure that it remains true at the moment he learns it; how can he command forces that he can neither see nor hear? These questions are philosophical. The solutions must be technical.

The technology developed to manage this expanded battlefield does not merely support the old system of warfare, it also makes it redundant and paves the way for a dramatically new system of intelligence that permits global warfare. The same technology—sensors, guidance systems, satellite communications, and so on—that permits us to see targets thousands of miles away is the same technology that permits us to strike at that target without using conventional weapons platforms. If we can see an enemy target, we can now strike it unerringly without putting troops in harm's way.

We are entering a time when weapons fired from one continent will guide themselves to targets on another continent in a matter of minutes and at minimal cost. The hyperintelligent, hypersonic, long-range, low-cost projectile is almost here. Born of the air age, it will destroy the old way of making war, while securing the new geopolitical system for generations. Where guns were inaccurate, these projectiles are extraordinarily precise. Where guns must travel to within miles of a target before firing, precision munitions can devastate an enemy from any distance. Where gun/petrochemical technology requires total commitment of resources and mass production, precision munitions require technical skill. The new weaponry places inherent limits on war, both in terms of scope and in terms of damage to unintended targets. The age of total war is at an end and a more limited type of war is at hand.

Like any new great power, the United States has appeared to be unsure and clumsy in many things, not the least of which is military power. The emergence of unparalleled American military power in the sands of Kuwait was prefaced by the ignominious failure of the United States in Vietnam, the massive uncertainties of the Cold War, and the naive triumph of World War II. Throughout this entire period, as the new weapons culture slowly emerged, the United States has been at war with itself over the nature of its military power and the principles that ought to guide that power. At its core, this has been a struggle to define the proper place of technology—and of technologists—in American military doctrine. More precisely, it has been an agonizing search for a doctrine that could define and control the overwhelming claims that technology makes on all aspects of American life—a search for principles of waging war that transcend mere weaponry.

2

Soldiers and Scientists

THE ORIGINS OF AMERICAN MILITARY FAILURE

Contemporary military thought was born on December 7, 1941. On that day, Lt. Kermit A. Tyler, an air-defense officer in Hawaii, replied to radar reports of incoming aircraft with the memorable phrase, "Well, don't worry abut it."[1] Tyler dismissed the reports because his common sense told him that the Japanese could not be attacking Pearl Harbor. No alarm was sounded and the American Pacific fleet was shattered. As the American fleet sank, an entire way of thinking about war sank with it. This was not merely a matter of aircraft carriers eclipsing battleships. A more fundamental change was taking place as well, a change in America's understanding of war's relation to politics and the relation of common sense to technical expertise. Common sense was inextricably linked with complacency. Planners turned to technology to overcome this.

Pearl Harbor taught American military thinkers that war might come at any moment and at any place. An attack on Pearl Harbor had been improbable, but it nevertheless happened. Therefore, calculations about the likelihood of events could no longer define and limit military preparations. A nation that defended itself only against expected enemies would be destroyed by the enemy who was unexpected. Military preparations now had to be measured against a vaster, less controllable universe of possibilities—the worst-case scenario. War became distinct from politics. It became a technical enterprise prepared for anything, divorced from the comforts of political prudence, a prudence that showed itself to be wholly inadequate on December 7, 1941.

Curtis LeMay, the creator of the Strategic Air Command, explained the principle that guided his thinking in the early days of the Cold War: "I remembered a couple of guys called Kimmel and Short—at Pearl Harbor. So I got busy and did what I could."[2] What LeMay could do was to create a system in which technology would dominate the common sense of future Lieutenant Tylers. The era of the permanent military emergency, the endless, perpetual surveillance of the skies after World War II, was a direct result of Pearl Harbor. American defense policy became twenty-four-hour vigilance, twenty-four-hour readiness, an unabating awareness that catastrophe might come at any moment.

The United States was not alone in entering an age of military obsession ungentled by the common sense of political calculation. Both American and Soviet military cultures were shaped by similar events—catastrophic failures to anticipate the worst—a mere six months apart. Like the Americans with the Japanese, the Soviets misconstrued German intentions. Like the Americans, they were unprepared for war and suffered a catastrophe infinitely greater. But like the Americans, the Soviet conviction that this must never happen again led them to create a war machine prepared for any contingency, regardless of the political situation. As the Cold War waxed and waned, both the Americans and the Soviets relied on their "national technical means"—their satellites, radar stations, and communications networks—to warn them of threat, rather than relying on the political instincts of their leaders. So, for about forty years both nations remained on permanent alert, deeply committed to the principle that the next moment could be the beginning of war.

Stunning, hidden surprises are rare in war. As Karl von Clausewitz put it:

> The preparations for a war usually occupy several months; the assembly of an Army at its principal positions requires generally the formation of depots and magazines and long marches, the object of which can be guessed soon enough. It

therefore rarely happens that one state surprises another by
War, or by the direction which it gives the mass of its forces.[3]

Of course, Clausewitz assumes that statesmen will use their common sense to interpret the intentions of the enemy. So, in traditional
warfare there were two lines of warning: the first was the sheer magnitude of the preparation for war; the second was the statesman's ability to
understand and take action.

Prior to World War I, all of the great European powers were on
the lookout not for flying weapons but for political signals and events
that were the natural precursors to war. After Pearl Harbor, war dis-
engaged itself from politics and common sense and wedded itself to
technology.

Ever since the rise of the nation-state, the traditional soldier had
given way to a new class of warrior, capable of managing the huge standing armies that modern weaponry—from early artillery to the jet aircraft—and the logic of the nation-state made necessary. Modern
weapons required vast numbers of men to fire them and vast numbers of
workers to produce them. And the esoteric skills they required had to be
constantly honed and frequently had no counterpart in nonmilitary life.
The production, delivery, and utilization of these weapons required a
class of officers who excelled in the management of the war-making
process, rather than who inspired by personal example. Along with this,
a class of trained soldiers, distinguished from the rest of the population,
came into being, if for no other reason than to serve as a core cadre for
the training of conscripts in time of war. This professionalization of the
soldier was the first step in the separation of war from the ordinary experience of men.

The professional officer who emerged differed from the traditional
warrior in several ways:

- Commanding a standing army, he remained on duty in peace as
  well as in war. Thus, his tasks included not only the fighting of
  wars but the administration of vast numbers of men in peacetime. Frequently the officer needed skills more suited to business than to war.

- The modern soldier was defined by learning and training rather
  than by birth or soul. His skills constituted a corpus of knowledge that, while difficult to learn, was knowable by anyone with
  sufficient wit. The military became a sphere of merit rather than
  birth.

■ As with all professions, the soldier's body of knowledge became codified by the great military strategist Karl von Clausewitz. Ironically, he was a descendant of the Prussian aristocracy, a warrior by birth and soul, who, by professionalizing it, undermined the Prussian warrior class.

Clausewitz distinguished between hereditary warriors and professional soldiers by turning war into a method that was transnational in scope and fundamentally apolitical. The Clausewitzian model applies to Russians, Chinese, Americans, and Israelis quite as much as it did to Prussians. It applies to fascists, communists, liberals, Christians, Jews, and Moslems. The rules of war—as defined by Clausewitz—transcend nation, religion, or ideology. His insights turned soldiering into a profession like any other—medicine, engineering, or law—and, in the spirit of enlightenment, applied to it the principles of science.

Clausewitz further de-aristocratized soldiering with the introduction of a radically democratic notion—genius:

Every special calling in life, if it is to be followed with success, requires peculiar qualifications of understanding and soul. Where these are of a high order, and manifest themselves by extraordinary achievements, the mind to which they belong is termed *genius*.[4]

Genius is due, perhaps, to birth, but it falls randomly. The teaching of the enlightenment on equality is not that all men are equal in ability but that no man is excluded from the possibility of excellence because of social class. Genius has been the great leveler of modernity, assaulting social class, political order, and tradition. Genius was celebrated among all other values in a society in which innovation, creativity, and the conquest of nature were held to be natural ends.

The genius of the warrior coupled with the learned expertise of the professional soldier integrated the military into the rest of society, creating a system of shared values. Yet, even as the military democratized and professionalized, even as it became dependent on the randomly occurring genius, and even as it became a science, the military understood itself to be a radically different sort of science, requiring a set of radically different virtues from the conventional scientist. In the end, the soldier had to be prepared to risk and even give his life. The moment for this might never come in a long career, or it might arise at any moment.[5]

This physical courage separated the soldier and his profession from other professions and was more in keeping with the traditional warrior class than with other modern professions. The scientist did not need physical courage to do his work. The modern soldier needed it only rarely, but he still needed it. The soldier's life was separated from the contemplative life of the scientist. Both were professionals, but the soldier, unlike the scientist, might be called on to face death. This was the soldier's badge of honor, and in his mind, this made him the rightful ruler of the battlefield.

The Rise of the Technicians

The emergence of technicians as the dominant class in the military and the decisive force on the battlefield could be seen, with stunning clarity, during the Battle of Britain. After the fall of France in 1940, Hitler mounted a massive air offensive against Great Britain. The Royal Air Force, battling for its life, scored stunning successes against the Luftwaffe. In large part the RAF's successes were due less to the skill or bravery of its pilots—after all, the Luftwaffe's men were skilled and brave as well—than to a new and unprecedented device: radar.

The development of radar revolutionized warfare in two ways. In the lesser sense, it made it possible to develop efficient defenses against aircraft approaching from long distances at relatively high speeds. Men could see things at superhuman distances and through fog and clouds. The RAF could see German planes taking off from French air fields, forming up over France, and heading out toward British targets. The RAF could scramble interceptors and effectively vector the fighters to the intercept point because of radar. In this sense, radar was militarily significant and plays a key role in the endless vigilance that marks modern American strategy.[6]

But radar revolutionized warfare in a deeper sociological sense. As a result of radar's success, a class of men previously only tangentially connected to warfare—the basic scientists—became as essential to war as the military men.

Modern warfare has always borrowed from the technology of the modern scientist in such fields as explosives, civil engineering, and aeronautical engineering, but such enterprises were not seen as central to military success, nor were scientists as a class considered uniquely important. Even during World War I, the British sent distinguished physicists to the front as ordinary soldiers rather than treating them as rare national assets.[7] All this changed between the World Wars as modern science's insights into nature deepened.

Enter the boffins, a name British officers gave scientists, a name combining affection and contempt for the thoroughly unmilitary and unwarlike thinkers for whom common sense was an inadequate way in which to manage reality.[8] The discovery of radar was not something that bicycle-shop tinkerers could come up with. It required a profound and radical insight into the nature of radiation and of the electromagnetic spectrum. It required an insight into a world hidden from ordinary human vision and in opposition to common sense. The theoretical physics that underlay radar and the technologies that came after it required a mode of thought that was inaccessible to both the traditional warrior, who depended on his courage and discipline for victory, and the modern professional soldier, who relied on his skill at organizing the vast enterprise of modern war.

The power of the scientist came from his ability to understand the workings of the physical world. It arose from disciplined contemplation rather than from physical courage. The issue for men such as Watson-Watt, the inventor of radar, was how the electromagnetic spectrum worked, what its possibilities were and how those possibilities could be manipulated. Radar, and all of the military hardware that derived from it, were, for the scientist, useful but peripheral outcomes of the core project, understanding the hidden truths of nature. Certainly not all scientists maintained such an Olympian perspective, but enough were of such a nature, and enough others adopted the pose, to drive the military to distraction. The professional soldier wanted to use the insights of science as the Romans had used engineering—to get things done. The down-to-earth world of the soldier was separated by an abyss from the wildly exotic world of the physicist or mathematician. This was all the more galling to the soldier as it was the impractical musings of the scientist that produced the tools that constantly revolutionized the military world.

Scientists constantly violated intuitive reality by discovering hidden truths about the universe that initially appeared absurd and preposterous. For example, germ theory, the idea that diseases were caused by microorganisms not visible to the naked eye, was held up to universal ridicule by the commonsense approach until it permitted the cure of disease. In time, the term *scientist* ceased to merely describe an activity. It described a priestly class, with the power to peer into the hidden dimensions of nature and, like a magician, conjure up extraordinary powers. The world of the scientist was esoteric, with barely comprehensible rules and its own special, secret language.

The name of this language was mathematics. Mathematics was a profoundly unnatural language; no child ever spoke this tongue at his mother's knee. Nevertheless, it provided a key for discovering the secret truths of nature and using them for the benefit of man. As Immanuel Kant put it in *The Critique of Pure Reason,* "Mathematics presents the most splendid example of the successful extension of pure reason, without the help of experience."[9] The question, for Kant and for us, was the limits of mathematical reason.

Clausewitz could answer easily: War had little to do with mathematics:

> We say therefore War belongs not to the province of Arts and Sciences, but to the province of social life. It is a conflict of great interests which is settled by bloodshed, and only in that is it different from others. It would be better, instead of comparing it with any Art, to liken it to business competition, which is also a conflict of human interests and activities; and it is still more like state policy, which again, on its part may be looked upon as a kind of business competition on a great scale. Besides, State policy is the womb in which War is developed, in which its outlines lie hidden in a rudimentary state, like the qualities of living creatures in their germs.[10]

For Clausewitz, summing up the traditional view, war cannot be codified into a science or art because it is too lively, too filled with interests, too competitive to have the predictability that science insists on.

The scientist, by contrast, had no sense of war or at least none that was inherent to his profession. He had no interest in knowing anything of the history of warfare nor any of the martial arts. Rather, he knew a small piece of nature and how to manipulate it. For example, scientists who understood the laws that governed the flow of radiation through the electromagnetic spectrum were able to manipulate that flow so that when it encountered an airplane, the radiation returned to the radar installation and the operator was informed of the presence of an aircraft. In turn, fighter pilots who were under the control of this warning mechanism were able to scramble and shoot down the intruder.

The victor in this combat appeared to be the warrior, but appearances were deceptive. The pilot had become a mere functionary in the system, delivered to the point of contact by a device created by a boffin, a whiz kid, a pencil-necked geek who probably got airsick. The life of the pilot, the success of the mission, and, ultimately, defeat or victory in the war were far more in the hands of the geek than of the warrior. A sea change was taking place not only in war but in the manner in which the warrior looked at himself. From time immemorial, there was never any doubt but that the warrior who risked his life, who put himself between home and war's desolation, was the hero of the battlefield. Now the hero was the scientist, hidden both by the difficulty of his language and the urgent secrecy of his task.

The warrior's dependency on the scientist was really only hinted at during the Battle of Britain. It reappeared endlessly during the war, in the development of submarine and antisubmarine warfare, proximity fuses, flashless gunpowder, the German V-1 and V-2 rockets, and the endless innovations that made Allied victory possible. However, the emerging hegemony of the scientist over the warrior was driven home with an absolute finality during the Manhattan Project with the development of the atomic bomb. The Manhattan Project was certainly a radical break in history, but it was particularly critical in redefining military thinking. In a sense, the creation of the atomic bomb went a long way toward abolishing the military and military thinking in general.

The Atomic Bomb: Physicist as Warrior

The atomic bomb appeared at first to have made the military as a whole, with its traditional values and virtues, obsolete. As soon as it exploded over Hiroshima, it seemed to some that war, understood in traditional terms, had been abolished. No nation could withstand the ability of another nation to obliterate its cities at will. No form of conventional warfare could be expected to challenge the atomic bomb.

But military processes did survive the bomb. After all, countless conventional wars have been fought since Hiroshima, and countless conventional armies have remained in existence. In terms of the military, the bomb's major aftereffects were psychological and cut to the heart of the military's sense of self.

According to the initial plan, the Manhattan Project was to be an Army endeavor, with Army personnel. Therefore, Robert Oppenheimer, who was selected to run the scientific aspect of the operation, was to be commissioned as a lieutenant colonel. However, a rebellion by key scientists made it necessary to transform the terms of the project, placing it under contract to the University of California, and putting scientists—not officers—in key positions of authority. This defeat of military culture by scientific culture was to set a pattern for postwar relationships.[11]

The institutional arrangements revealed the emerging power of the scientists as a class. Scientists were not merely useful, they were indispensable in the practical circumstance of war, for their arcane experiments held open the possibility of victory. The bomb was esoteric to the core. It could be understood only by recourse to a methodology that was not accessible to common sense and the common man, which in fact flew in the face of common sense. The physics of Newton, which gave rise to artillery and aircraft, was displaced by the physics of Einstein. Laws about the visible universe were replaced with theories about the nature of time and space, quantum mechanics, and radiation physics.

In the conventional view, the war against Japan had been won by the atomic bomb, which was seen as an irresistible weapon. By implication this meant that the creator of the atomic bomb held the real glory and was the true victor. Even if it was employed by conventional soldiers, it was understood that the creator of the bomb, who alone genuinely understood the weapon and who held out the promise of even more wonderful and terrible engines of war, was in some way the real power in the new kingdom. Much of the history of the Manhattan Project was the struggle for control between the conventional military and the scientists. However much Gen. Leslie Groves, the military commander, retained formal control of the operation, the actual creation of the weapons was beyond his power.

According to the traditionalists, there is certainly a place for science in war—in the design and construction of weapons and fortifications and other subsidiary elements. But the essential dimension of combat, determining who would win and lose battles, or how a battle should be fought, could no more be turned over to science and scientists than could the business of government or the business of business. Indeed, during World War II, field commanders persistently tended to exclude scientists from actual operations, in spite of the fact that their expertise might be useful. As the official history of the World War II Office of Research and Scientific Development put it, exclusion of scientists from operations was rooted in an "indoctrination in traditions of the Army and Navy, which tacitly seem to imply that an individual who is not literally and legally a part of the organization must be regarded with apprehension."[12] In other words, the gulf between warriors and scientists was seen as too great to bridge, even if a genuine emergency required it.

But what would happen if a scientist were to devise a weapon that would transform not merely the practice of war but the nature of war, indeed, the very possibility of war? This was precisely what the nuclear physicists of the Manhattan Project appeared to have done. Nuclear physics had originated in the expansion of mathematics' "construction of concepts" in the hidden realms of nature, a realm so distant from us in size and speed and energy that it defied not only our eyesight but our very imagination. Nevertheless, a team of scientists, led by Einstein, Bohr, Lawrence, and the rest, constructed mathematical models of this universe and, using their strange, unnatural, hidden language, conjured up forces that dwarfed everything else around them.

The atomic bomb appeared to transform the rules of war. From the very beginning, it destroyed whatever it touched, from emplacements to divisions. Precision deployment of conventional forces, no matter how well they were trained and equipped, appeared to be irrelevant in the face of atomic weapons. What was the point of deploying a force of even a million men, if a handful of weapons could destroy them instantaneously? What was the point of invading another nation with a vast army, when a handful of bombers could wreak infinitely more damage? What indeed was the point of studying the history of warfare at all, when nothing like it had ever been seen before?[13]

Following World War II, J. F. C. Fuller, the great military historian and analyst, envisioned the future of war:

> Miles above the surface of the earth, noiseless battles will be fought between blast and counterblast. Now and again an invader will get through, and up will go London, Paris, or New York in a 40,000-foot-high mushroom of smoke and dust; and as nobody will know what is happening above or beyond or be certain who is fighting whom—let alone what for—the war will be a kind of bellicose perpetual motion until the last laboratory blows up.[14]

In a war where none of the traditional martial virtues have meaning, where the laboratory is the scene of military power, it followed that the men who ruled the laboratory would take over the art of war.

The scientists, the class that had delivered the ultimate weapon, suddenly emerged as a center of power, power based on knowledge. Physicist S. K. Allison, who chanted the countdown on the first atomic bomb test, recalled, "The scientists quickly discovered, to their embarrassment, that *atom* was a magic word in Washington and that they, the only ones who fully understood its meaning, were looked upon as glamour boys. Some people, dimly conscious of the fact that they were up against a new kind of problem, regarded them as the men who knew all the answers."[15]

As the extent of the problem became clearer to the political and military leadership, the search for a workable solution became increasingly urgent.

The Soldier Falls in Love with the Scientist

The problem was simple: the atomic bomb was a weapon, or at least gave every appearance of being a weapon. How was it to be used? What military strategy was to be followed? The military understood the sort of warfare that had dominated World War II until Hiroshima and was painfully aware of how little it knew about the bomb. How could it devise a strategy for deploying something it did not really understand? Even as it devised operational principles for handling what was to be called conventional warfare, the pressing need to bring the atomic bomb under the control of strategy seemed unreachable.

If the military was at a loss, then surely the creators of the bomb ought to know how it was to be used. On May 31, 1945, a meeting of the Interim Committee governing the Manhattan Project was convened. In addition to Secretary of War Henry Stimson, who chaired the committee, and Gen. George Marshall, army chief of staff, and Gen. Leslie Groves, commander of the Manhattan Project, present by special invitation were a group of scientists, including Robert Oppenheimer, Enrico Fermi, Arthur Compton, and E. O. Lawrence. According to the minutes, Secretary of War Stimson addressed them:

> The Secretary explained that General Marshall shared responsibility with him for making recommendations to the President on this project with particular reference to its military aspects; therefore, it was considered highly desirable that General Marshall be present at this meeting to secure at first hand the views of the scientists.
>
> The Secretary addressed the view, a view shared by General Marshall, that this project should not be considered simply in terms of military weapons, but as a new relationship of man to the universe. This discovery might be compared to the discoveries of the Copernican theory and of the laws of

gravity, but far more important than these is its effect on the lives of men.[16]

The meeting went on to discuss the use of the bomb against Japan and the possible political and diplomatic repercussions of its use.

What was extraordinary about this meeting was that it took place at all. The secretary of war and the army chief of staff were consulting with a group of physicists not only on the mechanics of the weapon but on the strategy for its use. The physicists were advising the warriors on the art of war. This was the beginning of a process that would see the military transferring more and more intellectual, if not operational, control of war to scientists.

The immediate problem was the creation of a conceptual framework for using the new atomic bomb. Before the advent of aircraft, war was about destroying enemy forces and taking and holding enemy terrain. All military questions were posed in spatial terms. Now, the problem was not one of space but of time. The length of time necessary for takeoff or launch, for warning and for targeting, became the variables that determined the fate of nations. Mountains, rivers, even oceans were no longer politically relevant. Time was the crucial factor.

Politico-military questions such as when to launch a preemptive strike and what targets to fire on mingled with purely military questions concerning the basic size of warheads, strategies, and so on. Sorting through these issues with an urgency imposed by a deteriorating postwar situation and an awareness of the temporary nature of the American nuclear monopoly placed tremendous pressure on the military.

The Air Force in particular was keenly aware of the need for a strategy, even excluding the atomic question. In 1944, Hap Arnold, commander of the Army Air Corps, wrote a memo to Theodore Von Karman, a Hungarian refugee physicist: "I believe the security of the United States of America will continue to rest in part in developments instituted by our educational and professional scientists. I am anxious that the Air Force's post war and next war research and development be placed on a sound and continuing footing."[17]

Through this memo, Arnold created the Army Air Force Scientific Advisory Board. Within a year, Gen. Curtis LeMay was named head of the air force's new Directorate of Research and Development.

These two occurrences were fateful. In the first, the Army air force's highest-ranking general placed his prestige not merely behind the idea of technical research and development but squarely behind the idea of depending on civilians to provide it. The appointment of LeMay was equally significant. In Stanley Kubrick's film *Dr. Strangelove* and similar movies, he has been portrayed as the prototype of the war-mad, anti-intellectual flyboy: a tough, profane warrior with a stogie stuck in his face. With him in the head slot, research and development lost its sense of effeteness and became inextricably linked to the machismo of warfare.

It was LeMay who, casting about for a way to implement his mission, decided to take Hap Arnold's idea of civilian research one step further by creating a nonprofit corporation, under contract to the Air Force, named the Research and Development Corporation, or Rand Corporation. The Rand Corporation epitomized the new relationship between science and the military. It was formally independent, but was under exclusive contract to the Air Force.[18] It was termed a think tank, a military installation whose primary product was ideas.

It was part of the changing of the guard. LeMay charged Rand to develop a continuing "program of study and research on the broad subject of Aerospace Power with the object of recommending to the United States Air Force preferred methods, techniques and instrumentalities for the development and employment of Aerospace Power."[19]

By using the term *preferred,* LeMay put the scientists in the position of judging.[20] Obviously he intended that these scientists remain fully subordinate to himself and the Air Force, and in a strictly formal sense this remained true. But in a deeper sense, the military inexorably lost authority and control to the scientists. Bernard Brodie, one of the great analysts of Rand's early years and a father of contemporary nuclear strategy, said of this period:

> One of the most remarkable changes in the intellectual landscape over the last dozen years, especially in the area of public affairs, is the entry of civilian scientists into the field of military strategy. . . . Most of the distinctively modern concepts of military strategy that have been embraced by the military services themselves have been evolved by them.[21]

Brodie did not exaggerate about the importance of the civilian scientists. They shaped and reshaped the entire field of modern strategy and, of course, nuclear strategy.

The scientists, having no intrinsic understanding of a military analysis, naturally turned to their own frame of reference—scientific method and mathematics. For scientists, reality is divided between the pure, empirical phenomena and the mathematical restatement of the event, which allows the scientist to manipulate and analyze the phenomena symbolically. Newton's physics and his development of calculus was perhaps the purest representation of this relationship.

Of course Newton's job was fairly easy, since material objects are predictable. Whenever the same forces and masses are present, the same result will occur. But human activities—and military strategy is certainly one of these—are not so predictable. In dealing with large groups of people, it is impossible to say what any single individual will do in a given circumstance. For example, if enough potential voters are sampled, with a sophisticated enough mathematical model to analyze the results, we can predict with a high degree of accuracy what the mass of voters will do; in other words, we can know early on election night who the winner is. But we could not say with any reliability how a particular voter would cast his or her ballot.

War, like voting, is a human activity in which predicting the outcome and analyzing the reasons for the outcome *beforehand* is crucial. Traditionally, strategy is argued by historical analogy and common sense—but such methods are thoroughly alien to scientists. At the same time, their own methods—which require mathematical predictability—are useless when applied to the randomness of human activities. Gradually, a third way became apparent.

Operations Research and the Imperial Scientist

A discipline named operations research had begun to develop prior to World War II that aspired to use quantitative methodologies to develop a science of management. As an early text sponsored by the U.S. Navy defined it, "Operations research is a scientific method of providing executive departments with a quantitative basis for decisions regarding the operations under their control."[22]

The key term here was "quantitative." For the physicists and mathematicians of the Rand Corporation, the intuitions of common sense were utterly insufficient as to a guide to management. Mathematical precision was necessary, and operations research promised to supply that precision.

Operations research had already been used extensively in World War II to solve fundamental management problems that had previously been the exclusive responsibility of the field commander. During the war, a key participant in these affairs, Sir Solly Zuckerman, described it: " 'Operational research' emerged as a new procedure in military affairs, as an intellectual tool for the military, so as to help the proper exercise of judgement and control. Without it, warfare was already becoming too difficult technically for those in whose hands its conduct rested."[23]

An example might be seen in the problems the British had in dealing with German submarines in 1940. Pilots had difficulty judging how far below the surface depth charges ought to be set for submarines that were diving. Through statistical analysis of available combat data, operations researchers determined the optimal depth for maximum damage to the maximum number of submarines. As a result, the kill rate dramatically increased. The use of radar, the basing of fighter aircraft, and the structuring of convoys were among the endless applications of operations research carried out by the British, and later by the Americans when the operational principles of the vast bombing campaigns had to be developed.[24]

Military affairs lent themselves particularly well to operations research because of the vast numbers involved. Statistical analysis requires a large enough universe of events so that meaningful predictions can be made. Military life, with its swarms of personnel and constantly repeatable events, from endless reconnaissance flights to endless firing of rifles, is the perfect arena for the use of statistics. And indeed, operations research flourished in the military.

With the prestige of successes in the field behind it and the thunderous success of the Manhattan Project as well, physicists using operations-research techniques were in a unique position to influence and even form America's postwar military strategy. They could even promise the analytic precision necessary to manage the extraordinary atomic bomb, for which neither experience nor common sense was sufficient.

A group of physicists and mathematicians—including Edward Teller, Herman Kahn, Theodore Von Karman, and John Von Neumann—were brought together at Rand and other places to create the new strategy. Each individual was marked by two characteristics: brilliance in mathematics and physics, and no real understanding of the political.

In developing a pure, mathematical model of nuclear warfare, the strategy they developed was essentially independent of the realities of political life and even common sense. Many questions were posed, such as: What is the optimum response to a sudden attack on the United States? Where should U.S. weapons be located? What is the most efficient targeting procedure? The end of the conflict was defined as destroying the enemy's ability to inflict further damage or deterring him from striking at all. Left out of the equation was motive: Why would the enemy choose to strike in the first place? That question was answered not in terms of realpolitik but only in terms of nuclear weapons: he would strike if his weapons appeared about to be destroyed or if he felt he could destroy ours.

The models generated behaved as if the confrontation were not between two nation-states but between two complex weapons systems, each having as its end its own survival or the destruction of the other. From the standpoint of an operations-research model, such a sketchy rationality might be useful for making determinations about the control and management of weapons, but it did not actually constitute a political strategy.

Consider this example from the movie *Fail-Safe*. An accidental attack has been launched on the Soviet Union and the fear is that the Soviets will retaliate. Dr. Grotteschel, a caricature drawn from the annals of the Rand Corporation, declares that the Soviets will simply surrender. Since the purpose of the Soviet system was deterrence and since deterrence has failed, a counterattack would achieve nothing. Thus surrender was the only rational move. In the world of modeling, this might be correct. In the real world of nations, where the thirst for vengeance is real but cannot be cleanly included in the mathematical models, surrender is usually not an option.

Herman Kahn, one of the original team at Rand, produced a classic work, *On Thermonuclear War*, which in many ways both summarized and caught the spirit of mathematical modeling in warfare. Kahn, a physicist, was in effect attempting to sum up the nature of modern warfare, a task that had never before been undertaken by a scientist from a scientific standpoint. It had previously been treated by historians such as Thucydides and warriors such as Clausewitz.

The essential problem for Kahn was the quantitative analysis of nuclear war. In this pursuit, he hypothesized a set of scenarios of possible warfare built around the number and type of weapons exchanged. His intention was to develop a theory of the optimal strategy to be pursued in each circumstance. Of the eight possible cases, only one—deterioration of international relations—acknowledges the existence of a dynamic outside of the nuclear relationship itself. Curiously missing from his work was any systematic consideration of the political circumstances that would lead a country to use its missiles. In other words, in discussing strategy, he was not concerned with why the United States and the Soviet Union would engage in conflict in the first place.

Implicit in Kahn's thinking was the idea that the protection of a nation's nuclear arsenal was a prime strategic goal and that the possibility of losing that arsenal warranted going to war. Obviously, the Soviet's ability to destroy the American arsenal (and vice versa) depended on their having a nuclear arsenal of their own. So, in Kahn's model, it was the existence of nuclear weapons that threatened the United States rather than any geopolitical problem. The political relations among nations become marginal as a trigger in nuclear war. Technical shifts and problems within the nuclear relationship are far more destabilizing.

It is interesting to note that later advocates of disarmament, who believed that the elimination of weapons would eradicate the possibility of war, took their bearings not from classical military theorists but from the physicist strategists. Both groups ignored the problems of geopolitics and grand strategy, focusing on the technical and quantifiable.

For Kahn and the other physicists, the key to making strategy was to quantify it. As Kahn said of *On Thermonuclear War:* "The major quality that distinguishes this book from most of the other works in this field is the adoption of the Systems Analysis point of view—the use of quantitative analysis where possible, and the setting up of a clear line of demarcation showing where quantitative analysis was not found relevant in whole or in part."[25]

Obviously, much of international relations is neither quantifiable nor subject to systems analysis. For example, the degree of German insecurity before World War II, the collapse of the Soviet Union, and American moral fatigue in the Vietnam War were all real factors in international relations. Indeed, to a large extent they determined the structure of international power. None of them was amenable to quantitative analysis.

This reminds us of an old Yiddish story.

> Seeing a rich man in a restaurant eating a blintze, Herschel rushes home to his wife and exclaims, "Mama, I must have blintzes. The rich look as if they are in paradise when they eat them."
>
> His wife replies, "We're poor, Herschel, we haven't the ingredients for making blintzes. I have no eggs."
>
> "So make it without eggs."
>
> "I have no cream!"
>
> "Make it without cream."
>
> "I have no sugar!"
>
> "Make it without sugar."
>
> And so she did.

When she was done, Herschel sat down to taste the blintze, anticipation setting his face aglow. He took a bite and chewed slowly. He took another bite and leaned back. "Mama," he said, "for the life of me, I can't understand what the rich see in blintzes."

Kahn had set out to quantify war. He had thrown all of the non-quantitative dimensions of strategic policy out of the recipe, declaring them to be inessential, and tried to make a blintze out of what was left. The problem was that he liked what he had cooked up. He was quite proud of having depoliticized military strategy and turned it into a quantitative science that could be modeled and managed with precision. But in the end, a blintze is a blintze, a strategy is a strategy, and all pretense aside, nothing can change hard reality.

The Fall of the Scientist Strategists

The hegemony of operations research over nuclear strategic planning created a vacuum in the area of conventional weapons, a vacuum that can be seen in nuclear warfare's usurpation of the very term *strategic*. If nuclear weapons had come to occupy the realm of the strategic—meaning, something that by itself determined the outcome of conflicts—then what was to be the place of nonnuclear military power, of which there were still enormous quantities? This was not merely a semantic issue. If nuclear warfare was strategic and fundamental to national security, it followed that conventional warfare, by definition nonstrategic, was less than essential. If strategic planning was necessary for nuclear warfare, then what sort of planning was necessary for "conventional" warfare? How was a commander to think about conventional warfare? The latter was not merely a question of morale, although there was certainly a crisis of confidence in the ranks of infantry officers, between World War II and the Korean War, about the meaning of their lives and efforts. Even after the Korean War, there was a problem. It was established that subnuclear warfare would continue on a limited scale. But the Korean War ended inconclusively because the United States chose not to use its atomic weapons to bring the conflict to a close. The sort of total war America had gotten used to appeared to have been made impossible by the atom bomb. More precisely, nuclear weapons had usurped the sphere of total war. Total war now meant simply nuclear war.

A doctrinal issue began to emerge concerning the principles that should govern the development of strategy for the use of nonnuclear forces in limited war. The traditional methodologies used during and prior to World War II seemed anachronistic. For example, in 1941, when General Marshall ordered General Wedemeyer to draw up the Victory Plan for executing World War II, Wedemeyer observed that "the specific operations necessary to accomplish the defeat of the Axis powers cannot be predicted at this time. Irrespective of the scope and nature of these operations, we must prepare to fight Germany by actually coming to grips with and defeating her ground forces and definitely breaking her will to combat."[26]

A perfectly commonsensical plan, it faced the unpredictability of war squarely and stated the obvious: victory required the defeat of the enemy by breaking his will to fight. The Victory Plan did not attempt a precise calibration of forces. Rather, it began demographically: What is the maximum number of men that the United States could place under arms without harming industrial wartime production? The answer was based on the assumption that the enemy could probably field a greater force and that given the stakes, more on our side was better than less. There was no consideration of cost-effectiveness, appropriate response, or any other subtlety: just the use of everything available. Rather than an abstract mathematical model of force structure, the Victory Plan contained a detailed *political* sketch, trying to predict requirements from the political order.

Given his reading habits, Wedemeyer's decisions were understandable. He cited as his most important texts Clausewitz, Sun Tzu, Mackinder, Fuller.[27] All of these emphasized the primacy of the political over the technical, of the qualitative over the quantitative, of geography over time.

The Victory Plan was thus a rough and imprecise document. The mere fact that it was successful in leading American troops to victory did not satisfy strategists in the postwar period. More to the point, given the mathematical precision of nuclear strategic planning, reliance on common sense and Clausewitz, on the subnuclear level, appeared counterintuitive. There had to be a better way, a more accurate way, to deal with the use of force, if for no other reason than to economize both destructiveness and cost.

The extension of mathematical methods from the nuclear level to the subnuclear was actually a return to operations research, for, after all, its origins were in submarine warfare, aerial combat, and logistics. It had not jumped from the management of particular, limited areas of warfare to the structuring of entire campaigns and wars. Operations research had not penetrated to the very marrow of conventional warfare, that is, not until an attempt was made in 1961 to revolutionize the idea of war. This was done by an industrialist named Robert McNamara, who had been president at Ford Motor Company.

McNamara's revolution built on an idea that was central to operations research and propounded by many nuclear strategists, such as Bernard Brodie, that war was not methodologically distinguishable from economics. The process whereby you analyzed, managed, and controlled an economy was not essentially different from the way you managed a war, except that one was an economy of production and the other was an economy of force. The principle underlying both was the doctrine of efficiency: maximizing the benefits received from the efforts and expenditures—a cost-benefit analysis.

Since quantitative economic and military analysis had common roots, a statement such as this by McNamara was not surprising: "I equate planning and budgeting and consider the terms almost synonymous, the budget being simply a quantitative expression of the operating plans."[28]

In a way, this notion made sense, since in economics, one might represent power in terms of money, and money purchases the weapons that create military power. Its weakness was in assuming that expenditures equaled power. More important, without substantial political input, the effectiveness side of the equation was difficult to gauge.

Operations research extended itself into a new variation: systems analysis. E. S. Quade, one of the giants of mathematical modeling at the Rand Corporation, described the difference between systems analysis and operations research:

> In a sense, the main difference between systems analysis and operations research may well lie just in emphasis. A good deal of the earlier work tended to emphasize mathematical models and optimization techniques. Honors went to practitioners who used or improved mathematical techniques like linear programming or queuing theory and found new applications for them. These people were usually associated with decision makers who knew what their objectives were and how to compute their costs, largely in terms of some single, clear-cut criterion. On the other hand, systems analysis— while it does make use of much of the same mathematics—is associated with that class of problems where the difficulty lies in deciding what ought to be done—not simply how to do it. The total analysis is thus likely to be a more complex and less

neat and tidy procedure, one seldom suitable for quantitative optimization.[29]

In other words, the complex problems formerly dealt with in the Victory Program through a political discussion would now be absorbed in systems analysis. This represented a vast claim for authority over the entire sphere of politico-military action.

Still, quantitative abilities distinguished systems analysis from other sorts of thought. As the economists taught, the simplest and most direct measure of almost all things was money. So money—the defense budget—became the central focus of analysis, and in many ways it was the defense budget that supplanted strategy as the military's central concern.[30] Making the budget preeminent was far more than simply a case of bureaucratic self-interest. The budget was understood to reflect the commitments and capabilities of the American military. Strategy and the budget were seen as different ways of looking at the same thing.

The purpose of McNamara's revolution was to bring scientific analysis and management techniques to the Pentagon. The assumption was that an increase in efficiency would increase both the military power of the United States as well as the controls over the military. In short, McNamara was searching for means of improving the *operational* efficiency of the military and assumed that this would yield *strategy*. This conceptual problem was compounded by another: a lack of agreement about nuclear weapons and their uses.

The need for some sort of redefinition was clear. Since the Korean War, the Army had constantly rejected the idea that nuclear weapons made conventional warfare impossible. Quite the contrary, it argued that nuclear weapons were unusable and that deterrence was an illusion. Indeed, it argued against the very doctrine of civil destruction on moral and practical grounds.[31] The problem was that a coherent conventional doctrine was missing. If nuclear weapons were not used, then how would a conventional war be fought? Lacking an alternative, McNamara wanted to use the insights of nuclear modeling for the development of a doctrine and conventional weapons.

Instead of strategic goals, McNamara generated analytic principles and operational norms, the latter being summed up in the term *crisis management*. The concept of crisis management originated out of nuclear strategy and systems analysis, in which the efficient employment of power was at a premium. The problem with nuclear weapons was time: there was so little time in which to make decisions that the process had to be both precise and efficient. Moreover, the movement from decision to action had to be both swift and certain. These principles were transferred to conventional weapons.

The great success story of the Kennedy administration was the Cuban missile crisis. Using the doctrine of escalation to assure economy of force, Kennedy and McNamara employed precise increments of force to achieve the desired end—the withdrawal of Soviet missiles from Cuba. Of course, this had to be done without triggering actual violence or a nuclear exchange. Kennedy also used a management system that avoided the cumbersome formal chain of command in favor of the more informal "Excomm," which included people whom Kennedy trusted, such as Dean Acheson, who did not hold positions in the government.

From Kennedy's point of view, the extraordinary success of the missile crisis confirmed the logic of McNamara's revolution. From a more traditional point of view, there were other reasons for the American success:

- The United States had forced a naval confrontation, and the Soviets were in no position to challenge the United States at sea.
- The confrontation took place within range of U.S. ground-based aircraft.
- The Soviets could not reinforce Cuba.
- The United States had vast nuclear superiority.
- The U.S. action, rather than being viewed by the Soviets as an act of economy of force, was actually an overwhelming psychological blow, as the United States simultaneously organized an enormous invasion force and with its strategic forces went to DEFCON 3—two notches above peace and two notches below war.

In other words, overwhelming strength, favorable geography, and a convincing display of the will to employ force overwhelmed the Soviet will to resist. Though a great deal of attention was paid to the precise calibration of the American response and the careful, slow, methodical movement up the ladder of escalation, the U.S. response was not modulated. Rather, potentially limitless, it compelled the Soviets to withdraw.

So great was the relief that war had been avoided that no one really noticed that war had never been a Soviet option and that the installation of missiles in Cuba had been designed to *create* such an option.

The United States succeeded in the Cuban missile crisis because it threatened the Soviets' fundamental interest with overwhelming power, forcing them to capitulate. But this intuitive understanding of the crisis was rejected in favor of a counterintuitive reading—that the United States had succeeded because fundamental Soviet interests had *not* been threatened.

Emulating natural science, the new strategists rejected the obvious explanation in favor of the hidden reading of events. Traditional commonsense doctrines, such as the idea that you cannot deploy too much force, doctrines concerning geography and its influence on policy-making, and so on, went by the boards. The doctrine of escalation became coincidental with the doctrine of economy of force, and both became tied up with effective decision-making. It was all extremely sophisticated, and it did not lead to catastrophe until several years later in Vietnam.

Vietnam was a test of Kahn's systems analysis of war, with its principle of escalation, economy of force, and measured response. It was the nuclear strategy of the physicists applied to jungle warfare. Vietnam was also the arena for a test of another dimension of the scientific method: social science. If the first way modern science came to control foreign policy was through its domination of nuclear strategy, the second was through the creation of a class of imitators—social scientists—who sought to apply natural science's methods to social processes. In many ways, the indirect influence of the social sciences was even more damaging to American military strategy in Vietnam than the direct influence of the physicists.

In one sense, Vietnam was a classic geopolitical conflict. The United States wished to prevent the domination by communism of all of mainland Asia and therefore sought to prevent the collapse of Laos and South Vietnam.[32] This was one justification for the war. The other, and apparently more sophisticated, justification was unprecedented and gave rise to bizarre strategic and operational principles. The war was understood to be primarily a sociological phenomenon instigated from the outside. The Kennedy and Johnson administrations were persuaded that the Vietnam War was the result of inequality, poverty, and social unrest. Interestingly, this was also the enemy's justification for the war. The United States, in adopting a social scientific reading of the war's causes, accepted the communist doctrine that the military victory depended on social rectification—a social revolution. Thus, the United States sought, in the midst of war, to foment a social revolution, thereby making communism irrelevant.

The argument that victory in Vietnam was impossible without land reform, a decline in poverty, and the creation of a class of modern civil administrators who would replace the corrupt ancient regime was part of a more general social scientific theory. The collapse of the European empires had spawned a large number of states without real foundations on which to rest. The problem of how to create nations out of the debris of colonialism became politically urgent, and sociology, economics, and political science were all forced to focus on it. Together, they generated a new approach—development theory, as it was formally called. Development theory was designed to explain how postcolonial states could create nations with rapidly developing economies—not incidentally preventing seduction by communism. Vietnam became, in many ways, the classic case of nation building—a postcolonial state trying to create a coherent nation and modern economy in the midst of the inevitable social unrest, taken advantage of by unscrupulous communists.

One of the most important figures in this project was Walt Whitman Rostow, author of a work called *Stages of Economic Growth*.[33] Published in the early 1950s, this work postulated that Western approaches to economic and political organization were more likely than the communist model to yield the economic "takeoff" stage desired by nation builders. Written very much in the tradition of military operations research, Rostow's book was also an important contribution to the argument between Marxist and non-Marxist developmental economists.[34] It was even more significant as it became one of the key intellectual cornerstones of the Vietnam War.

Rostow replaced McGeorge Bundy as Johnson's national security adviser (and preceded Henry Kissinger) and was one of the key architects of the execution of the war after Kennedy's assassination. But even before his appointment to the National Security Council, Rostow was special assistant to Bundy. In June 1961, Rostow made a speech at Fort Bragg, cleared by Kennedy, entitled "Guerrilla Warfare in the Underdeveloped Areas."[35] The speech set the basic approach of combining a geopolitical and ideological concern, the spread of communism in Asia, with a sociological analysis of the problem. The real problem was the existence of social instability, which the communists could exploit. The task of the military was to create the conditions for solving the social crisis, building the nation, and undermining the attractiveness of communism for third-world peasants. It was appropriate that Rostow made this speech at Fort Bragg—home of the U.S. Special Forces, Kennedy's favorite military force and the key instrument of this understanding of war as a social process.

The military was impotent in the face of this social-scientific analysis. Years of obsession with nuclear strategy had left a generation of military officers without a clear understanding of how to fight in Vietnam.

During the Korean War, the military had devised an effective strategy: to fight a conventional war against regular communist units. The task of political warfare—pacifying and controlling the rear, containing communist irregulars, and creating infrastructure—was to be carried out by the Koreans. Unfortunately, the U.S. Army did not recall Korea happily and, instead of drawing on its relative successes there, cast about for a new understanding, which it found in the doctrine of counterinsurgency, deeply rooted in development theory.

The basic premise of the war was set: Vietnam was not a conventional war, but a struggle for "the hearts and minds" of the peasant masses. U.S. leaders believed the hearts and minds could be captured through economic growth, the creation of a stable environment, and so on. The problem they faced was how to create this stable environment when it was communist strategy to disrupt peasant life to breed distrust of the government's ability to protect them and create dependency on communist cadres.

In Rostow's analysis, all military operations had to have as their ultimate goal securing the peasants. Thus, the doctrine of counterinsurgency was imposed on the Army. Already trapped from above by the overwhelming presence of nuclear strategic planning, the Army was now trapped from below by the elevation of a previously marginal form of warfare, irregular operations, to the center of American strategy in Vietnam.

To a large degree counterinsurgency took its bearings from the communists' own interpretation of what they were doing. The Leninist idea of revolution, when extended to the third world by Mao Tse-tung and Ho Chi Minh, was to modernize societies by taking advantage of the ongoing unrest among peasants. Thus the focus was on issues like land ownership, official corruption, and inadequate prices for agricultural products. By simultaneously intensifying these problems by small-scale military activity and creating better living conditions in secured areas, the communists intended to build a peasant-based army, isolate the cities, and ultimately overwhelm them.

Rostow's model accepted this basic analysis as valid. It conceived of the guerrillas as indigenous fighters, led by some outside cadre, using locally levied supplies and captured weapons and ammunition. If this was true, then conventional warfare, short of annihilating the entire population, would not work. The key would be to separate the communist cadres from the population by increasing the risks to the population of collaboration and increasing the advantages for cooperation with the government.

To carry out this strategy, the regular army had to be replaced by the counterinsurgency specialists, who would serve as advisers to friendly indigenous soldiers. The function of the Special Forces teams was to separate the population from the communists using both a carrot and a stick. Indeed, the entire advisory process, central to American strategy in Vietnam, was intended to strengthen the administrative function of the Vietnamese, so that an efficient government apparatus could both prosecute the war and administer the nation efficiently. In June 1962, a school to train an American anticommunist cadre was opened.[36]

Quantitative methodologies permeated the war. The infamous body counts of enemy killed and wounded rested on an attrition model based on the assumption that when the casualty rate reached a sufficient level (how high, no one knew) the communists, as rational actors, would capitulate. The hypothesis was that the cost-benefit and risk-reward ratios that dominated American strategic planning also governed the communists. In other words, it was assumed that the American value structure and those of any rational nation were the same. The numbers of dead on each side were scrupulously collected, analyzed, searched for a hint of the future.

Rural pacification, the attempt to bring the vast areas of the Vietnamese countryside under secure government control, was similarly operated by devotees of nation building and systems analysis. William A. Nighswonger, in a study published in 1966, said of the pacification program:

> The rural revolution would involve the establishment of peace and order to displace the insecurity of war and terror. It would give the peasant free and open political participation in local village affairs and eventually in all levels of political life. He would increase his standard of living by learning how to grow more and more food and by receiving or being able to buy better seed, livestock and fertilizer. His children would be assured a primary school education and possibly more. His health care would be improved through better trained staffs and closer facilities. He would be better protected from the extremes of his own government through guarantees of equality before the law and he would be protected from corruption and other abuses by having direct access, through elected representatives and/or grievance and redress systems, to the highest levels of government. A land reform program would provide him with more equitable land distribution and

legal title of ownership in areas where serious inequities now exist.[37]

Rostow's theory of nation building was transformed into a rural utopia in the midst of all-out war.

The counterinsurgency program was unable to defeat the Viet Cong on the village level or separate them from the peasant. In response to the failure of rural pacification, the social planners became even more creative. If the communists could not be removed from the peasants, they concluded, then the peasants would be removed from the communists. The result was the "strategic hamlets program," which resulted in the creation of artificial villages, usually without an economic and social infrastructure, where peasants were forcibly resettled. Vast statistics were collected, proving the program to be successful, but the reality was far different.[38] The rest of the war consisted of attempts at implementing the basic lunacy: the idea that the social structure of a country could be manipulated by outsiders with minimal understanding of the nation or culture.

Social reform and nation building continued to be basic goals, and U.S. planners continued to use sophisticated socioeconomic theories and advanced mathematical models to monitor and plan. At the same time, the actual military situation deteriorated. While the Americans were obsessed with social-science methodology, the war had shifted from a guerrilla war to a conventional war, albeit on discontinuous fronts. What this meant was that the communist forces were no longer operating on food taken from villagers and weapons captured from the Americans. A complex and sophisticated logistical system had been created, bringing supplies in from China and the Soviet Union by rail and sea, then moving them down a complex system of roads and trails—the Ho Chi Minh Trail—through the Laotian panhandle, into Cambodia, and then, on a vast system of lesser supply trails, into South Vietnam proper. There they were used by conventional units, up to regiment size, who engaged the Americans at irregular points in the country.

A conventional approach to this war would immediately have suggested the following:

- Place the enemy's fundamental interests at risk. Therefore, have as a strategic objective the invasion and occupation of North Vietnam.
- Cutting lines of supply is essential to an effective strategy.

Therefore, having sealed the demilitarized zone effectively, extend this line westward to cut the Ho Chi Minh Trail in Laos.

- Having cut off North Vietnamese forces in the South, let their fighting strength wither as their supplies run out and allow pacification to fall to the South Vietnamese Army.
- Do not place a direct political role on the military, which is neither trained to deal with the subtleties of cross-cultural exchange nor is of so little value as to be wasted on a hopeless mission.

As a result of Rostow's theories and operations research and systems analysis, a completely different strategy was followed:

- Engage in direct rural pacification through the Civil Operations and Rural Development Support program (CORDS), using military personnel to function as village-level security as well as social revolutionaries, supplanting both the Vietcong and the North Vietnamese Army in this role.
- Stop the North Vietnamese invasion of the South by initiating a strategic bombardment campaign designed to escalate the war gradually, so as to follow the principles of economy of force and using models developed during World War II for the effective allocation of forces. Force the North Vietnamese to cease hostilities in the South because of the pain caused by aerial bombardment.
- Attempt to cut the flow of supplies to the South through air strikes along the Ho Chi Minh Trail rather than through ground interdiction, using sophisticated sensing devices and mathematical models to guide interdiction.
- Use large-scale ground operations to harass and destroy the enemy, using superior firepower and mobility to draw the enemy into disadvantageous exchanges.

Advanced planning techniques helped develop these operations. Sophisticated studies abounded, carrying titles such as *An Interdiction Model for Sparsely Traveled Networks*.[39] Complex mathematical and computer models were used to strategize and help execute the air war as well as many aspects of the ground operation. As a result, **the Vietnam War was operationally brilliant and strategically demented.**

- Asking an army to serve as an instrument for social revolution and reconstruction was asking it to do something it was not trained for. Only by abstracting them, treating them as an unreal quantum, could U.S. Army Rangers be transformed into warrior/social workers.

■ During World War II, strategic bombing of the civilian population had failed as a deterrent in Britain, Germany, and probably Japan. Thus, there was no historical basis for the idea that a gradual increase in the suffering of noncombatants would influence North Vietnam's war strategy or drive a wedge between the leadership and the masses. It was wishful thinking masquerading as social science, cost-benefit analysis incapable of evaluating either the benefit or the cost.

■ Failing to make the cutting of the Ho Chi Minh Trail an absolute imperative and accepting operational criteria and battle objectives from operations-research and systems-analysis models led to massive defeat.

■ Accepting war on the enemy's terms, a series of small-unit engagements in which the attrition rate for U.S. forces quickly reached unacceptable levels, was a strategic blunder of the first order.

■ Above all, by claiming that traditional models of warfare were irrelevant in the Vietnam War and substituting for them methods developed for nuclear warfare and social-scientific models of nation building, the United States lost the essence of strategy—a clear evaluation of one's own strength. The primary strength of the United States in Vietnam was the ability of the Army to succeed in any engagements it chose to attempt.

Conclusion

One of the keys to the United States defeat in Vietnam was the failure to define the war geographically. It has been said that the war in Vietnam resembled naval warfare more than land combat, as units moved around each other simply with the intent of inflicting harm, rarely with the intent of taking and holding terrain. Since the war had no geographical goal, terrain had no meaning in the overall scheme. This meant that the U.S. Army had no strategic goal other than outwaiting and outenduring the enemy.

Caught between the guerrilla and the atomic bomb, the Army was utterly without a self-conception. From the end of the Vietnam War until the beginning of the Reagan administration, it lived in semidisgrace, with a strong sense that it no longer had a mission, that it could no longer shape or influence a world full of unruly social forces.

The central problem in military strategy, of course, was the awesome, brooding presence of nuclear weapons—weapons that had paralyzed military thinking. But despite the potential for massive destruction, wars continued, unabated, and wars—even empires—were won and lost. History continued. All of this raises a possibility—that while nuclear weapons overwhelmed military planners, their actual importance, their real power, was far less than was thought.

3

False Dawn

THE FAILURE OF NUCLEAR WEAPONS

Ever since the atom bomb was dropped on Japan in 1945, it has been alleged that the introduction of nuclear weapons has revolutionized warfare—indeed, that nuclear weapons have rendered warfare obsolete by making war so terrible that no sane nation would attempt it.

There is an obvious sense in which this claim is untrue—wars continue to be fought, undeterred by nuclear weapons. A refinement of this argument has been more persuasive: war between nuclear powers is no longer possible, since each side knows that the other is unwilling to lose and, therefore, would resort to nuclear weapons to prevent defeat. Thus, neither side would initiate war in the first place, since no gain would be worth the risk of mutual annihilation.

The problem with this position is that it only partially answers the question of the nonuse of nuclear weapons. It does not explain why nuclear powers have not used their weapons against non-nuclear powers. Nuclear powers have consistently permitted themselves to be either stalemated or defeated in war without recourse to nuclear weapons: the United States permitted a stalemate in Korea and a defeat in Vietnam; the French also permitted themselves to be defeated in Indochina as did the Soviet Union in Afghanistan.

Furthermore, nations have not refrained from attacking a nuclear power even if they themselves did not have this capacity. For example, Israel was known to be a nuclear power when the Egyptians and Syrians attacked it in 1973, but this did not serve as a deterrent. Nor did Vietnam refrain from combat with China in 1977, even though China, by then, had a substantial nuclear force and the ability to deliver it. For some reason, the possession of nuclear weapons neither prevents attacks by a nonnuclear foe nor guarantees victory in war.

It is, of course, true that the United States and the Soviet Union never directly went to war—although they engaged in countless, dangerous proxy wars. Indeed, many theorists have contended that this was because each feared the possible use of nuclear weapons in a direct confrontation. While this argument has much justification, it is important to realize that the issues between the Soviet Union and the United States, although intensely significant geopolitically and ideologically, were not sufficiently threatening in any immediate sense as to make war of any sort, let alone nuclear war, a rational choice.

The Mighty Weirdness of Nuclear Weapons

There is, then, something exceedingly strange about nuclear devices. They are said to be the ultimate weapons, able by themselves to overwhelm any correlation of forces on the conventional battlefield. Several nations now possess them, yet not once, in more than half a century, has any nation found it prudent to use them. It may well be that we have merely been fortunate, but it may also be that some characteristic of the weapons makes their use generally improbable.

Another claim is that the world is a much more dangerous place because of the existence of nuclear weapons. In the early 1980s, during the nuclear freeze movement, the argument was made that the elimination or dramatic limitation of these weapons would lessen the chances of nuclear war. Indeed, this was the premise of the entire disarmament movement. There had been disarmament movements in the past. In the 1920s, there was a serious attempt made to limit the size of navies in order to balance the power of nations. But this time, the problem was not seen as political, part of the clash of national interests, but as technical: the physical existence of nuclear weapons was what threatened the world. If these weapons did not exist, the threat of nuclear war would decline.

So, instead of arguing that a fundamental political settlement between the United States and Soviet Union had to be reached, the debate constantly focused on the reduction of nuclear weapons. For example, a great imperial agreement could have been reached between the two superpowers, defining the spheres of influence of each, assigning nations to each and agreeing to keep hands off the other's nations. After all, this was the basic outcome of the Yalta Accords at the end of World War II. We could simply have extended to the rest of the world the political agreements that maintained the peace in Europe. This, of course, cut across the anti-imperial grain of both powers, and the focus remained on the hardware rather than on politics.

The point missed by advocates of nuclear disarmament is that the disappearance of every last nuclear weapon in the world would have no effect on the possibility of nuclear war; it would, at most, delay its onset. The danger in nuclear weapons does not lie in their physical existence but in the knowledge of how they are constructed. So long as that knowledge exists, the weapons' absence or presence in the world means nothing. In times of turmoil those nations that know how would be able to build nuclear weapons, this time not in the relatively open and slow atmosphere of the peacetime arms race of 1950–1990 but in the overheated, secretive, and profoundly insecure atmosphere of a crisis—an infinitely more dangerous circumstance for an arms race.

The danger of nuclear war, like the danger of any war, is political. Consider the following: From 1950 onward, two nations were each armed with sophisticated nuclear weapons. These nations, virtually bordering on each other, had a history of rivalry and warfare going back centuries and a deep-seated culture of mutual distrust and even dislike. In previous wars, each had felt betrayed by the other or felt that the other had failed to fulfill a commitment. Yet, although each had sufficient nuclear weapons to devastate the other, not a single radar on either side watched for an attack. Had a missile been accidentally launched, the results, however tragic, would have had no military repercussions. The existence of nuclear weapons had no effect on the security of either nation.

This is not a fanciful example—it refers to two real countries, Britain and France. Both were armed with nuclear weapons of substantial sophistication; both had a history of conflict and mistrust. But nuclear war was an impossibility between the two, regardless of the existence of nuclear weapons. This is because nuclear war, like all war, is *political* in nature, and the mere *physical* existence of these powerful weapons on either side has little significance. Abolishing them means nothing if the underlying political conflict remains, just as retaining them when there is no conflict means little as well.

Consider the relative positions of the United States and the Soviet Union during the Persian Gulf crisis. In 1980, the Soviet Union had fewer nuclear weapons than in 1991, yet, during the Iranian crisis, no action could be taken by the United States without profound concern for the Soviet response. In 1991, however, the United States acted with almost complete indifference toward Soviet power. The fact that the Soviets had more missiles, more warheads, and greater accuracy did not really matter. The *political* equation had shifted—a fact that the possession of nuclear weapons could not change.

The argument we are making is not that nuclear weapons are not frightening, but rather that they are not particularly useful as an instrument of national power. Moreover, their utility is limited in any real political application, even when held in the vast numbers in which the Soviets held them. You might say that nuclear weapons are simply overrated.

To describe the atom bomb, or the considerably more powerful hydrogen bomb, as overrated is, obviously, an odd and seemingly preposterous contention. Yet if the claim is true, then the shape of the military world shifts dramatically, and with it, the geopolitics of the twenty-first century. Certainly, precision-guided munitions were born from the womb of the nuclear culture. The search for nuclear-tipped missiles created the technology of precision that precision munitions drew on. The surveillance system created to guard against nuclear war was the foundation of their targeting systems. Most important, precision-guided munitions were created to solve the same problem of inaccuracy that nuclear weapons tried to solve—although the former solved the problem with the subtlety of a stiletto while the latter solved it with the clumsiness of a sledgehammer. In a sense, therefore, precision-guided munitions and nuclear weapons are mirror opposites.

Therefore, the future is incomprehensible without a careful consideration of nuclear weapons: whether or not they are usable components of a military strategy, when they are appropriate as a solution, why they have not been used in the past. We must begin, however, by considering why nuclear weapons were produced in the first place.

Atomic Weapons and Strategic Bombardment

Nuclear weapons were made possible by advances in nuclear physics and were made necessary by the evolution of military strategy after the failure of the airplane to be decisive in war. By 1943, when the Manhattan Project went into high gear, the atomic bomb and nuclear weapons were seen as a means of fulfilling airpower's promise and overcoming its inherent weaknesses, which we might summarize as:

- The large number of sorties needed to bring sufficient explosive power to bear on the target.
- The large number of support personnel needed to produce and maintain the aircraft, as well as the number needed to maintain the logistical system.
- The large consumption of petroleum, oil, and lubricants necessary for the large number of sorties.
- The inability of bombers to be precise enough to destroy a target if the area did not consist of flimsy wooden buildings susceptible to incendiary bombs—as was the case with Tokyo and other Japanese cities.

But if these problems could be overcome—if sorties were held to a minimum and precision or devastation dramatically increased—and if the number of personnel required could be dramatically lowered, then the strategic bomber might well prove effective. The atomic bomb solved the problem of resource allocation (once the massive costs of the Manhattan Project were discounted) by reducing to one the number of delivery vehicles needed. It solved the problem of the inefficiency of conventional airpower by introducing an explosive power unprecedented in history.

However, the atomic bomb also posed a series of new and unanticipated questions. First, it raised the issue of the appropriate delivery vehicle—how to get the bomb to the target—triggering a debate between advocates of manned bombers and advocates of missiles, invoking technical questions far beyond the competence of the military. The second issue was how to defend planes and missiles from attack by enemy forces, an issue that was to evolve into an extraordinarily complex system of calculations and countercalculations, moves and countermoves.

It is extremely important to remember that, in its early stages, the atomic bomb was not a fundamental improvement on conventional strategic bombardment in terms of its destructive power. Sixty-six cities in Japan had been subjected to incendiary raids; these cities suffered between 25 and 90 percent destruction. On average, 50 percent of the city was destroyed. Substantial, if not decisive, damage was done to war facilities by these raids—production levels dropped from between 54 percent (for army ordnance) and 96 percent (for motor vehicles).[1] It is not altogether clear, of course, whether these drops were due to the air campaign or to the submarine blockade, which strangled the flow of raw materials from Southeast Asia.

To achieve these results, 29,153 sorties had to be carried out, in which a total of 176,059 tons of bombs were dropped, over a period of about 14 months. In one raid on March 10, 1945, 334 B-29s destroyed 16 square miles of Tokyo, 267,171 buildings (a quarter of the city), killed 83,793 people, wounded 40,918, and burned down the homes of over 1 million people.[2] By contrast, 78,000 people died in Hiroshima, 51,000 were injured, and 48,000 buildings—two-thirds of the structures—were destroyed or damaged. About 4 square miles of the city was destroyed, less than the area destroyed in some of the fire-bombing raids. In Nagasaki, 35,000 were killed and 60,000 wounded, with about 40 percent of the structures damaged.

The low level of damage in Nagasaki pointed out one of the inefficiencies of an atomic explosion: the concentration of the explosive. Had the twenty thousand tons of TNT equivalent of the Hiroshima blast been distributed more widely, as a massed bombing attack would have done, the destruction would have vastly increased. A line of hills surrounding the city enclosed the blast, creating far fewer casualties than would have been produced by a large bombing raid.

Therefore, the actual quantity of damage done at Nagasaki was not in any way greater than the damage from conventional bombing. The virtue of the atomic bomb was the relative ease of delivery. In the raid on Tokyo, 334 B-29s were used; only three planes (the *Enola Gay,* a chase plane, and a weather plane) were necessary in Hiroshima. On the surface, this was a great advantage—but only on the surface.

The world was stunned by Hiroshima. Yet many in the military were not entirely impressed; Maj. Alexander P. de Seversky, of the U.S. Army, claimed that the bomb dropped on Hiroshima would do no more damage than a ten-ton bomb of TNT had it been dropped on New York City's financial district, with its more robust construction.[3] The head of the Navy's Aviation Ordnance Branch said that if someone stood at one end of the runway at Washington's National Airport and an atomic bomb was dropped at the other end, no major harm would come to him.[4]

Such nonsense missed the point. Atomic bombs were enormously powerful weapons, but so were conventional bombs. Terrible damage was done to both Tokyo and Hiroshima, yet both survived as cities, and indeed, both were functioning within days of the bombing. The results were similar—devastating without being decisive. What was not clear was whether the nuclear delivery system (taken as a whole), from research to explosion, was more efficient than conventional bombardment.

Delivering the bombs on Nagasaki and Hiroshima carried with it a $2 billion price tag ($1 billion per city) for the research, development, and production costs of the Manhattan Project. The Manhattan Project diverted massive resources to achieve a goal that conventional weapons could have achieved, though not as easily, at a much lower cost. A single B-29 bomber cost $639,188;[5] a fleet of the size that devastated Tokyo cost $213,488,792. That was 10 percent of the cost of the Manhattan Project, and it was reusable. While the maintenance crews for an air wing appeared to be much larger (the Twentieth Bomber Command had 11,144 personnel)[6] than that for an atomic squadron, the number of individuals needed to build, deliver, and maintain an atomic device was much greater, not only in terms of numbers but in terms of technical skill as well. The cream of American science and engineering were devoted to the Manhattan Project, and important aspects of the war effort were dangerously delayed because of the Manhattan Project's priority claim on key material. Indeed, amphibious operations in the Pacific as well as D day in Europe were jeopardized because of this.

It would be difficult to argue that the Manhattan Project was essential for victory in World War II. And this is not a statement made entirely without hindsight. Less exuberant calculations of the explosive power of the weapon and the amount of time necessary to produce bombs would have shown that the advantages offered by the bomb in the context of the war were limited. But such calculations had little bearing in the context of the war. The bomb was built.

Painfully aware of the various criticisms of the bomb, Bernard Brodie responded in 1946 by arguing:

> It should be obvious that there is much more than a logistic difference involved between a situation where a single plane sortie can cause the destruction of a city like Hiroshima and one in which at least 500 bomber sorties are required to do the same job. Nevertheless, certain officers of the United States Army Air Forces, in an effort to "deflate" the atomic bomb, have observed publicly enough to have their comments reported in the press that the destruction wrought at Hiroshima could have been effected by two days of routine bombing with ordinary bombs. Undoubtedly so, but the 500 or more bombers needed to do the job under those circumstances would, if they were loaded with atomic bombs, be

physically capable of destroying 500 or more Hiroshimas in the same interval of time.[7]

This was the heart of the matter. The Hiroshima-style atomic bomb did not produce unprecedented damage, but its damage had an extremely low logistical cost. While this was true only if the logistical cost of developing and producing the weapon was ignored, it leaves another question unanswered. **If the virtue of the bomb was that it allowed five hundred cities to be destroyed simultaneously, then this is a benefit only if circumstances were to arise in which it was useful, politically and militarily, to destroy five hundred of the enemy's cities simultaneously.**

This is not weakheartedness. Twenty million Soviets died defending their nation against the Nazis. It was worth that and twice the price, but their deaths had a political point, their deaths had meaning. By dying, they helped defeat Hitler—a definable and desirable goal. Thus, a nation threatened with the death of tens of millions of its citizens would not necessarily submit to an enemy. The assumption, therefore, that the ability to destroy five hundred cities would compel an enemy nation to submit to unbearable political demands is far from obvious. Moreover, the power to destroy an enemy's cities may or may not mean the ability to destroy an enemy's military force, which, particularly in the short run, might be able to continue operating, dispersed in the countryside.

How Nuclear Weapons Became Irrelevant

Psychological power is not trivial, but it is not by itself decisive. Over time, a bluff will be revealed to be hollow. Underlying the psychological power that came from having the bomb was a grimmer reality. At the beginning of the Cold War, the United States did not have enough atomic bombs to make a difference. The United States had only nine bombs in 1946, thirteen in 1947, fifty in 1948.[8] At the rate the United States was producing these weapons, it would be decades before five hundred Soviet cities could be threatened. An even worse problem was that the United States had no way to deliver them. The United States had only one atomic bombing group, the 509th Composite Group with twenty-seven B-29s. Available crews could prepare only two bombs a day for use.[9] Much of the eastern Soviet Union was out of American reach. The distances involved in reaching the western part were extreme and required a willingness to place these devastating weapons in such places as Turkey or Iran.

This meant that the United States actually had no way to threaten the Soviets with nuclear attack, and the Soviets clearly behaved as if they knew that this was the case. The 1945 atomic bombs fell on a nation that had been stripped of its defenses by months of aerial warfare. Still, the B-29 *Enola Gay* had to be stripped of all weapons and had to fly to the limits of its range—about three thousand miles.[10] Adding armaments would have dropped the range sharply. Against a vigorous, fully armed enemy such as the Soviet Union, the ability to penetrate thousands of miles to a target would have depended more on luck than anything else. The fact was that the Americans and their allies were more impressed with the power of the atomic bomb than were their enemies, an oddity apparent from the beginning of airpower.

American doctrine was that the atom bomb was to be a city killer, as in World War II, used against civilian populations rather than against purely military assets. Its function, if it could be delivered, would be as a terror weapon, to frighten the enemy into acquiescence. During the Cold War its key purpose was to prevent communist aggression. The fact is that the U.S. atomic arsenal did not deter the North Koreans from invading the South, prevent to Soviets from imposing the Berlin blockade, nor force them to stop supporting the Greek communists. During the Truman years massive retaliation against Soviet cities in the event of war was implicit in American military doctrine. Under Eisenhower massive retaliation became explicit. As Field Marshal Montgomery, then deputy supreme allied commander Europe, put it in 1954:

"I want to make it absolutely clear that we in SHAPE [Supreme Headquarters Allied Powers Europe] are basing all our planning on using atomic and thermonuclear weapons in our defense. With us it is no longer: 'They may possibly be used.' It is very definitely: 'They will be used, if we are attacked.' "[11]

A clear politico-military function for nuclear weapons had emerged: to deter conventional attacks with the threat of overwhelming massive retaliation. This doctrine immediately posed two questions. First, what would the Soviets do if the United States massively retaliated against them? Second, could the United States in fact achieve massive retaliation? Could enough weapons, with enough explosive power, be deployed and delivered to targets in a reasonable length of time to deter an attack? As we saw in the last chapter, the political question rapidly became a highly technical one that the research and development staffs of the armed services urgently tried to solve.

Their efforts proceeded in two directions. One was toward a more effective explosive device, while the other was toward a more efficient delivery system. The more efficient explosive device was, of course, the hydrogen bomb. The power of the hydrogen bomb was more than just a quantitative improvement over the atomic bomb, it was a qualitative one as well. A single atomic bomb could not cause a city to cease functioning. A hydrogen bomb, at least theoretically, had no limit on its size and power. True, it suffered from the same inefficiency as its predecessors— its explosive power was too concentrated. But with its power exponentially expanded into the megaton range, the H-bomb compensated for this deficiency.

An early test, the "Mike" explosion at Eniwetok in the Pacific on November 1, 1952, equaled 10.2 megatons. The fireball alone had a diameter of 3.5 miles, and it left a crater measuring well over one mile in diameter.[12] A single H-bomb could obviously devastate a medium-size city, such as Washington, and destroy the heart of a megalopolis, such as New York. And the Mike shots were far from the largest. The Soviets exploded a fifty-three-megaton monster in the late 1950s. A genuine city killer existed; the question was, what to do with it?

The central technical problem was, how to deliver the weapon to its target? An early operational device was the M-17, weighing twenty-one tons and having a yield of about twenty-five megatons. In 1954, the only plane capable of even lifting this weapon was the Convair B-36, a propeller-driven plane operating well below the speed of sound.[13] But it could not possibly deliver the M-17. Nothing could.

Two long-term solutions soon became apparent. One was the manned bomber, the logical extension of the theory of strategic bombardment embraced by American air planners during the 1930s and executed during World War II. The other was the unmanned, ballistic missile.

The manned bomber had certain obvious advantages. No fundamental scientific or engineering breakthroughs were required; in a sense, it was ready to go. American planners had a great deal of experience in the construction and deployment of manned bombers as a result of deep-penetration raids into Germany and Japan. They understood the logistical and maintenance issues as well as the command and control problems.

The real issue was whether the manned bomber, such as the B-29, could penetrate with the high degree of certainty necessary for a nuclear strike. Nothing would be worse than to launch a nuclear strike with scarce weapons only to have the attacking aircraft shot down. During World War II, the ability of the Allies to outproduce the Axis in airframes and engines meant that attritional warfare was to their advantage. The strategy was to constantly engage and exhaust German and Japanese air defenses, relying on vast numbers of aircrews and aircraft to gradually wear the enemy down and gain superiority in the skies.

The delivery of a nuclear weapon cannot wait for the culmination of long-term operations to establish air superiority. Quite the contrary. It must be the opening move in a war. It takes place while the enemy's air defenses are at their freshest and strongest. Since the war, the Soviets had turned their attention to constructing air defenses to deter an American attack from bases in Europe and Turkey. Their Air Defense Command was regarded as so important that it was separated from their army into a separate service and, by 1953, was equipped with over two thousand advanced jet interceptors, thousands of air defense artillery pieces, and a complex electronic surveillance system.[14]

The Soviets felt that they had properly analyzed the strategic threat created by the bomber, and they demonstrated this by not building bombers themselves. In part this represented a historic distrust of manned strategic bombers—the Soviets had no experience with them during World War II.[15] In part the decision had to do with geopolitical reality. Unlike the Americans, the Soviets had no bases within range of the enemy. This meant that Soviet bombers simply could not reach the United States. The Soviets did build one intercontinental bomber in the 1950s—the Bear—but the likelihood of its being able to cross the oceans or traverse the Arctic and then penetrate the American heartland was near zero.

It is also important to remember that in the Soviet Union no built-in political apparatus defended the interests of a strategic bomber command, as most assuredly was the case in the United States. The Soviets were free to solve the problem in a more radical way, more in keeping with their own historical interest in artillery and rocketry. The Soviets, therefore, sought to leapfrog the manned-bomber stage by developing an intercontinental missile that could deliver nuclear weapons in a matter of minutes. The Soviet decision frightened the Americans into speeding up their own missile program, particularly after the success of the *Sputnik* launch.[16]

The idea of an intercontinental missile had first been raised by the Germans, dreaming of vengeance on the United States. The V-1 air-breathing missile and the V-2 rocket had been intended as preludes to a missile that could reach New York. This vision of an intercontinental rocket had transfixed the Soviets as well as the Americans, as both sides captured plans and scientists. Missiles became even more interesting when the H-bomb was invented. A single warhead would now be sufficient to destroy the center of a city, so an expensive and expendable delivery system would be cost-effective. At the same time, the mature H-bomb was much lighter than the early A-bomb, making it more practical for the limited thrust of the early rockets.

The Soviet decision to go to missiles caused the United States to pursue that path every bit as vigorously—giving it another dimension to its nuclear threat. The result was the creation of a Soviet intercontinental ballistic missile force while the Americans built intermediate-range ballistic missiles. The United States did not initially need long-range missiles because of its ring of bases surrounding the Soviet Union.

Now the problem shifted from the ability to strike enemy targets to the ability to withstand an enemy surprise attack, or at least to have enough forces survive an attack so that they could counterattack.

Assuming that an intercontinental missile could hit a Strategic Air Command (SAC) bomber base in twenty minutes, it was doubtful that SAC could detect the launch, transmit a scramble order, scramble the planes, and get them out of the blast zone before impact. SAC, painfully aware of this, had to keep many of its bombers in the air at all times and others ready to scramble at a moment's notice. The United States and Canada created the North American Air Defense Command (NORAD) just to provide a few minutes extra warning to the bombers. One of the enduring images of the Cold War are films of SAC bomber crews running to their aircraft with warning Klaxons blaring, practicing for the day they would have to beat a missile.

During the 1950s, the nuclear relationship between the United States and the Soviet Union moved to a stalemate, just a few scant years after the introduction of what was to be the most revolutionary weapon of all time. Robert Oppenheimer described the situation:

"We may anticipate a state of affairs in which the two Great Powers will each be in a position to put an end to the civilization and life of the other, though not without risking its own. We may be likened to two scorpions in a bottle, each capable of killing the other, but only at the risk of its own life."[17]

The essential fact was that nations and leaders were rational. The risk, which could never be eliminated, was that any retaliation by the other side would cause devastating damage. Neither side had any strategic objectives worth risking its survival as a nation.

The stalemate was rooted in geopolitics and not, as nuclear strategists fantasized, in the nuclear balance. Both the United States and Soviet Union were relatively self-contained powers. The Americans were more wealthy than the Soviets, but victory in nuclear war would not transfer any of this wealth to the Soviet Union. The disparity of wealth between the two *might* be rectified were the Soviets, for example, to seize Western Europe, but the risk and cost of the invasion would not be worth the prize, even if full value could be realized. The Soviets were relatively poor, but not so poor that winning a war against them would be easy. Most important, neither side could place the other in an untenable position. The Soviets were not dependent on the American alliance for their economic well-being, so the Americans could not force them to act out of desperation. The Soviets, on the other hand, were in no position to break their geographic encirclement. Thus, the United States was not put into a position of geopolitical desperation.

Wars begin in desperation—the fear that without war a nation would face catastrophe. This is what drove German and Japanese aggression in World War II. The Cold War was a confrontation between two vast powers, neither of which could be made to feel desperate, therefore neither side would risk war.

The balance of terror became a balance of ennui. In fact, the nuclear balance appeared to be a confrontation between two technical staffs floating in the subtle metaphysics of the nuclear calculus, a calculus utterly meaningless, since no political leader would translate his staff's calculations into action. The arms race and arms control talks proceeded, while history went on unheeding. The European empire disintegrated, Japan reemerged as a great power, the United States and the Soviet Union were both defeated in various wars by nonnuclear powers without once considering the use of nuclear weapons, and finally the Soviet Union, still a nuclear power, collapsed.

The deadlock between the two nations on the nuclear level was repeated on the political level. Throughout the former European colonies, the Americans and the Soviets struggled with each other for geopolitical advantage—the Soviets wanted to strike at the rear of the American encirclement in Africa and Asia and directly threaten American interests in Latin America. The Americans sought to stabilize their position by stopping Soviet influence. During the 1960s, the Cold War turned into a vast kaleidoscope of conflict between American and Soviet agents throughout the European imperium. As decolonialization advanced, each side sought to take full advantage of the situation. No one could find a use for nuclear weapons in all of this.

The problem was that while the struggle for global domination was important, it was not worth American or Soviet cities. So long as these were held hostage by nuclear weapons, the stalemate continued, and so did the vast, nonnuclear confrontation. Nuclear strategists, increasingly

isolated from the flow of history, sought for a means to break the grid-
lock and make nuclear weapons useful again. The solution was obvious,
if obscured by nuclear strategic jargon: each side was stopped from using
nuclear weapons out of fear of the enemy's response. If, however, the
enemy's missiles were destroyed, there would be nothing to fear, and
nuclear weapons could be used to bully others.

Therefore, the idea arose that maybe the weapons themselves
could be destroyed before they were launched—counterforce. City
killing shifted to missile killing. Since exposed missiles were too vulnerable to the enemy's missiles, even inaccurate ones, they were moved to
underground silos and submarines. Then these underground silos were
hardened to withstand closer and closer explosions, while the missiles
were made more accurate, bringing them closer and closer to the hardened silos.

An endlessly subtle game of hide-and-seek, culminating in the mar-
velous lunacy of the MX missile, ensued. Treaties limited the number of
missiles each side could have. If the location of a missile were known, it
could be destroyed. But if the location of the missile could not be discovered? Imagine a system of railroad tracks on which a missile would shuttle between three or four possible launch sites—all underground. The
Soviets would not know which to strike and would not have enough missiles to hit all the possibilities. A vast mathematical system came into
existence, calculating throw weight and kill probabilities, all incomprehensible and profoundly irrelevant. Submarines were built with nuclear
missiles on board, not accurate enough to destroy silos but accurate
enough to destroy cities. These nuclear submarines were difficult to
locate under the sea, and harder still to kill, and were the final guarantor
of national security. No matter what happened, enough nuclear submarines would survive to devastate enemy cities.

During the entire arms race between the United States and the
Soviet Union, the basic situation never shifted: neither side could launch
an attack with sufficient certainty that the other side would not have
enough missiles or bombers left to reach a few cities and inflict damage.
Given this reality, neither side would use its missiles. What either side
would win in a successful attack never approached in value what they
would lose in an unsuccessful or partially successful attack. During the
Cuban missile crisis, where U.S. interests were seen as imperiled, the
threat of nuclear war prompted the Soviets to accept defeat: the advantage they sought was not worth the risk.

Atomic bombs were not so powerful as to abolish the rules of war.
Hydrogen bombs were so powerful that they were relevant only to each

other and not to any real military conflict. The sole purpose of thermonuclear weapons was to prevent their own annihilation rather than to allow one nation to impose its political will on another. Imagine that the head of a military establishment came to a leader and announced that a system of attack had been devised that could destroy the enemy's nuclear arsenal. Would any leader place sufficient credence in that declaration to risk his nation's safety?

During the Cold War the nuclear balance became a metaphysician's delight, floating outside the real political arena, existing in its own set of relationships regardless of what was happening in the outside world. Horrible prophecies were made of what would happen when China developed the bomb; then China did get the bomb and nothing happened. And it did not stop there: India got the bomb and nothing happened; the Israelis got the bomb, and again nothing happened except that the Arabs attacked them, without caring at all about the fact that Israel had the nuclear bomb. The Vietnamese went to war with the Americans, and it did not matter that the Americans had the bomb. The same was true of the Afghans and the Soviets, except that the Afghans may not have known what a bomb was but were not going to miss a good fight with a halfway decent infidel.

The acquisition of nuclear weapons made almost no difference to the acquisition of power in the international system. China, India, Israel, all became nuclear powers. Their basic place in the politico-military order remained the same as it would have been had they not built nuclear weapons. France was a nuclear power, Germany was not. France's role in the international system did not dwarf Germany's. Britain was a serious nuclear power in the 1950s, but it did not help her hold her empire. South Africa became a nuclear power but gave up the weapons because they made absolutely no difference. Not once in half a century did the acquisition of nuclear weapons increase the power of a nation. In political terms—and these are the only real measure of military power—acquiring nuclear weapons had not made a bit of difference, except to the psychological mind-set of the owner.

In the meantime, the nuclear planners calculated warhead counts and throw weights, and the antinuclear protesters protested warhead counts and throw weights, and no one cared except the Americans and the Soviets, each of whom was terrified that the other would get the wrong idea that their weapons could all be destroyed—although it was never clear what either side thought it would do with such information. In the meantime, wars were waged, nations rose and fell, and men died. History went on, unmindful of the absolute weapon.

The Use and Abuse of Nuclear Weapons

It is an extraordinary fact that in the half century of the existence of nuclear weapons, no nation, except the United States in World War II, has ever found it in its national interest to use them. The United States lost the war in Vietnam, the Soviets lost the war in Afghanistan, France lost the war in Algeria, and so on. In each case, nations preferred defeat to the use of nuclear weapons. One is tempted to ascribe this decision to humanitarianism, but neither Richard Nixon nor Leonid Brezhnev reminds us of a humanitarian. They do remind us of superb politicians, committed to the national interest and their own power. Why were they unwilling to use nuclear weapons?

One answer is that they were afraid of retaliation from the other side. This is not altogether persuasive. It is difficult to imagine that, in return for an attack on Hanoi or Kabul, the United States or the Soviets would have unleashed a nuclear attack on the other. Yet it is possible that the slightest potential for miscalculation persuaded both sides that even victory in these marginal wars would not be worth the risk. If that were the case, and even the most minuscule chance of nuclear exchange would be enough to deter the use of nuclear weapons, then such weapons would be truly useless, except for deterring the other nation from using them.

Historically, nations have refrained from using all of their power in all of their wars, calibrating the use of force carefully so as to be in a position to meet other eventualities. Consequently, nations held forces in reserve, even accepting defeat, to prevent the exhaustion of all their forces. One might argue that this was the American strategy in Korea, where the United States was prepared to risk defeat twice and finally accept stalemate rather than use atomic bombs. The Americans had very few atomic bombs and did not want to squander them on what was essentially a sideshow.[18]

Yet, even when nuclear weapons had become available in both the Soviet and American arsenals, they were not used to prevent defeats. The explanation ultimately rests in the fact that they would not have achieved a decisive result. This is ironic, since the entire justification for nuclear weapons has been that it was the "absolute" weapon, an irresistible force that brought the Pacific war to a halt and that could, therefore, halt any war. Mao was ridiculed when, during the Korean War, he said, "The atom bomb is a paper tiger with which the U.S. reactionaries try to terrify people. It looks terrible, but in fact is not. Of course, the atom bomb is a weapon of mass destruction, but the outcome of war is decided by the people, not by one or two new weapons."[19]

What Mao grasped, and his detractors missed, was that the atom bomb was not a *decisive* weapon. It would inflict damage, but it would not make continued resistance impossible. Mao analyzed it as a problem of a lack of suitable targets and insufficient force. There was, however, another problem.

The initial idea for aerial bombardment was not to use the atom bomb as a terror weapon but, rather, to use it as a means of destroying the industrial base of a nation and therefore break the ability, rather than the will, to resist. Neither Vietnam nor Afghanistan, nor for that matter Korea, was waging war on its own industrial base. Support for the struggles was coming from the outside. The United States could use nuclear weapons to destroy the rail lines from China, but it must be recalled that reconstructing rail lines, even after an atomic blast, is not particularly difficult. After all, trains were running in Hiroshima twelve hours after the bomb exploded.

The industrial base of these countries could only be broken with attacks on third countries. This did not pose merely a military problem. A Soviet nuclear attack on the staging areas of Pakistan or an American atomic attack on China could have triggered some interesting nonnuclear military responses. But the essential problem was political: neither American interests in Korea or Vietnam nor the Soviet interests in Afghanistan were worth the political cost of alienating the supporting power. And since the strike would not have guaranteed victory, what was the point?

The possibility that nuclear weapons may be used in a particular conflict increases the risk of involvement. Even if the possibility is slight, the outcome is perceived to be so catastrophic that anyone without a direct and overriding interest immediately recoils. So, in Vietnam, American allies with more marginal interests in the conflict, such as Japan or Australia, would have protested America's use of nuclear weapons. The political fallout would have been so great, and the military payoff so small, that using the bomb would have been irrational. The United States preferred defeat to turning to nuclear weapons.

Since this pattern was confirmed by the Soviets in Afghanistan, we must now regard it as rather firm. There is, of course, the possibility of an "irrational" third-world leader using such weapons promiscuously. Interestingly, the usual examples are drawn from the Islamic world: Iran, Iraq, Libya. Sometimes the Pakistanis are thrown in with this lot and, more recently North Korea.

Of course, China is an example of an apparently irrational nation that was expected to behave aggressively once it obtained the bomb. In fact, it did not behave aggressively at all, but was quite rational. China found that the bomb, by deterring a Soviet attack, or at least contributing to such deterrence, allowed it to take a prudent course. Indeed, nations that lack a nuclear device may feel the need for a display of bravado, which is obviated by possession.

Under what circumstances are nuclear weapons considered a threat in the hands of "lunatics"? Consider the example of Saddam Hussein. Now, Saddam is certainly a ruthless and brutal man, but he is not a lunatic. Hussein fought a long, costly war with Iran to become the preeminent power in the region, clearly intending to recoup his losses in that war by conquering the weak and wealthy states on the western shore of the Persian Gulf. He began by conquering a small, weak, but very rich neighbor—Kuwait—and toyed with the possibility of seizing Saudi Arabia. Hussein calculated that the United States would not intervene. He was wrong, but not irrational. Indeed, he maneuvered his way to surviving a massive military defeat.

It is difficult to perceive any hint of irrationality in Saddam's behavior. Certainly one can see miscalculation, poor intelligence, and adventurism. But there is not the slightest sign of lunacy or fanaticism, which one would take to be a willingness to pursue to the end, regardless of consequence. Therefore, the vision of a leader who unleashes nuclear weapons for the sheer idea of doing so is rather hard to fathom. Men become political leaders because they have a large measure of prudence, understood as the will to survive. Such men are rarely suicidal. Still, the specter of some demented leader launching a nuclear attack at New York is a possibility, but it is not one on which anyone's strategic policy ought to turn. Though less concern is expressed over "rational nations" such as India, Brazil, or Israel, one could argue that they might find the use of nuclear weapons to be in their self-interest and, thus, pose more of a potential risk.

What then are the uses of nuclear weapons? Obviously, smaller, tactical nuclear weapons will have a place on the battlefield, although even here the mystery of nonuse in Vietnam and Afghanistan, where they would clearly have served a military purpose, remains to be answered. But, by themselves, they will not provide a nation with great power, any more than will strategic nuclear weapons. We are again reminded of the collapse of Soviet power regardless of its vast array of nuclear weapons, both tactical and strategic.

It would appear that the Franco-Israeli strategy provides the best clue to the use of nuclear weapons. For France, the key was the retention of its existence as a nation-state, not subsumed in the American alliance system. As de Gaulle put it:

Consequently, it is evident that what is necessary and what we must achieve during the coming years is a force capable of acting exclusively on our behalf, a force which has been conveniently called the *force de frappe* susceptible to deployment anywhere at any time. It goes without saying that the basis for such a force will be atomic armament—whether we manufacture it or buy it—but one which belongs to us; and since France could be destroyed on occasion from any point in the world, it is necessary that our force be designed so that it can act anywhere on the earth.[20]

For the French, the function of nuclear weapons was to preserve the integrity of the nation, to make certain that fundamental issues of war and peace affecting France remained in French hands. In so doing, France could pursue the interests of a normal nation-state, beneath the nuclear level.

In this sense, nuclear weaponry has only a negative function—to prevent the obliteration of the French nation-state under the nuclear pressure of other nations, friends or enemies. On its own, France cannot achieve any political goals such as compelling another nation to succumb to French interests. It could not win the war in Algeria nor prevent the emergence of a powerful Germany. However, merely by possessing nuclear weapons, France can prevent any other nation from forcing it to act against its own interest.

The number of weapons necessary to achieve this end is quite small. De Gaulle did not need to totally destroy his enemy. All he needed, in his famous phrase, was the ability to tear an enemy's arm off, under the reasonable assumption that most enemies would rather not have an arm torn off. In an all-out attack on France by the Soviet Union, for example, a small force, consisting of a few land-based bombers and nuclear submarines, would do significant damage. So any Soviet attack on Western Europe might halt on the Rhine.

This is similar to the policy adopted by Israel.[21] Obviously, the Israeli nuclear force was not capable of deterring an Arab attack in 1973, when, fully aware of Israel's nuclear capability, Egypt and Syria attacked. According to reports, Israel considered a nuclear response during the first twenty-four hours of the war, when it was not clear to them that the Golan could be held and that Syrian forces would not enter into the Galilee.

Two facts are relevant here. First, the Israeli nuclear deterrent was not enough to stop the Arab attack. Second, Arab armies, with a full head of steam and scenting victory, might not have been halted even by the direct threat of nuclear attacks on their capitals. The paradox of nuclear weapons is that unless they are credible as a deterrent, they might as well not be applied. Israel's purpose was to be subtly ferocious *before* the outbreak of war. Subtle in the sense of not flaunting the weapons too much; ferocious in making their presence absolutely clear.

The Franco-Israeli strategy is to employ nuclear weapons as a means of deterring an attack on the nation's very existence. For France, nuclear weapons were a means of limiting a Soviet adventure in Western Europe. For the Israelis, it was also a means of preserving the nation, not only by threatening the Arabs but by threatening their strategic backer, the Soviet Union, in extremis. North Korea's strategy seems to have taken its bearings from this. Isolated from both of its patrons after the end of the Cold War, suffering severe economic problems, the North Korean regime saw nuclear weapons as a guarantor that outside powers would think twice before seeking to undermine communist rule. In other words, nuclear power was seen as a lever for guaranteeing defensive interests. It was never tested for such a purpose in the French case, and it was insufficient to guarantee security in the Israeli case.

And so it seems that the best use of nuclear weapons is in limiting the nuclear and, more important, conventional threat to one's nation, under the untested assumption that no rational leader would absorb nuclear punishment to overwhelm an enemy. In this role, nuclear weapons would appear to be a limiter of war, making total war—in the sense of a struggle to the finish between two nations—impossible. Certainly such wars are rare indeed. World War I, as brutal as it was, was not a fight to the finish, and World War II was only apparently such a fight. All of the combatants continue to survive as independent states, tracing their lineage to the wartime regimes.

Nuclear weapons have not been used in fifty years because they are fairly useless. All weapons must relate to strategy, and all strategy must relate to politics. For fifty years there has been no connection between nuclear weapons and politics. For fifty years we have lived under the threat of a nuclear war that has never materialized. During this period, in not a single case did the existence of nuclear weapons influence the flow of conflicts.

As the new epoch opens, the preeminent role of nuclear weapons in both the making of strategy and the structure of the international system will decline dramatically. Wars in Korea, Vietnam, and Afghanistan mingle with images of a nuclear-potent, politically impotent Soviet Union, losing its international influence even before it collapsed. There will still be a place for nuclear weapons, as a block against other nuclear weapons and as a putative guarantor of the existence of the nation. But such a force is both small and of marginal politico-military importance. Of far greater significance will be the evolution of real military power, previously known as conventional military force. And for the United States, no form of conventional warfare is more important than naval warfare, for control of the sea is the foundation of American national security.

4

Fundamentals

THE SEA IN AMERICAN STRATEGY

Nuclear weapons have been called strategic weapons. This has been a misnomer. The fate of the United States has, since its founding, rested in the hands of the U.S. Navy, its true strategic force. For the United States, as for Britain before it, control of the sea was the foundation of the national interest. Controlling the sea meant that America was secure from attack. Control of the sea meant that goods would flow to and from the United States, assuring prosperity. Control of the sea made America into a global power. Losing control of the sea could mean that the United States would be torn apart. Thus, the Navy has been the key strategic weapon in the American arsenal since its founding.

It is no accident, therefore, that the U.S. Navy produced America's foremost strategist, Alfred Thayer Mahan. Nor, paradoxically, is it an accident that the U.S. Navy also produced America's narrowest advocate of technical proficiency, Hyman Rickover. A warship is the epitome of technical complexity. It demands absolute attention. At the same time, naval operations constantly interact with the fundamental interests of other maritime nations. Thus the study of naval warfare continuously moves between the broadest, global considerations and the narrowest technical matters. On land, vast battles can take place without overthrowing the political order. At sea, where a nation's maritime power is condensed into a few hundred or even a few dozen ships, the outcome of an engagement involving a handful of vessels can undermine the geopolitical foundations of a nation or elevate it to preeminence. The United States, at the height of its military expansion in the 1980s, had tens of thousands of tanks but only about six hundred ships. Thus, the loss of each ship is of much greater importance than the loss of a tank. Not only is the cost of each warship considerably larger than the cost of any land weapon, each is, in effect, responsible for a far greater portion of the earth's surface. Thus the smallest defect can have enormous consequences.

The defeat of the Spanish Armada, the battles of Trafalgar and Midway, were tiny engagements compared to the battles raging on land. They turned frequently on technical issues, arcane matters of engineering or even luck. With naval power consisting of such a small number of units, the failure of any system on any ship could undermine the fleet. Thus, good naval officers are less likely to focus on the strategic importance of their mission than on engineering.

Thus, we have a paradox. The U.S. Navy is the most strategically significant part of the American armed force yet it is usually the one that most neglects the broader strategic issues of which it is a central part. Assuming that the U.S. Navy does its job, there will be no land battles in North America in the foreseeable future. If the U.S. Navy holds the line, a lost battle in Eurasia would be painful but not devastating. The ability of the Navy to protect U.S. shipping or interdict enemy shipping is of fundamental importance to U.S. grand strategy. As Admiral Mahan understood, control of the sea must be the central concern of American strategists.

Strategic doctrine comes hard to the Navy. Indeed, operational doctrine in general has been intentionally neglected by the Navy, under the assumption that, in battle, tactics derive from the circumstance at the moment, rather than from some sort of theoretical model. It has been only in recent years, with the new emphasis on joint operations in the American defense community, that the Navy has paid attention to doctrine in an attempt to conform to the practices of the other services. The basic strategies of the U.S. Navy changed only once during the Cold War. During the 1980s, John Lehmann, then secretary of the navy, argued that rather than forming a line in the North Atlantic to hold out the

Soviets, the Navy should go north of the Greenland–Iceland–United Kingdom gap, to engage them before they broke into the Atlantic. This relatively modest change in strategy—it shifted to a question of where the Soviets would be contained, not whether they would be contained—caused tremendous uproar within and without the Navy.

There is good reason for such conservatism. Naval strategy is rooted in the fundamental, unalterable geography of the globe. While strategically significant changes in trade routes emerge over generations, the configurations of coasts and waterways change not at all. Thus, for example, ever since the emergence of the United States as a global power, American strategy has depended on the ability to transport troops and supplies from the east coast of the United States to Western Europe. World War I, World War II, and planning for World War III all revolved around keeping sea-lanes open. Throughout the century, the threat of closure from submarines and, later, from aircraft, was a central strategic concern that drove the operational and tactical developments in the Navy.

The strategic problem remained constant; technology evolved. Antisubmarine warfare destroyers were supplemented after World War II by hunter-killer submarines. The growing air threat meant that the primary screen in the North Atlantic would consist of carrier battle groups rather than battleship divisions. The replacement of the German threat by the Soviet threat meant that the screen would form around the Greenland–Iceland–United Kingdom gap rather than the Scottish-Norwegian gap. But the essential problem of the Atlantic remained the same, as did other issues driven by geography. As American power grew, its fleets expanded into the Indian Ocean and the Mediterranean. As technology evolved, U.S. submarines haunted the Arctic. Strategy changed along with technology, but always slowly. The underlying battlefield was eternal, and trade patterns changed as glacially as the social and economic patterns they were based on.

In spite of the reluctance to rethink or even to contemplate rethinking basic truths and broader strategic questions, the end of the Cold War forced the Navy to reconsider its role. The resulting doctrine, published in 1992, entitled "From the Sea," was in fact a fundamental break with the past and an important event in military history. "From the Sea" maintained that the end of the Cold War meant that the United States no longer faced a challenge from the Soviet Union over its control of the blue water—the vast ocean tracts through which global sea-lanes move. Rather, it argued, the problem facing the United States was one of "littoral warfare," warfare in the coastal areas of nations—particularly

Eurasian nations. Moreover it urged that rather than focus on confronting enemy fleets or protecting our own, the U.S. Navy should be concerned with projecting forces into the edges of Eurasia.

"From the Sea" implicitly argues that the central strategic challenges facing the United States are:

- To maintain the balance of power in Eurasia to assure that no Eurasian power could attempt to impose hegemony in the region or aspire to challenge U.S. global domination.
- To police the American empire to assure that the peace of the new, militarily unipolar world is not disturbed by internal unrest or regional imperialists. (A number of operations along this line have already taken place, in the Persian Gulf, Liberia, Somalia, Haiti, and, in a limited way, Bosnia. The point of all these "peacekeeping" missions has been to assure that the order necessary for imperial prosperity is maintained. In this function, the United States is no different from the countless procession of empires that have preceded it.)

The U.S. Navy is the rock on which American global power rests. Maintaining both the balance of power and world peace requires that the United States send its forces to foreign shores. Because the U.S. Navy will, of necessity, transport troops and supply them till they are there, U.S. warships will come close to the enemy shore. Now the issue is, simply put, whether an American warship can approach an enemy shore in the coming generation without being sunk.

Given the global nature of the battlefield and the fact that engagements can swirl over thousands of miles, and given the relatively small numbers of weapons platforms involved, the evolution of naval tactics is intimately tied up with the evolution of naval strategy. We must also consider in detail the strategic imperatives guiding the U.S. Navy, the technical issues facing the Navy in its execution of U.S. strategy, and whether any alternative strategies might be followed.

The Shaping of U.S. Naval Strategy

Ever since the United States completed its occupation and domination of the North American continent in the late nineteenth century, American security has depended on two factors: the maintenance of the Eurasian balance of power and the freedom to navigate the two great oceans separating it from Eurasia. With absolute sea control in 1944, the United States was able to launch simultaneous amphibious assaults against both European and Asian enemies. Were hostile powers able to dominate those oceans, an amphibious operation against the United States, at least as decisive as the one launched by it during World War II, would not be inconceivable.

U.S. grand strategy has, therefore, been predicated on achieving and maintaining control of at least the North Atlantic and of the Pacific as far west as Hawaii. Indeed, part of the history of the twentieth century has been the story of America's success in forcing Britain out of the Caribbean basin and the western Atlantic, and finally, during and after World War II, of placing the Royal Navy under the control of a U.S.-dominated alliance.

But the best strategy for naval supremacy is not to defeat rival navies in battle or to win a naval arms race. It is to have potential enemies so engrossed by land-based threats that they have no time to construct fleets. This permits the cheapest and most effective form of naval supremacy—supremacy by default. This was the foundation of British strategic policy from the Spanish Armada to World War II—to manipulate the European balance of power through political, economic, and minimal military means so as to maximize regional tension and mistrust. Not incidentally, Europe enjoyed a century of peace between the fall of Napoleon and World War I because of the success of Britain's policy.

Britain's naval preeminence was undermined by the rise of the United States. To be more precise, the success of the United States in achieving control over North America meant that it could divert resources from protecting its borders to patrolling its coasts and oceans. The inability of the British (in spite of several attempts) to play the balance-of-power game in North America as it was doing in Europe meant that the United States could harness its vast resources and challenge Britain's naval superiority. As Britain's control over the European balance of power declined, and it was drawn into a series of catastrophic land conflicts in Europe, it lost its ability to compete with the U.S. Navy. By the end of the Second World War, the United States had replaced Britain as the world's preeminent naval power.

The American strategic problem is the British problem multiplied many times over. The British were concerned with maintaining a balance of power in Europe. The American concern is with maintaining a balance of power in Eurasia as a whole and preventing the emergence of a Eurasian force able to project its will transoceanically and, thereby, threaten the security of American trade and territory.

Since 1917, the United States has consistently followed a policy designed to prevent the emergence of a rival power, preferably through the maintenance of a balance of power operating under its own dynamics or manipulated by the United States or, failing that, through direct political or military intervention.

- 1917—With the Anglo-French alliance under heavy pressure from Germany as Russia collapsed, the United States intervened to prevent a German victory and a German threat to Atlantic sea-lanes.
- 1919—At Versailles, the United States prevented France from dismembering Germany, thereby preventing France's emergence as a hegemonic land power and limiting Britain's ability to increase its naval forces.
- 1919–40—The United States permitted the internal dynamic of European politics to maintain the balance of power, minimizing American risk by its policy of isolationism—a policy that is fully rational when the balance of power is stable. In Asia, the United States intervened subtly and increasingly against Japanese expansionist interests in China.
- 1942—Following the collapse of the European balance of power after Germany's defeat of France, the United States intervened on the side of the weaker powers in both Europe and Asia, thereby preventing the victory of either Germany or Japan.
- 1945–91—After World War II, the United States created a system of alliances intended to encircle and contain the USSR, preventing its hegemonization of Eurasia.

In each of these cases, American foreign policy remained operationally consistent:

- To rely, as far as possible, on internal regional dynamics to maintain the regional balance of power. Improperly labeled isolationism, this is merely a recognition that under optimal circumstances, forces in Eurasia are sufficiently balanced so as to allow the United States to avoid the risk and burden of intervention.
- Where isolation fails, to use nonmilitary instruments, such as trade, aid, political support, and such, to support the weaker side, without entering into permanent alliance. At Versailles in 1919, the United States threw its weight behind Germany with-

out entering into alliance. In 1972, the United States supported China against the Soviet Union, without entanglements.

- Where informal support proves insufficient, to enter into formal alliances with powers resisting foreign hegemony. Ideally, the alliance should be based on the Eurasian power contributing the bulk of forces, and taking the maximal risk, while the United States limits itself to financial assistance, military and technical transfers, and expertise. The classic examples of this strategy are lend-lease to the Soviet Union in World War II and aid to Israel.

- Where essential to stabilize the regional situation, to intervene on the periphery of Eurasia using relatively small forces. Examples include Lebanon in 1959 and 1982, Korea in 1950, and Vietnam in 1964. These latter two interventions are considered major wars in American history. But when the relative size of forces involved is considered and when the potential U.S. deployment is measured against what was actually deployed, they remained relatively low-cost, low-risk operations designed to stabilize the balance of power rather than impose American hegemony.

- Where a major national effort is required to restore the balance of power, to operate only within the context of an alliance in which allies have much more at stake and much less room for maneuver than the United States and in which they can bear the primary burden of war. The classic case was World War II, where allied casualties dwarfed American casualties. One might add that the outcome of World War II was wholly beneficial to the United States. Germany and Japan were decimated, American allies were economically wounded. The United States emerged as the sole major economic power, fueling a generation of extraordinary economic growth.

- Under no circumstances should the United States bear the brunt of a major Eurasian war. The nearest we have come to going too far was during the Cold War, when the power of the Soviet military pushed the United States to station substantial forces along Europe's periphery, which would have been at risk in the event of all-out war. Of course, it should be remembered that while the United States had some forces at risk, the Germans and other allies had their entire populations in the line of fire. Thus, the Cold War maintained the American strategy of minimal exposure while maintaining the balance of power.

To reiterate: The fundamental principle of American grand strategy over the past century has been to manipulate the Eurasian balance of power to prevent the emergence of any hegemonic powers able to harness Eurasia's vast resources and challenge American naval supremacy. The implementation of this strategy turned on the twin principles of economy of force and indirection. Economy of force was realized by using allies to bear the strategic visions and the brunt of combat. Indirection involved never attacking an enemy frontally, only when he had been sufficiently weakened by internal or external opposition, and relying on economic power to produce the preconditions of victory rather than matching force to force prematurely.

The American victory over the Soviet Union illustrated both the strategy and the operational principles perfectly. The United States maintained the balance of power in Eurasia through a maximum economy of force. Rather than mobilizing vast numbers of American troops, the United States relied on nuclear weapons to increase the risk to Soviet interests in any direct attack, while using allied forces for the bulk of ground-based power. The United States supplemented this force with minimal ground formations, massive tactical air support in the various theaters, and an overwhelming naval presence able to keep the sea-lanes open for reinforcement and resupply, while also threatening maritime counteroffensives against Soviet regions.

Soviet attempts to break free of encirclement, as in Vietnam or Korea, were met with minimal force and political countermaneuver (the entente with China). Meanwhile, the United States relied on the inherent superiority of its maritime-based economic system to seduce potential Soviet allies into the American fold and to impose an arms race on the Soviet Union that it could not afford to win or lose. In the end, the strategy proved fabulously successful.

But in many ways, this strategy was deceptive. In refusing to engage Soviet power directly, by relying on an indirect strategy, the United States appeared, to the unaided eye, to lack the will to resist Soviet aggression. What was happening in fact was that the United States was permitting the Soviets to dissipate their strength in attacks on inessential targets while maintaining a policy of encirclement and economy of force. The United States had to walk a fine line between the appearance of weakness—an appearance that could have deadly consequences—and showing its hand. Thus, the United States was forced into occasional unfortunate episodes—such as the Vietnam War, where major resources were devoted to peripheral issues—simply so as not to appear weak. The willingness to intervene in Vietnam was not the measure of American commitment to take on the Soviet Union. Rather, it marked the U.S. desire to deter the Soviet Union from dominating Eurasia, while using minimal power to guarantee its allies against attack.

Old Strategies, New Geopolitics

The collapse of the Soviet Union has been of enormous importance to the United States because it has meant that for the first time since the American emergence as a global power, no one is threatening to impose hegemony on Eurasia. In part, this is a blessing to the United States, since it means that no single power is capable of utilizing Eurasia's resources to challenge American maritime preeminence. But while this new configuration eliminates a direct threat to American dominance, it does not, by itself, secure it.

The basic American interest in Eurasia is a stable balance of power that forces regional adversaries to devote their resources to defending themselves against one another—making it impossible for them to devote resources to challenging American interests. While the collapse of the Soviet Union has alleviated the threat that a single power will upset the political status quo and usurp Eurasia, it also opens the door to lesser powers to secure vital regions of Eurasia without opposition. The creation of stable, regional hegemons poses the same basic threat to the United States that a single Eurasian power did: free from challenges from other land powers, they will be in a position to challenge American preeminence at sea.

The current fragmentation of Eurasia appears to have insulated key powers and given them room for maneuver within a relatively narrow range. We have already seen Iraq's attempt to dominate the Persian Gulf. Having upset the regional balance of power with the defeat of Iran and enjoying the freedom provided by the breakup of the Eurasian power balance, Iraq sought to create a large, wealthy, geopolitically secure empire by moving down the western shore of the Persian Gulf. This would have allowed it to use oil revenues to impose an arms race on Iran, which Iran could not possibly win. Iran would have been forced to seek outside interventions and, failing that, accommodate itself to Iraqi preeminence. Then, protected in the west by the Jordanian desert, in the north by rugged Turkish mountains, and in the east by a prostrate Iran, Iraq would have been free to pursue broader regional interests in the Indian Ocean basin.

Following Iraq's defeat by the United States, Iran has sought to create a regional hegemony. Freed from an Iraqi threat in the west, secure in the north because of the collapse of the Soviet Union, facing a marginal threat from the Kurds, it has aggressively developed its naval forces, until they pose real problems for the United States. Indeed, it was to prevent precisely this scenario that the United States refused to topple Saddam's government and dismember Iraq. Even so, the United States was unable to avert the rise of Iranian maritime power, and it may have to confront this problem directly in the coming years.

The issue is not merely oil, although the presence of petroleum reserves in the region certainly raises the stakes. Leaving oil aside, it is not in the American interest for any single regional power to consolidate a large area of Eurasia. Where such consolidation exists, the United States seeks to counterbalance it with another regional force, but, too often, this is not possible due to geopolitical isolation. For example, the collapse of the Soviet Union has insulated China regionally. Similarly, the Persian Gulf region is fairly insulated. It is difficult for Eurasian land powers to cross the mountains and deserts that surround the region. Thus a potential hegemon may come from within the region itself or from the sea. Once established, however, a Persian Gulf power is an inherently maritime power. History abounds with examples of nations that have projected themselves into the Indian Ocean. The idea of a wealthy, secure regional power turning its attention to the oceans is something that is clearly not in the American interest.

Potentially insulated maritime powers can be found all over Eurasia:

- A four-power system in the northwest Pacific—Japan, China, Korea, Russia—becomes an insulated region should any one power emerge as dominant.
- The collapse of China would make Vietnam a dominant, insulated hegemon in Southeast Asia.
- The collapse of Pakistan in the west and China in the north would create a fully insulated regional power out of India. Indeed, India already enjoys a substantial insulation and is emerging as a maritime power in South Asia.
- A further weakening of Russia would create the opportunity for Turkey's emergence as a regional power.
- Both Iran and Iraq threaten to emerge as insulated hegemons.
- Peace in the Middle East could upset the tenuous balance of power in the eastern Mediterranean, allowing Israel to become a regional center of power.

▪ The collapse of the Soviet Union and the withdrawal of the United States from Europe open the door for a traditional competition between Atlantic Europe and Germany, the outcome of which could well create an insulated hegemon.

Throughout Eurasia, therefore, in these and possibly other cases, the regional balance of power prevents the emergence of insulated powers able to marshal scarce resources for maritime expansion. In the post–Cold War world, maintaining these regional power balances is the foundation of U.S. foreign policy. Ideally, the United States can achieve its geopolitical ends without direct intervention. Yet, throughout the twentieth century, this has not always been possible. The United States has had to project its force into Eurasia on numerous occasions in an effort to roll back or contain an ascendant Eurasian nation. At times, as with D day in June 1944, this entailed a lunge at the very heart of a Eurasian power. At other times, as in Korea or Vietnam, it was part of an attempt to contain a power using a sustained force projection. At other times, as in Lebanon in 1959 or Grenada in 1982, the invasion was a small, marginal effort aimed at maintaining the stability of a particular region not directly involved with the major confrontation.

In the long run, the United States must maintain a dual capability: to confront contenders for Eurasian hegemony and to contain, limit, or destroy contenders for regional hegemony. In one sense, this is the same capability. The only difference is the extent and sustainability of the effort.

In containing Iraq, which as we have seen was not a contender for Eurasian hegemony, the United States had to be in a position to transport troops and matériel from the continental United States to a particular location in Eurasia, land those troops even in the face of enemy hostility, engage and destroy the enemy's armed forces to the extent that geopolitical reality required, and sustain its troops there for a short period of time—months rather than years. In Vietnam, on the other hand, sustainment lasted for years. However, Vietnam and Iraq were similar in this respect—the bulk of the operations were carried out without significantly expanding the standing forces of the United States, as had been the case in World War I and World War II.

In a conflict with a Eurasian hegemon, as in World War I and World War II, U.S. forces would have to be expanded dramatically, and the Navy would have to guarantee the sea-lanes (as well as the transport capacity) for an extraordinarily larger force. It would also need the ability to carry out amphibious operations—to kick in the door—if necessary at the point of intervention and to supply forces for sustained combat in the area of conflict following the intervention.

Control of the sea, amphibious warfare, and logistical systems appear to be radically different problems. In fact, they are part of the same geopolitical system—the ability to influence the balance of power in Eurasia. Should any part of this system be lacking, should any part be vulnerable or insufficient for the mission, then the entire geopolitical imperative collapses.

The dilemma faced by the United States is that the evolution of weapons has raised the possibility that none of these three missions can be carried out by the U.S. Navy in the near future. All three missions depend on surface vessels—from antisubmarine warfare to intercontinental transport to landing troops on hostile beaches. Surface vessels are, of course, extremely vulnerable to a variety of explosives—a fact amply demonstrated at Pearl Harbor and Midway. Today, the threat of destruction no longer comes from iron bombs strewn with relative randomness by aircraft but from self-guiding missiles launched outside the ship's range and able to guide themselves to their targets.

With the advent of these missiles, the ability of a surface vessel to operate, and of the U.S. Navy to carry out its mission, becomes increasingly dubious. If this is true, then the geopolitics of the twenty-first century will be far different from those of the twentieth century, as U.S. isolation will be a matter of technology rather than ideology. But the truth of this matter depends most of all on technology, and therefore we must address ourselves carefully to the technological question posed by precision-guided munitions at sea.

From Strategy to Operations

The ability of the United States to influence events in Eurasian regions depends, obviously, on control of strategic sea-lanes connecting the United States and Eurasia. Indeed, the entire reason for an American presence in Eurasia is because of the need to control strategic sea-lanes. Ends and means are the same, as they ought to be in any effective strategy.

Yet—and this is the critical politico-military fact of our time—there is no immediate challenge to American maritime control. At this moment, no Eurasian nation deploys a fleet able to challenge American control even regionally, let alone globally. This is not to say that some nations aren't contemplating such challenges. India has an ambitious program under way to create a powerful regional navy, including aircraft carriers, in the Indian Ocean.[1] China appears to be on the verge of a powerful shipbuilding program as well, which would include the construction of a carrier force. In China's case, as Yihong Zhang, writing in the *International Defense Review,* put it: "Partial resolution of the eastern border conflict with Russia has enabled the People's Republic of China to shift its front line from the north to the south and its interests from the interior to the 'blue waters' of the South and East China seas."[2]

This crisp geopolitical analysis points out one of the dangers faced by the United States in the event that insulation spreads in Eurasia. Yet there is something futile in both India's and China's efforts. Even without threats from their land neighbors, the costs involved in creating a true blue-water navy that could threaten American interests far outstrips each country's respective resources. No matter how ambitious their dreams or how legitimate their geopolitical concerns, neither the Indians nor the Chinese have the wherewithal to create the carrier battle groups that would be required to challenge the American maritime hegemony. Since neither can hope to match American power at sea in conventional forces, each hopes to use its naval power to create coastal buffer zones against American encroachment and thus affect the regional balance of power. Yet even this more modest goal is beyond their means, as is the cost of matching weapon system against weapon system. The industrial and intellectual capacity of the U.S. Navy dwarfs Chinese or Indian capabilities and will continue to do so for the next generation.[3]

Japan, on the other hand, has the intellectual and industrial capability to challenge American naval hegemony. Indeed, alone among the great powers, Japan has a long-term interest in doing so.[4] Geopolitically, Japan is an anomaly. A Eurasian power separated by a narrow channel from the mainland, Japan, like England, operates in a very special way. Because of its geography, it is either a totally enclosed nation, without foreign contacts, or it is a naval power. Unlike England, however, Japan is almost completely devoid of industrial resources. This means that its ability to manipulate the political balance on mainland Asia is much more limited than Britain's ability to manipulate the European balance in the nineteenth century. This fact further intensifies Japan's interest in becoming a maritime power—an even stronger imperative than England's.

Japan's pursuit of maritime power is, of course, conditioned by political and economic realities. It is dependent on the United States for trade, and however aggressive Japan might be, it is loathe to challenge

American power. Still, Japan has emerged as the greatest naval power in Asia, second only to the United States in the region.[5] It is thus instructive to consider Japan's approach to maritime issues.

The United States has based its contemporary naval power on the carrier battle group, an effective but expensive weapons system. Japan has chosen what appears to be a more modest approach, building destroyers. Historically, destroyers, which are smaller ships, have had as their primary mission escorting merchantmen and defending them against enemy submarines. Given Japan's need to import raw materials from around the world by sea, honing this convoying capability appears prudent.

Of course the Japanese approach to the problem of convoying may involve more than mere prudence. We find a hint of it in their decision to mount the U.S.-built Aegis defense system on some of their destroyers. Aegis is a complex radar system designed to protect carrier battle groups against air-launched missiles. It is extremely expensive, and no nation other than Japan has chosen to use it.

The Japanese appear to have grasped a critically important point—that large surface vessels are extremely vulnerable to precision-guided munitions. In developing a more dispersed force structure—in which the platforms are increased in number and decreased in size, with key vessels armed with sophisticated antiair systems—they have clearly located the problem. The more targets there are and the smaller each target is, the greater the probability that at least some will survive to protect the convoy and carry out their antisubmarine warfare mission.

For maritime trading powers such as the United States and Japan, the advent of antiship precision-guided munitions represents both a threat and a challenge. These countries cannot abandon the seas, so each must find a solution to the problem. For nations less dependent on the sea, or nations much weaker than the United States and Japan, these same missiles represent a tremendous opportunity. At the very least, for minimal cost, secondary and tertiary powers that own these missiles can impose risks on maritime trading nations that cannot be borne—even if they cannot deny them use of the seas altogether.

The risk of having a cruiser or a carrier hit by a missile means that a commander must ask how much risk he is willing to take for each mission with a $20 billion carrier battle group.[6] Effective antiship missiles can impose a potential risk that makes many, if not most, operations too risky to undertake. Similarly, even if antiship missiles cannot sink all or even most merchantmen, increased insurance costs can easily destroy the economic viability of trade—which is the same as stopping it altogether. Lloyd's of London is as prudent a risk manager as the chief of naval operations.

Antiship missiles are, therefore, a low-cost, highly effective means for weaker powers to deny sea-lane control to great powers. And no nation is as vulnerable to this maneuver as the United States. As the world's leading naval power, it is exposed globally. More important, with its naval strategy built around a small number of extremely large, powerful, and expensive ships organized into battle groups, the threat posed by these missiles is unusually high. One or two leaking through the Aegis shield could cause havoc, rendering the battle group inoperative during a crisis.

As we have seen, secondary powers such as China and India, and tertiary powers such as Iran and Iraq, are at a distinct disadvantage if control of the sea is understood to derive from traditional surface vessels and aircraft carriers. None of them has the wherewithal to deploy forces that can challenge the United States even regionally. But, as we have suggested, challenging the United States would not necessarily require this sort of deployment. There may now be other weapons at hand that could drive the U.S. Navy out of a region at prices easily affordable to even tertiary powers.

Once the United States or Russia transfers missiles to countries like Israel or India, the technology can be adapted and expanded and resold to other countries like Iran or Iraq. R&D is expensive; production is not. There is a very real threat that Chinese Silkworm missiles, mimicking U.S. and Soviet technology, will be sold to the Iranians and threaten access to the Persian Gulf. The importance of rooting out missiles hidden on land—the problem faced by the Scud hunters during Desert Storm—is doubled when the stakes are not a few city blocks but the maritime choke points that the world's oil flows through.

The threat posed by missiles is compounded by the diffusion of submarines. The Iranian Navy, for example, has acquired several Russian diesel-powered submarines that are quiet to operate and difficult to detect by sonar. In the event of a conflict inside the Strait of Hormuz, the United States would be unwilling to move its carrier battle groups inside the narrow strait unless they had the absolute assurance that no hostile submarines were in position to launch a torpedo or missile. Barring absolute certainty, which Russian submarines may well deny the U.S. Navy, the battle groups would remain outside the strait, making air strikes deep into Iraq or northern Iraq difficult, if not impossible. Thus, in the event of conflict, Iranian submarines, along with judiciously strewn mines, would negate the aircraft carrier.[7]

The surface vessel, which, in various forms, has been the mainstay of naval warfare since the beginning of time, is under tremendous pressure from two directions: torpedoes and rockets launched from submarines and cruise missiles and rockets launched from the air. Potentially extremely intelligent, able to target the most vulnerable parts of a vessel, able to travel longer and longer distances, these projectiles promise to become a new form of sea control. At the very least, they will become a dramatically new way to influence the balance of power at sea.

The situation at sea is analogous to the evolution of battle on land and in the air. As accuracy increases, the number of projectiles that have to be fired to hit a vessel with dramatically decline. Antimissile missiles will be more of an issue than finding and hitting targets. As important, with the advent of long-range missiles, the need to bring naval vessels into close proximity with each other will disappear.

This process actually began with the introduction of aircraft carriers. The Battle of Coral Sea, in 1942, was the first sea battle in history in which no ship saw the other. Thereafter, throughout the great naval struggle between the United States and Japan, the range limits imposed by guns were replaced by range and time limits of aircraft—it took longer for a plane than for a shell to reach its target. With the beginning of midair refueling after the war, the ranges increased again. One should note that accuracy had declined as bombs and torpedoes replaced guns. Naval aircraft were no match for naval gunfire in terms of accuracy and time—only in range.

With the introduction of high-speed precision-guided munitions, increases in range will not necessarily mean increases in time. More important, they will not mean decreases in accuracy. As the range of projectiles increases, the high seas come into the range of land-based munitions. Where ships were needed to carry guns within range of the enemy and planes were needed to bring bombs and torpedoes, it will become possible for shore-based missiles to bring the entire ocean into range—and every fleet and merchantman into danger.

At that point, the naval battle will no longer be the confrontation of ship and ship, but rather, the confrontation of ship and missile. It will be uneven combat, as the stationary ship (stationary relative to a supersonic or hypersonic missile) confronts a tiny, extremely fast, agile, intelligent projectile attacking from any and all directions. Under these circumstances, it is not at all clear that control of the sea will rest with any power. Seaborne trade—that is, the bulk of international trade—will be hostage to any nation with the wherewithal to field a cluster of antiship missiles. Many nations will have more to gain from disrupting global trade patterns than from maintaining them. One is reminded of the days of the Roman Empire's decline, when the Roman roads were at the mercy of brigands whose extortion brought trade to a halt and ushered in the Dark Ages.

We already had a taste of this in the 1980s, when both Iran and Iraq found it in their interest to use missiles to attack oil tankers. The American response to this relatively primitive threat was to organize convoys of tankers protected by Aegis cruisers—one of which was hit by an Iraqi missile (USS *Stark*), another of which shot down an Iranian passenger jet (USS *Vincennes*). The tanker wars were a prelude to a new sort of naval warfare. The missiles fired by aircraft do not have to be launched from the air. They have cousins that can also be launched from land. As range increases, ships on the high seas will be in danger, as well as in coastal waters and straits.

This evolution will be particularly disturbing to the United States, which has built its global power on a ship-based navy. As those ships become more and more vulnerable, and as they spend more and more of their resources merely defending themselves, the ability of the U.S. Navy to control events on the high seas and deliver U.S. forces anywhere in the world will decline. No one will replace the U.S. Navy. No one else will be able to exert operational control over the seas. Rather, everyone will have a veto power over who will use the sea.

As precision-strike technology diffuses and the cost and accessibility of missiles allows them to become commonplace, the price of defending warships will soar. The cost of putting missile defense systems on merchantmen will make them completely uneconomical.

Such apocalyptic thoughts about the consequences of naval precision-guided weapons need not be correct for us to be able to state unequivocally that these weapons pose a substantial problem for the U.S. Navy and for U.S. foreign policy. If, as we have argued, U.S. policy requires that we maintain the Eurasian balance of power, then the United States must be able to continue intervening in Eurasia. These interventions will be of three basic varieties:

- Interventions other than war, in which the primary motivation is either humanitarian, to stabilize a friendly regime, or a peacekeeping operation designed to halt a conflict whose continuation is not in the interest of the United States. Such operations are less likely in Eurasia than elsewhere because, as we have seen in the former Yugoslavia, Eurasian force structures make such interventions costly and potentially risky. Somalia and Haiti are the prototype of operations other than war. Such operations, occurring in peripheral, isolated nations where the United States can bring overwhelming power to bear to achieve limited ends, do not tax the logistical capabilities of the United States.

- Small-scale interventions of relatively short duration intended to stabilize situations or deter attacks by other powers. Lebanon in 1959 is the classical example in Eurasia of such a case. Size and duration limits the stress on American logistical capabilities.

- Large-scale interventions at the divisional level, intended to engage enemy forces directly in middle- or high-intensity conflict. Examples of this include both World Wars, Korea, Vietnam, and Desert Storm.

This last form of warfare is the most rare, the most important, and the most dangerous. While it does not tend to occur more frequently than once a generation, its outcome defines both the generation and the balance of power in Eurasia—or at least of a region in Eurasia. American planners have tended to downplay the likelihood of this type of conflict in the near future, focusing instead on low-intensity confrontations and peacekeeping operations outside of Europe. This is a fundamental mistake. Maintaining the balance of power in Eurasia is critical to American interests, and large-scale conflicts are the fundamental mission of the American armed forces. All other actions are peripheral.

If we have correctly assessed both the potential power of precision weapons and the current geopolitical situation, it is not clear that the United States will be able to continue to intervene in Eurasia. As precision-guided munitions mature in range, speed, and intelligence, the long sea-lanes traversed by American merchantmen during the wars of the twentieth century will become increasingly hazardous. Just as German submarines threatened to cut the sea-lanes between North America and Britain during the two World Wars, and just as U.S. submarines *did* cut Japan's lines of supply, these weapons have the potential to threaten U.S. lines of supply into Eurasian theaters *without* necessarily using the submarine as a platform.

Destroying even a portion of American supply vessels could so disrupt the tempo of a logistical buildup as to delay offensive operations indefinitely. This would be worth enormous expenditures from nations expecting to engage American ground forces.

Standard solutions do not apply here. The United States is already hard-pressed to field a sufficient logistical capability. Stealthy supply ships are not in the cards. The same can be said for Aegis. Aegis was not designed to defend large convoys of merchantmen but, rather, carrier battle groups. The latter were to be used along the Greenland–Iceland–United Kingdom gap, preventing Soviet aircraft from penetrating into the North Atlantic. The forward deployment permitted the supply ships passage from secure port to secure port. If the supply ships were deployed anywhere but the North Atlantic, they would have to sail close to hostile shores. Even if they were not attacked by long-range missiles, merchantmen resupply forces in the Persian Gulf or in Korea could well come within range of land-based antiship missiles only a few hundred or even dozen miles away.

U.S. strategic planners can no longer guarantee that its supply ships on the high seas or in the areas closer to shore can avoid being hit by missiles. The model we need to bear in mind today is not Korea, Vietnam, or Desert Storm, where the enemy lacked the ability to interdict or even constrain American supply lines. Rather, we should have in mind the North Atlantic in 1942–43 and the western Pacific from 1943–45, where submarines were able to control the tempo of military operations and limit strategic buildup—in one case, substantially delaying operations, and in another, actually choking off supply lines.

During Desert Storm, the United States was fortunate that the enemy lacked the will or the capability to force the tempo of war. Had the Iraqis been able to continue to mount offensives between August and October 1990, the supply situation might well have become a severe limitation on American strategy, as it was during World War II and the first stage of the Korean War. As it was, American logistical planners were hard-pressed to build up sufficient theater reserves to mount the air operation followed by a massive ground assault. When the war concluded, the supply of aviation-grade fuel was severely depleted, almost at the crisis point.

Now, imagine Desert Storm with intelligent missiles able to hit and sink or slow a substantial portion of the American logistical flotilla. Such a capability in Iraqi hands might well have postponed the U.S. offensive indefinitely, at the very least. As it was, Iraqi mines in the Persian Gulf raised serious operational problems for the few U.S. ships that were involved.

During all of our wars, the key to successful U.S. operations has been forward logistical bases on which to build up supplies to sustain extended operations. Britain served this function during World War II, Japan during the Korean War, logistical bases such as Cam Ranh Bay during the Vietnam War. In Desert Storm, prepositioned equipment and supplies in Saudi Arabia and prepositioned ships and supplies at Diego Garcia, in the Indian Ocean, served this function. During Korea, lack of sufficient prepositioning and the enemy's control of the tempo of war nearly led to disaster. In World War II, Vietnam, and Desert Storm, prepositioning coupled with the tempo of war permitted stabilization and a relatively extended buildup.

Our next war may, like Korea, deny us control of the appropriate timing of operations, while political circumstances may preclude access to prepositioned supplies. For example, an Iranian attack on Iraq that threatens to upset the Eurasian balance of power might require intervention of effective (not symbolic) forces in a matter of days, while the politics of the region might cause the Saudis and others in the region to deny us access to our logistical bases. Indeed, if the United States lacks the means to intervene, it might be eliminated as a deterrent force and embolden the Iranians. This situation could be multiplied from Serbia to Korea—throughout the southern rim of Eurasia.

The choice at this point is between abandoning American grand strategy and devising novel solutions for carrying it out. In a world of regional hegemons with massed armies able to move unexpectedly and at low risk from neighboring powers, the ability of the United States to protect its amphibious force is absolutely essential. But depending on the transport of equipment and supplies from the United States to initiate American operations is not a viable alternative. Therefore, we must turn to the blue-water issue of intercontinental sea-lane control in an age of intelligent weapons.

Indeed, we must consider carefully and in detail the effect of these weapons on warfare in general, and not simply on naval warfare. To do that, we must look at the historical process that has brought the ballistic technology of Europe to senility, paving the way for both precision-guided munitions and the American epoch.

PART TWO

THE SENILITY OF EUROPEAN WEAPONRY

Introduction

FROM BALLISTICS TO BRILLIANCE

World War II represented the apex of Europe's weaponry. Masses of ballistic weapons, mounted on masses of hydrocarbon-powered chassis, battled one another throughout the Eastern Hemisphere, wreaking unimaginable slaughter. World War II marked the end of Europe's domination of the world. Forced by the logic of the ballistic projectile to field ever larger armies, build ever larger war industries, spend ever larger sums of money, and endure ever larger casualties merely to hold on to what it had won for a mere pittance in the preceding centuries, Europe exhausted itself. Caught between the emerging vigor of the new global maritime power, the United States, and the united strength of the Eurasian heartland, the Soviet Union, Europe finally succumbed.

A generation later, Europe's weapons culture began to crumble—an event still barely noticed. Over a period of a few years, from 1967 to 1973, first at sea, then in the air and on land, a new species of weapons showed its worth in battle. Still in their most primitive forms, the new weapons—smart, guided munitions—faced and defeated the

113

old weapons platforms, or allowed those platforms to carry out missions that would have been impossible before. At three points and times—Port Said in 1967, North Vietnam in 1972, and in the Sinai Desert in 1973—a new weapons culture showed itself. A new epoch was born.

- On October 21, 1967, at about 5:20 P.M., two Styx missiles were fired by a Soviet-built Komar-class patrol boat belonging to the Egyptian Navy. The target was the Israeli destroyer *Eilat*, steaming off Port Said. The missiles were launched from inside the harbor at Port Said, about fifteen miles from the destroyer.[1] They were fired in the general direction of the Israeli warship, climbing to around 475 feet. At about seven miles from the target, the Styx missiles turned on their radar and located the *Eilat*. The destroyer burst into flames upon being hit, and three hours later, after receiving another Styx, it finally sank. For the first time in history, a ship was destroyed solely by surface-to-surface missiles.

- On April 27, 1972, twelve F-4 Phantoms of the Eighth Tactical Fighter Wing in Thailand were ordered to strike at the Thanh Hoa Bridge. Located south of Hanoi, the Thanh Hoa connected the ports and rail lines of the North with South Vietnam. Were it to fall, North Vietnam would find itself unable to resupply its forces in the South. Since 1967, hundreds of aircraft had attempted to destroy the bridge. Still it stood. U.S. aircraft simply lacked the necessary accuracy.

  Then on April 27, a mere twelve aircraft attacked and destroyed the Thanh Hoa. Eight aircraft carried two-thousand-pound bombs, four carried chaff designed to foil North Vietnamese radar and protect the attacking planes from surface-to-air missiles and antiaircraft guns. The explosives in the bombs were no different from explosives used in other raids. What was different was that the bombs, once released, were not merely subject to gravity. Rather, they were "smart" bombs that could correct their course as they fell. Using laser and electro-optical guidance, the weapons officers on the aircraft directed the weapons to their targets. This time, twelve planes succeeded where hundreds had failed. The Thanh Hoa came down—too late to affect the outcome of the Vietnam War, but a clear pointer to the future.

- The revolution in land warfare began in the midafternoon of October 8, 1973. It started in the northwestern corner of the

Sinai, at a point where invading armies had passed for thousands of years. Egyptian forces had crossed the Suez Canal on the previous day, and by the next day they had occupied most of Israel's Bar-Lev Line in the Sinai. The Egyptian attack was a brilliant success, but the general feeling around the world was that it was doomed. The Egyptians, after all, were engaging the most impressive military in the postwar world: Zahal, the Israeli Defense Forces.

The result was described by Chaim Herzog:

By midday Gaby's [a brigade commander] forces were approaching the Canal and were engaged by Egyptian tanks and infantry, which they could see clearly on the ramparts on the Egyptian side of Firdan. Gaby's left flank battalion attacked along the Firdan route and almost reached the Israeli ramparts along the Canal. Suddenly hundreds of Egyptian infantry appeared out of the sand dunes around him, all firing antitank weapons at short range. The battalion commander was wounded and the battalion withdrew, leaving twelve tanks aflame in the area. . . . After fifteen minutes another unit was under antitank fire. The leading battalion reported that two tanks had been knocked out and that the second in command of the battalion had been killed. Concentrated katyusha fire blocked their advance and they could not see more than a distance of a yard ahead because of the black smoke and dust covering the area. When they were 800 yards from the Suez Canal, a hail of antitank fire descended on them. Natke looked around him as the smoke lifted and saw tanks exploding on his right and left. What he saw convinced him that they must pull out. Of the force he had attacked with, only four tanks were capable of withdrawing from the inferno into which they had charged.[2]

The Israeli 190th Armored Brigade had attacked an *infantry* position and had been destroyed. A new age was born.

All three events marked the end of the age of ballistic warfare on which the European global system was built. At the heart of the ballistic age was the projectile-firing tube—the gun. Now the gun had many virtues, but the greatest of these was its ability to kill at greater distances than muscle-driven weapons—spears, arrows and bows. The gun also had a great defect—inaccuracy. The solution to this problem was the massed armies of the nineteenth and twentieth centuries. At Port Said, the Thanh Hoa Bridge, and in the Sinai Desert, the gun mounted on a weapons platform encountered a new species of weapon—a weapon that was so stunningly accurate that it ushered in an entirely new military culture and a new way of waging war.

War had become an enterprise that required the complete commitment of a society—its entire economy drawing on the wealth of entire continents. Nations smaller than the United States or the Soviet Union were simply not large enough—not enough men, minerals, or factories—to field the forces necessary to wage modern war. Indeed, modern warfare had become so vast and all-consuming that the mere act of feeding and fueling troops had become not only overwhelming but frequently impossible, dwarfing the war-fighting capabilities of armies. The European military culture had become senile. It still functioned, but only at a ruinous cost to everything around it.

Suddenly, inexpensive missiles carried by a small ship, a few planes, or by infantrymen proved themselves capable of shattering the behemoths of the reigning military culture. Giant warships, massive tanks, invulnerable bridges, all suddenly fell before a handful of simple and relatively inexpensive weapons. The age of total war had ended.

Desert Storm will be remembered less for its strategic significance than as the first war in which precision-guided munitions were decisive. Iraq, equipped with the latest weapons from the end of the European epoch, encountered the first, still-primitive weapons of the American epoch. What was most striking was not how thoroughly the Iraqi Army was crushed nor even how quickly the defeat occurred. What was striking was how many Iraqi soldiers died and how few Iraqi civilians. The image that will always be with us from that war—the TV camera in the nose of a precision-guided munition recording the unerring strike on a particular part of a particular building—told only half the story. The other half was the story of what was not hit—the civilian population. These new smart weapons made it possible to strike at an enemy's war-making capability without having to strike at noncombatants.

The gun dehumanized war because of its range and its inherent inaccuracy. A soldier could fire a weapon without being able to see who or what was being hit. The victims were legion and invisible to the warrior. With precision weapons, they will remain invisible, but they will be fewer. This is as important an event in human history as one can imagine—in light of the human devastation wrought by the wars of the twentieth century.

In thinking about the future of war, it is vital to understand its past, to understand how we have arrived at this moment, and why the transcendence of the old war-fighting culture was both possible and essential. Revolutions in military culture are not new. For example, the invention of the composite bow—consisting of both wood and sinew—about 2000 B.C. allowed soldiers to fire projectiles farther and more frequently than ever before. Armed arches in chariots or on horses brought an irresistible force to the battlefield and became the military foundation of the Persian empire.[3] It took another revolution in warfare—which occurred when the heavily armored Greek warrior (the hoplite) organized into the phalanx—to check the Persians at Marathon.[4]

The inherent limits of the horse/archer combination—the difficulty in training mounted archers, the cost of the chariot, the terrain limitations on horse-based fighting—revealed themselves openly only when a more advanced form of warfare was encountered. So too the inherent limits of modern combat—the fantastic cost, the logistical difficulties, the pressures of tank production on the civilian economy— reveal themselves only when the ballistic weapon confronts its negation, the precision-guided munition. We need, therefore, to understand the manner in which Europe's weapons systems—on land, sea, and air— came to be, and why they are now passing away.

5

From Gunpowder to Petroleum

TANKS TRIUMPHANT

The first serious use of the gun on land was not as a personal weapon, but as a means of firing heavy projectiles designed to breach fortifications. From biblical times, the fortification has been the great barrier to attacking forces. During the age of the city-state—that is, through most of human history—the wall represented the foundation of human life. The Greek word *polis,* the root of the word *political,* literally means walled city. Walls made it possible to be fully human. In *The Iliad,* Homer recounts the complex special operations that were required for breaching the Trojan walls. In the Bible, the books of Judges and Joshua are filled with examples of breaching the walls of cities, sometimes through divine intervention, sometimes through covert operations, and sometimes through direct assault. The need for divine intervention at Jericho is a measure of the difficulties involved in military operations against walled cities. So, the fortified walls of the city-state represented the central military problem: it was much easier to hold a fortification than to take it.

It is important to bear in mind that the earliest deployments of firearms did not immediately abolish the power of walled cities nor immediately give rise to the nation-state. Firearms were deployed in battle as early as the beginning of the fourteenth century. But they were not at all decisive in determining the course of battles, even in the case of sieges. Early weapons had many weaknesses: they were so weighty and unwieldy that they could not be brought to the battlefield in time to be useful unless the siege was lengthy. Artillery rounds were so heavy and firing procedures so complex that guns were often fired only once in a battle. Handheld firearms were inaccurate and unreliable. Lack of metallurgical skill and impurities in powder frequently led to failure and occasionally to disaster.[1]

Yet the handwriting was on the wall for premodern forms of warfare. The medieval knight, already heavily armored against muscle-driven weapons, had dramatically increased his armor to defend against the crossbow. A fallen knight, wearing between three hundred and four hundred pounds of armor, could not regain a standing position without the help of servants.[2]

When firearms were first used, only the smaller fortifications fell to them. Not until later, during the fifteenth century, did massive fortifications capitulate. For example, Constantinople fell to seventy Turkish artillery pieces, including one monster firing a stone ball weighing eight hundred pounds.[3] New technologies are frequently less efficient initially in performing given tasks than old technologies. **A revolutionary weapon type does not, when first introduced, need to have a decisive or even significant effect in battle. In the case of firearms, the impact did not even begin to emerge for two or more centuries.**

The Problem with Guns

All firearms suffer from two basic defects—inaccuracy and a low rate of fire. The eye can easily see a target, but the complex muscular coordination necessary to accurately fire a weapon is rare. Even with expert training and all the time in the world, most men cannot learn to hit a distant target, even when not in danger themselves. This fact, compounded by imperfections in the weapons themselves, makes most gunfire random, especially at great distances. The problem with getting close to the target is that the closer you get, the more likely you are to be shot yourself.

Cannon posed somewhat less of a problem in terms of accuracy, if for no other reason than that their large size made the exquisite finesse of small-arms firing unnecessary. The other advantage of cannons was that they could fire a projectile that exploded on impact, shredding into smaller, deadly shards that would fly a substantial distance from the impact point, killing and wounding. In some cases, large projectiles would contain smaller projectiles. In other cases, the cannon itself would act like a shotgun, spewing a mass of bullets outward in a field that expanded as distance from the muzzle expanded. In the twentieth century, in the most common form of cannon, the casing itself would shred, causing *shrapnel* to fly.[4]

In the late eighteenth century, the probability of a single twelve-pound shot hitting a single target at 750 yards was about 34 percent. To assure complete destruction at a given point, three rounds would have to be fired. It took about 7.5 minutes to fire a round. If a line of troops was advancing toward an artillery piece at the rate of a yard a second, it would take 12.5 minutes for the troops to reach an artillery position 750 yards away. An artillery piece could fire no more than two shots in this period, providing only a 68 percent probability of destruction, assuming no misfires—and not taking into account increased accuracy at closer ranges or decreased accuracy if the target was moving.[5] To increase the probability of a hit, the obvious solution was to increase the number of shots fired. But with early muzzle-loading firearms, this was impossible. Waiting for the barrel to cool down; then cleaning out the debris from the last round; loading the powder, the round; firing—all of this took time. The next step was to increase the number of weapons firing to have at least two guns at each point of an advancing line—a larger number of artillery pieces.

Used as a mass of weapons rather than as a series of single shots, artillery worked extremely well. However, in working well artillery posed a new problem. The greater the increase in the rate of fire, the greater the logistical burden. In the British Army around 1800, an artillery piece was provided with 180 rounds per gun.[6] At twelve pounds per round, this meant that the gunners had to carry around 2,160 pounds of ordnance per gun. This amount had to be resupplied as it was consumed—along with powder and replacement barrels.

As with artillery, personal firearms suffered from low rates of fire and accuracy. Many weapons were extremely cumbersome. The British Short Land musket of 1768 was nearly five feet long, weighed over ten pounds, and fired a .75-caliber round without a sight. It took twelve separate motions to load and fire, with a maximum range of only 125 yards. For all practical purposes, the weapon could not be fired more than twice a minute and tended to misfire.[7]

A British study conducted in 1823 described the probability of inflicting a hit on a line of infantrymen at various distances. The extreme range was 250 yards, where the probability of a hit was about 20 percent. At 50 yards, the probability of a hit rose to about 65 percent. One should recall that this was not a hit on a single man, but on some point of a line of men—a target-rich environment!

Assume that you were shooting at a line of infantry, marching in formation, advancing at about two yards per second. The effective rate of fire of a muzzle-loading weapon was between two and three rounds a minute; therefore assume that one shot was fired every thirty seconds. The probability of a hit would be:

Shot	Range in Yards	Hit Probability
1	250	20.0%
2	190	33.5
3	130	47.0
4	70	60.5

In this extreme example of rapidly moving troops and low rates of fire, the probability of a single soldier hitting someone in a line of soldiers approaches certainty by the fourth shot. Thus, with sufficient defenders, an advancing line could be decimated.

There are three possible solutions for the attacker. One is to increase the rate of movement, which would reduce the time at risk and the number of rounds exposed to. But this is not possible over extended marches. The second is to increase the number of soldiers advancing. The third is to have the advancing soldiers fire back, thus reducing the defenders' rate of fire. Massed armies and massive casualties were born out of this logic. Losses could reach as high as 40 percent for both victor and vanquished, with the number of dead and wounded ranging in the 10,000–25,000 range per battle. In one conflict, a single volley disabled seven hundred men.[8]

A soldier could not simultaneously advance and fire a muzzle-loading weapon; he had to stop for about thirty seconds to load. But to maximize the probability of a hit on the enemy, *simultaneous* firing was necessary—in that way the horizon could be saturated. Indeed, if a commander wanted to keep up a continual hail of fire, he needed three lines—one kneeling and firing then falling to the rear to load, another standing and preparing to kneel and fire, the third line loading. This ballet could be combined with a slow, methodical advance—the origin of the close-order drill.

Increasing the rate of fire was the obvious next step in the development of guns and artillery. For example, instead of having the soldier assemble the round while under fire, prepackaged powder charges were developed, cutting loading time dramatically. But the most important advance was the complete cartridge—charge and projectile together—which could be inserted into the breech of the gun. This increased the rate of fire to five or six rounds a minute. When multiple cartridges could be inserted in a revolver or a repeater rifle, it became possible to fire a round every few seconds.

Gunpowder Meets Petroleum: The Origins of Armored Warfare

Even with these dramatic improvements, the probability of a hit remained low. To overcome this problem, it was necessary to saturate a target area—which is precisely what the machine gun did. Ammunition was automatically loaded, and depressing the trigger fired not one but a series of rounds. The original Maxim gun could fire six hundred rounds per minute—an excessive amount as it tended to oversaturate areas while melting the gun barrel.[9] The Vickers Mark I, a derivative of the early Maxim machine gun in wide use during World War I, could fire two hundred rounds a minute, or over three rounds a second. Because the barrel was not rifled, the rounds would disperse, which was better than having multiple hits at the same point. Like the Maxim, it was cooled by circulating water, but because of its lower rate of fire, and therefore lower heat, it could fire for an hour before its barrel had to be replaced.[10]

The weight of machine guns, particularly the early water-cooled models, made them difficult to move—which reversed the trend toward infantry mobility. The advantage shifted to the defense as the massive firepower of the machine gun shut down movement on the battlefield. In retrospect, it is ironic that the planners of World War I saw the essential problem of warfare as managing the rapid mobilization of men to the front and providing them with the power to move to the offensive.[11]

The machine gun and massed artillery made it impossible for an infantry and cavalry army to attack by making the cost of approaching an enemy position unbearable. On September 25, 1915, for example, two British divisions were ordered to attack the German lines at Loos. Following an ineffective twenty-minute bombardment, 10,000 British soldiers advanced. When the British were 1,500 yards out, the Germans opened fire with their machine guns, resulting in a carnage of 385 officers and 7,861 enlisted men. The Germans suffered no losses.[12] There had to be a way to protect the infantryman against machine-gun and artillery fire that would also allow him to move around the battlefield and fire his weapon.

The result of the effort to reinvent the offense—and to end the war—was a tank, a name originating as a clever code used by the British to disguise what was actually being produced in their factories. But *tank* was a good description. An armored tank, large enough to contain infantrymen, was mounted on wheels of some sort and moved by an internal combustion engine that was powerful enough to propel the entire load. Its armor was thick enough to provide protection against both machine-gun and artillery fire, and it could deliver men to critical points on the battlefield. More important, when armed with a machine gun or cannon itself, it could move around the battlefield and wreak havoc. As the wheels concentrated pressure at four small points, causing the heavy tank to sink into the ground, the decision was made to mount the tank on treads. Thus, the modern armored vehicle was born.[13]

The British created the first workable tank, the Mark I. Manned by a crew of eight, it had a maximum speed of 5.9 kmh, a range of 37 kilometers, and could cross a 1.5-meter-wide trench with parapets 1.4 meters high.[14] The tank was tall, so adding a turret and gun would have made it top-heavy. Instead, guns were mounted on the sides. As impressive as tanks were technically, the British were not completely clear what to do with them. At first, they were used to add fire support to infantry operations.[15]

On November 20, 1917, at the Battle of Cambrai, the history of warfare took a dramatic turn. The British Third Army, in an attempt to break the stalemate of trench warfare, launched an attack on the French town of Cambrai. All of the characteristics of a World War I offensive were present: careful planning, artillery bombardment, infantry assault. The only added element was the use of 378 tanks—attacking in a mass. To everyone's amazement, the seasoned men of the German Second Army broke and ran. In six hours the British penetrated four miles into German lines—extraordinary in a war where successes had been measured in yards. Because the British had not really expected success and had thus made no serious plans to exploit the breakthrough, the Germans managed to close the line, and the offensive failed. Still, it demonstrated things: first, when operating as a massed force, the tank could defeat infantry; second, in addition to breaking through enemy lines, the tank had to exploit that breakthrough, encircling the enemy and cutting his lines of supply and communication.

The tank was not immediately recognized as a dramatic step forward. Some strategists contended that it was merely a response to an idiosyncratic problem—trench warfare—which was unlikely to occur again. Thus, it was argued, the tank would not be necessary in the future.[16] Others, supporting the new orthodoxy that the defense was always superior to the offense—a view held as vehemently as the old orthodoxy that offense was always superior to defense—maintained that the tank was unimportant.

Interestingly, this latter position was held more strongly by the winners than the losers; the losers were forced by events to reexamine everything. Military losers are intellectual radicals; the winners, complacent in victory, feel the need for self-examination far less. Thus, for the French, the lesson of World War I was that offensive warfare could not succeed.

Even so, the French had built a large number of excellent tanks. They terms them "assault artillery" and distributed them among infantry formations to serve as mobile artillery in tactically precarious situations.[17] On the eve of World War II, the French had 3,245 tanks compared to only 2,574 tanks for the Germans. French tanks were armed with 37-mm and 47-mm guns and had 40-mm-thick armor. Fifteen hundred of the Germans' tanks were armed only with machine guns. Those that were armed with 37-mm cannon could not penetrate French armor at anything beyond point-blank range.[18] The French outgunned, outarmored, and outnumbered the German tanks.

Yet the Germans defeated the French. German tanks had three decisive characteristics: speed, range, and radios—and a doctrine for using these effectively. The most common German tank, the PzKmpw III, had a top speed of about 35 kmh, while the top-of-the-line PzKmpw IV could reach 40 kmh. This was more than double the speed of French tanks—which were not built for maneuvering but for holding a defensive position. Furthermore, French tanks had extremely limited ranges, due to both the limited amount of fuel they carried and their mechanical unreliability. The Germans expected their tanks to move up to 200 km without breaking down and without refueling. Each side built their tanks according to doctrine.

Perhaps the greatest innovation by the Germans was marrying the radio to the tank. The Germans envisioned fluid warfare, swirling over hundreds of kilometers and involving thousands of tanks. They needed to control the movement of tanks from the platoon to the army level, otherwise tank warfare degenerate into mechanized mob violence. By placing receivers in all tanks and transmitters in all command vehicles, the Germans created both a system of control and a system for intelligence gathering. This was the key to their success against France in 1940.

The doctrine that bound all of this together was the blitzkrieg—lightning war—a doctrine rooted in geography. Fighting from an inferior material base and faced with threats from several directions, the Germans had few natural barriers between themselves and their enemies to the east and west. Outnumbered and outflanked, the Germans hoped to deal with one threat at a time, initiating hostilities at the time and place most favorable to them. This had been their strategy in World War I, and it failed when their great opening offensive bogged down on the Marne and in trench warfare. In World War II, the tank became a technical expression for their geopolitical imperative.

Blitzkrieg is a form of risk-taking imposed by weakness, a game for the numerically inferior side.[19] The weaker power concentrates his forces at a certain point along the enemy front, crafting a localized superiority, and ruptures it. He then moves to the rear of enemy formations, cutting lines of supply and communications and seizing strategic points. The larger force is immobilized by the shock and effectiveness of the maneuver—and the result is victory. However, should the initial blow fail to break the enemy, the danger of an extended war of attrition looms.

During the Franco-German war of 1940, the Germans were inferior along the entire front but succeeded in the Ardennes-Sedan region by secretly concentrating their forces there. Using superb intelligence, concentrated power, and tanks' mobility, they broke the French line and defeated a numerically superior enemy. The Germans, thus, became the first great practitioners of tank warfare.

At the beginning of World War II, the tank's primary role was the traditional anti-infantry, antifortification, and antilogistical one designed for it during World War I. In the course of the war, however, it increasingly took on an antitank function, owing to the logic of the battlefield. Few infantry or artillery weapons could deal with a tank. The American bazooka, with its tube-fired rocket with an explosive warhead, had little success against the well-armored German tanks, nor did the German Panzerfaust against the more lightly armored American weapons. Hand-thrown grenades could, with luck, dislodge a tread on a tank or set fire to an external fuel tank.

Artillery used in a direct-fire mode could be more effective. Rifled, high-velocity antitank guns, such as the German 88-mm weapon, could penetrate armor with devastating effectiveness. But an antitank gun that was itself mobile could pursue and destroy a tank. German tanks, for example, were armed mostly with machine guns. By the end of the war, however, the German Tiger carried an 88-mm tank killer. At the beginning of the war, the American M-3 carried a 37-mm gun. By the end of the war, the Sherman and Chaffee tanks had 75-mm guns, while the Pershing was armed with a 90-mm gun. In the British case, the early tanks had two-pounders, while the later Cromwell had a six-pounder.[20]

A sort of arms race took place as each side built tanks with more efficient armor and bigger guns. Armor thickness soared. At the beginning of the war, the German Panzer I had 13 mm of rolled-steel armor on its front hull. By the end of the war, the German Tiger II had 185 mm, the American M-3 had 44.5 mm, while the Pershing had 102 mm and the Sherman 100 mm.

What is important here is that while absolutes surged—armor got thicker and guns bigger—*relatively,* there was little change. German tanks remained capable of killing enemies at ranges of up to 1,500 yards; American tanks remained unable to kill German tanks from the front— until the end of the war. There was an arms race in which relative advantages remained fixed.[21]

McNair grasped the problem more clearly than his contemporaries. He understood with Guderian and Patton the possibilities of tank warfare. But he also understood the dangers. What he did not understand was that the logic of armored warfare compelled a tank-tank confrontation, in which tanks would grow more powerful, expensive, and draining, without changing the relative balance at all.

In 1940, the best French tank carried 30 mm of armor on its front hull. The German 50-mm gun could penetrate that at less than 1,500 yards. At the end of the war, the heaviest Soviet tank, the IS-2, had 95 mm of front-hull armor. The best German tank firing an 88-mm cannon could penetrate that at less than 1,500 yards. The American Grant, firing its low-velocity 75-mm gun in North Africa in 1942 against the Panzer IVF with 80-mm front-hull armor, could not penetrate its armor at 1,500 yards. In 1945, a Sherman firing a high-velocity 75-mm gun at a Tiger with 100 mm of front-hull armor could still not penetrate the German target. It took the introduction of the 76-mm high-velocity gun to allow penetration at ranges closer than 1,000 yards.[22]

What did shift was the quantity of resources that had to be deducted to maintain the same levels of effectiveness. Increased weight meant larger engines consuming more fuel; heavier armor meant more steel; larger guns meant more metal and explosives. The result was greater strain on the logistical and industrial system. But at the end of the day, the same relationship existed—the Germans outgunned their enemies. But they could not outproduce and outdeliver them. All that had changed was that the amount of resources necessary for carrying out the same mission was beginning to soar.

Tanks Take Over

The tank was invented in World War I and was forged in the crucible of World War II. It reached its technical and strategic heights during the Cold War. The Cold War was organized around the operational principles of World War II: the primacy of the offensive. The great generals of World War II, Zhukov, Rommel, Guderian, Patton, Montgomery (indeed, all the European generals and none of the Pacific generals), understood that armor was the key to offensive power. Thus, tanks and other armored vehicles were at the heart of strategic planning and operations.

During the Cold War, the conceptual centrality of armor turned into practical centrality. The Soviets constructed a massive tank army and concentrated it in Central Europe.[23] The sole military purpose of that force was to strike westward, at an opportune moment, and seize Atlantic Europe. The sole political purpose was to convince Western European leaders of the inevitability of such a seizure so that they would seek an accommodation with the Soviets and expel the United States from the continent.

Infantry could not expel the United States from Europe, but an armored force could. Therefore, the American Cold War obsession was with stopping that armor from rolling westward. Nuclear weapons were built to deter the Soviet Army from moving. Other nuclear devices, from tactical nuclear weapons to the neutron bomb, were designed to destroy Soviet armored concentrations. But in the end, the United States and NATO accepted the conclusion they had reached by the end of World War II, which was that the best way to stop tanks was with other tanks. So, from very early in the Cold War, both the Americans and Soviets became obsessed with designing tanks that could destroy other tanks—guns that could penetrate armor, and armor that could stop guns. This obsession would, in the end, lead to massive extremes in the design of increasingly endangered armored vehicles.

Basic armor design was set in World War II. The size of the tank's main gun would be determined by the thickness of enemy armor but would be at least 75 mm, with as high a velocity as possible with rifling. It would be mounted on a turret that could traverse 360 degrees independent of the main chassis and could be elevated for indirect fire or depressed. The normal crew size would vary: three, if there was an auto-matic loader; four—driver, gunner, loader, commander; or five, with a machine gunner. A variety of ammunition would be used, from high explosive to antitank to smoke. Armor thicker than 100 mm and tanks weighing 50 tons and more would become common, although lighter (20 to 30 tons) medium tanks would predominate until the 1960s. The engine would be diesel or gasoline powered.

Larger guns meant thicker armor. Thicker armor meant more pow-erful engines. More powerful engines meant costlier tanks consuming more gasoline. It was obviously important to reduce the weight of tanks as much as possible—without making them vulnerable. But because armor had to encase a fairly fixed volume of space—a space large enough to hold a crew, engine, and ammunition—reducing size was dif-ficult.

An additional problem was the weight of armor. One way to control weight was to design armor only for its most likely uses. Lt. Col. J. M. Whittaker, a British tank specialist, developed what was called Whittaker's Directional Probability Variation (DPV), which was simply a study of where rounds were likely to strike a tank. Whittaker's DPV showed that the majority of hits would be on a frontal arc measuring sixty degrees.[24] Thus, if the front hull was thickened, other areas could be thinned, and the weight of the tank maintained or reduced. A certain number of tanks would be destroyed by hits in thinner areas—a calculated and acceptable cost. It should be noted that, in this cold-blooded analysis, we see one of the first uses of mathematical modeling in weapons design.

Another way to keep weight down was by sloping armor. Assume that a piece of armor is 50 mm thick and placed in a full upright posi-tion. A projectile striking it on an angle parallel to the ground would have to penetrate 50 mm of armor. Now, assume that the armor was inclined at an angle of sixty degrees. The same projectile moving parallel to the ground would have to penetrate *double* the thickness.[25] It was diffi-cult to angle rolled homogeneous armor (RHA) because it was extremely brittle and difficult to shape. Skilled—and expensive—labor was needed. Cast iron was easier and cheaper to work with, but it was even heavier than RHA. Thus, every step taken to make production more efficient made the tank heavier and less efficient.

The Cold War tank, therefore, retained the basic armor design of the World War II tank. It was made from RHA. The front hull was extremely thick, but less so at the sides and rear. Least armored were the top and sides, although the turret was well armored. In addition, the hull was shaped so that the part most at risk—the front—was angled to present maximum thickness to enemy rounds. The central concern was with line-of-sight attacks. Tanks lived in a two-dimensional world. Little thought was given to aircraft and land mines and almost none to artillery shells.

Armor developed incrementally, without any real breakthroughs during most of the Cold War. The same was true for tank guns. From World War II onward, the tank gun fired two types of projectile—kinetic rounds and high-explosive rounds. Kinetic rounds were solid masses fired at a tank at the highest possible speed. High explosives depended on explosive chemical energy to penetrate armor. Both needed a gun that could deliver the round accurately—and in the proper condition to penetrate the armor, kill the crew, and destroy the tank.

In a sense, *kinetic rounds* are the simplest and more logical projectile. They resemble bullets, using weight and velocity to provide the energy to penetrate their target. They ought to be cheaper and easier to manufacture and handle than high explosives—and in their earlier incarnations they were. As armor thickened and sloped, the caliber of guns had to be increased, along with muzzle velocities. The problem was that there seemed to be a limit as to how fast the round could be fired— about one kilometer a second—a limit imposed by the rate of explosive expansion of gunpowder-generated gasses. The only way to increase the lethality of the tank was to increase the weight of the projectile and, therefore, the size of the gun. Toward the end of the war the Germans were proposing guns of 150 and 170 mm and equally monstrous platforms to carry about the guns and their ammunition.[26]

One possibility was to create smaller and heavier rounds that could penetrate armor by bringing unbearable pressure on it. A small round extruding an even smaller penetrator rod made of a hard metal such as tungsten would, in a high-velocity impact, cause the armor to flow away from the point of impact, allowing the round to penetrate. But such small rounds could not be fired from the large guns that tanks carried to fire high-explosive rounds.

A sort of Rube Goldberg solution was found. The small antitank round was retained, but it was surrounded by a lightweight shoe, or *sabot,* which could be adjusted to fit any size gun. When fired, the saboted round would leave the rifled gun barrel spinning. The sabot would fall away, leaving a small, dartlike projectile, stabilized in flight by spin. The small, extruding probe would slam into the armor, concentrating all the energy of weight and velocity on a single tiny point, causing the armor to flow away as if it were liquid, allowing penetration and destruction. This was the *armor-piercing, discarding-sabot (APDS)* round.

As always, there was a problem: the thicker the armor, the longer and narrower the extrusion had to be. Spin stabilizes a projectile so long as its length is no more than seven times its diameter. If the ratio is greater than 7:1, the projectile becomes unstable and useless.[27] As tank armor was thickened and better angled, the extrusion had to get longer and longer, until it passed the 7:1 ratio. At that point, another means of stabilizing the round had to be found, something other than rifling. The solution was to place stabilizing fins on the round, creating an *armor-piercing, discarding-sabot, fin-stabilized (APDSFS)* round. Where the APDS *had* to be fired from a rifled gun, this round *could not be,* since a spinning finned projectile would behave even more erratically than a long, thin round.

Another solution was the *high-explosive antitank (HEAT)* round. Chemical explosives disperse their energy in all directions, with insufficient concentration for penetrating armor. But if one could devise a means for *containing* the explosion and *shaping* it, its energy could be concentrated enough to penetrate armor.

The HEAT round is peculiar in that it detonates *prior* to hitting the target, while still in flight. The explosion is shaped by a metallic liner in the shell, usually copper.[28] A shaped jet of liquefied metal extrudes, becoming long and narrow (in shape, not unlike the probe of the kinetic charge). While the round as a whole moves at about 1 kmps, the extrusion extends itself ahead of the main projectile at a fantastic speed— about 8 or 9 kmps, about twenty-five times the speed of sound.[29] Traveling faster than its own shock, the jet arrives completely undistorted and penetrates, slicing through the armor like a cutting torch. A general rule of thumb is that the round can penetrate armor six times the diameter of the base of the cone.[30]

On the one side, the APDSFS round argued for a smaller barrel. Pressure also grew to develop the means of delivering a large-caliber, low-velocity HEAT round, with at least the same accuracy as the laser-sighted tank gun had achieved. Antitank weapons thus began to take on two different tracks. One was the traditional gun; the other idea, brewing for a long time, was the rocket-propelled, guided missile armed with a HEAT warhead. Together, these tracks would conspire to push the tank toward senility and obsolescence.

The Snake Meets the Mongoose: Tanks and Antitank Missiles

The idea of antitank rockets dates back to World War II, when military planners sought a way to provide infantry with some protection against tanks. They designed small rockets carrying shaped charges whose purpose was to penetrate armor on the sides and rear of tanks. One such rocket, the American bazooka, consisted of a long tube, open in the rear to allow for exhaust. Bazooka-type weapons had several weaknesses. First, it took a lot of guts for an infantryman to get close enough to a tank to use one. Second, because the bazooka was so unwieldy and its projectile so unpredictable, it required superb hand-eye coordination. And third, its warhead was too small to do much damage.

The Germans, whose own Panzerfaust was only marginally superior to the bazooka, searched for a better means of empowering infantry.[31] Late in the war, they developed the X-7 Rotkäppchen. The X-7 was a winged missile powered by a solid-propellant rocket, which allowed it to reach a maximum speed of 98 mps and a range of 1,200 meters. What was most important about the X-7 was that it could be controlled *after* firing. The X-7 trailed wires behind it, through which signals could be transmitted by a gunner, who could guide the nine-kilogram mass to its target.[32]

The Germans placed the X-7 into production in 1945, far too late to influence the war. However, the French continued developing the wire-guided system and six years later produced the SS-10, a weapon that was produced until 1963, when it was replaced by the ENTAC. Meanwhile, the SS-11, another variant, was produced during the mid-1950s and became part of the arsenal of over twenty countries. The SS-11 had a higher velocity than the SS-10 and a longer range (160 mps and 3,000 meters, respectively) and could penetrate over 600 mm of armor under test conditions.[33]

The SS-10, ENTAC, and SS-11 were important because they reintroduced the possibility that tanks were not the only way to stop tanks. Essentially man-portable weapons, they allowed infantry to engage tanks from ambush positions from distances of up to nearly three kilometers. In addition, they could be mounted on thin-skinned vehicles, such as jeeps. There was a certain irony in all this. The tank had started as the ultimate anti-infantry weapon and the basic question had always been, How could it kill other tanks? Now the question began to be, How could the tank protect itself against infantry? With the SS-10 and SS-11, the first glimmer of the revenge of the infantry could be seen.

A large number of wire-guided antitank weapons were developed, ranging in size from the heavy American TOW (accurately called a "tube-launched, optically tracked, wire-guided" missile) to the smaller, man-portable Soviet AT-3 Sagger. All had the virtue of not being dependent on pure ballistics for trajectory. But all had a common defect: the velocity of the missiles was low, generally below 200 mps. The missile was guided to its target by a gunner—in early versions, such as the SS-10, the gunner had to command via a joystick, much as in today's video games. In later versions an optical element was included. The gunner peered through a sight and zeroed in on an enemy tank. A computer translated that sighting into commands to the missile to home in on the target.

The maximum range of a TOW missile varied from 2,000 to 3,750 meters—depending on which version was used.[34] Its maximum speed was 300 mps. So, it would take a shot ten seconds to hit a tank from 3,000 meters. The blast of a rocket is clearly visible to the naked eye, and even more visible to instruments scanning in the infrared portion of the spectrum. With the TOW and all other optically guided missiles, whether trailing wires or using some other means of command, the gunner has to keep his eye on the target—and therefore remain exposed—for what can seem like a very, very long time.

However, what is gained is an extremely high probability of a hit—somewhere between 83 and 96 percent of 3,000 meters. The HOT missile, a shoulder-fired infantry missile built by the French, had a hit probability of 86.7 percent of 500 meters. By contrast, rifled guns firing APDS rounds and using simple optical sighting techniques had about a 20 percent probability of hitting their target.[35] As important, it was found that the HEAT rounds being fired by shoulder-held missiles were able to penetrate even the tank's thick front armor. As the new antitank weapons came on line during the 1960s, a maverick school of thought emerged holding that the tank was finished.

The encounter between the Soviet AT-3 Sagger and Israel's Western tanks in 1973 stunned the West. The AT-3 could be mounted on a vehicle or carried around in a suitcase by a team of infantrymen.[36] The lid of the suitcase had a rail on it on which the rocket was mounted, along with its warhead. What was most significant was that it could be launched up to fifteen meters from the gunner. Although this standoff capability was important, as it protected the gunner from return fire while he held his position for the crucial seconds, it posed serious problems in the real-time computing of guidance commands. Solving these difficulties represented a breakthrough in control technology. Able to penetrate up to 400 mm of armor with a launch weight of only 11.3 kilos, the AT-3, along with its Western equivalents, promised to revolutionize warfare.[37]

Even before 1973, tank designers were scrambling to find a counter to antitank missiles, all of which carried HEAT warheads. The first impulse, of course, was to thicken the armor again. However, with Western tanks already about twenty tons heavier than their Soviet counterparts, further thickening would significantly increase weight and reduce performance—or require a radically larger, costlier, and logistically demanding engine. Instead, the focus turned to finding new types of high-efficiency, low-weight armor.

The density of steel makes it hard and durable, but it also means increased conductivity, making it more vulnerable to the effects of superheated explosives. As the effectiveness of antitank missiles with shaped charges became apparent, an urgent search was commenced to find materials that were less dense than steel yet extremely strong and able to resist kinetic-energy rounds.

Aluminum, in several forms, was seen as the most promising material, particularly when alloyed with magnesium. But it tended to melt at high temperatures, and shaped antitank charges would pass through it like a hot knife through butter.

During the 1970s, ceramics—produced from minerals such as boron carbide, silicon carbide, and alumina—seemed to provide a breakthrough in armor. Ceramics had a number of very real advantages: first, because of their relative lack of density, they could defeat shaped explosive charges; second, they were hard enough to defeat kinetic-energy rounds; third, they were light enough that they could be thickened as necessary, without imposing undue weight penalties. But ceramics also had several drawbacks. They were brittle.[38] This meant that while they could survive a single round, they would be shattered, leaving the tank unprotected at that point. They were also inflexible, which meant that they had to be applied as individual tiles to the tank, much as they were eventually applied to the Space Shuttle, thus increasing its vulnerability.[39] An additional problem was that ceramics were most efficient against projectiles striking them at right angles—an angled hull would decrease ceramics' effectiveness.[40]

Ceramics by themselves could not provide a tank that could survive both kinetic and shaped rounds, any more than could steel. The solution, therefore, was sought in composite armor, consisting of layers of different materials. Chobham armor, first designed by the British and named after the design facility in England, combined the advantages of high-grade steel with those of ceramics.

Chobham armor's prime disadvantage is that, like any combination of laminates, it has specific weaknesses as well as specific strengths. By knowing the precise makeup of the armor—the particular mix of steel layers and ceramic layers, the gaps between each, and so on—an enemy can customize his attack. For example, by striking at a particular angle, he can take advantage of the inevitable weaknesses to break through and destroy the tank.

The M-1 Abrams, the mainstay of the American armor force, has Chobham armor, and its precise makeup is one of the U.S. military's most closely held secrets. Of course, it takes only one engineer—or one custodian—to give the secret away. Moreover, tanks are destroyed and recovered by the enemy on the battlefield. In a protracted war, a particular laminate becomes obsolete after the first battle.

One of the most important military operations in the U.S. Army has been the development of a computerized database that contains information about the capabilities of various armors, which when fully implemented will enable the designer to identify and compare armor materials and systems that may meet his requirements.[41] Obviously, information on how to destroy particular laminates will be used by gun designers seeking to defeat enemy tanks. That the Tank-Automotive Command (TACOM) should place an emphasis on the development of a database provides us with a hint of the bewildering profusion and complexity facing the tank engineer.

The cost inherent in designing such complex armoring is compounded by the constant risk of becoming obsolete or, worse, of becoming obsolete without being aware of it. Other problems exist as well. Laminated armor is bulky. The larger the tank, the more visible it is on the battlefield—particularly at a time when advanced target-acquisition and fire-control systems are making the tank more and more a target.

A simpler approach to defeating the HEAT round was pioneered by the Israelis. Called Blazer, it consists of appliqués applied to the hull of tanks. These appliqués contain an explosive—the first versions were merely gasoline—that is set off by the superheat of a HEAT round's extruding jet. The appliqué explodes on contact with the extrusion, disrupting the structure of the jet and making it impossible for it to penetrate the armor.[42] This approach has been used by others, including the U.S. Marines, who adopted it as an interim solution to protect their aging M-60 tanks.[43]

Blazer was part of a class of armor called Explosive Reactive Armor (ERA). ERA has evolved in the last decade to combine lamination with an explosive. Armor is covered with a layer of explosive, which is then covered with a ceramic plate and finally by more explosives. Moreover, the explosive contained in the armor is less likely to explode when hit by a bullet or shrapnel. It is certainly effective. One estimate is that ERA now reduces the effectiveness of HEAT rounds by 75 percent. Its drawback is that it is so expensive that most countries cannot afford it, and, worse, it is not very effective against kinetic rounds.[44]

As armor got more sophisticated in blocking the HEAT round, gun designers went to greater and greater lengths to find exotic ways to destroy tanks. The German-made 120-mm gun currently used by the M-1A2 Abrams fires kinetic rounds made from depleted uranium, the material left after the refining process for weapons-grade uranium is finished.[45] Uranium is much heavier than tungsten, which means that using it increases the total kinetic energy of the round without requiring an increase in the size of the gun firing it. Its drawback is that it is mildly radioactive, and it might be remembered that depleted-uranium rounds caused some controversy during Desert Storm, when it was discovered that the war had left a residue of low-level radiation.[46]

We had reached the stage where exotic ceramic laminates covered by exploding appliqués were facing shaped plasma jets and depleted-uranium rounds. Things were getting out of hand.

Weird Science: Keeping the Tank Going

But the future began to look even wilder as the West contemplated future Soviet tank development in the 1990s and beyond. Everything discussed so far had been intended to deal with *existing* Soviet weapons, such as the feared T-80.[47] Before the breakup of the Soviet Union, NATO also feared the next generation of Soviet tank, reasonably called FST, the future Soviet tank. The FST 1 was expected to be produced with a 135-mm main gun mounted on a turretless chassis, while the FST 2, which was to be fielded in the mid- to late 1990s, was to have a 140- or 145-mm gun that used new firing methods and advanced, even more exotic armor.[48]

The prospect of facing a 140-mm gun caused the United States to make a massive commitment, on the order of *$60 billion*, to upgrade its own armored force. It was not until the fall of 1992, when it became apparent that the Soviets would not be able to field the FST 1 or 2, let alone a mysterious FST 3, that the American counter, the Armored Systems Modernization plan, was discontinued.[49] But while the plan to field six thousand new vehicles at an average cost of about $10 million apiece was dropped, the research and development on armor and anti-armor went on, including the production of prototypes. Ideas such as modularized armor, which could be changed for each mission, automated systems and computers reducing crew size, abandoning the turret to minimize height, were among dozens of ideas bruited about. But the most intriguing discussion concerned the future of the tank gun.

The best way to break through increasingly sophisticated armor is with brute force—faster and heavier bullets. But it had become obvious that the grizzled explosive-powder gun had reached its limits. A chemical explosion drives a projectile up a tube by creating a mass of expanding gas. No matter how much powder is placed inside the breech, the gas cannot expand faster than its chemical properties allow. Thus, if guns are to keep up with armor, new forms of energy must be found.[50] As Greg Ferdinand, a senior researcher in exotic gun technologies, put it, "Our program is based on the premise that conventional munitions have just about peaked out. We've reached the point of diminishing returns. We're looking to new technologies to give us quantum leaps."[51]

But what sort of energy ought to power a gun? The obvious answer is the purest and most efficient sort possible. Barring nuclear energy, only electricity is left. One possibility would be to generate a powerful plasma jet inside a superheated fluid.[52] The plasma would expand dramatically, creating powerful waves in the fluid, forcing a projectile from a tube at a velocity equal to that of the wave—a velocity much greater than what could be achieved with explosive powder. Thus, we would have an electrothermal gun.

But there were problems. First, for the gun to be accurate and consistent, events in the firing chamber have to be absolutely predictable. This was one of the nice things about gunpowder—the same quantity would explode with the same force time after time. This was not the case with the electrothermal gun. When the fluid is excited by electricity, a cavity is formed, to be filled with the electrically generated plasma. Since no two patterns repeat, no two cavities are identical. The wave behaves differently each time, and therefore the speed and trajectory of the projectile become unpredictable.[53]

The second problem involved the use of inert fluids, such as water, which resist energy. Two companies, Lockheed and FMC Corp., have found that when an energetic fluid is used—something that itself produces energy when stimulated (think of gasoline when a match touches its fumes)—the wave becomes both more powerful and more orderly. Thus the issue of the inert fluid is resolved.[54] The theory of the electrothermal (ET) gun or electrothermal-chemical (ETC) gun becomes more compelling. But reality intrudes.

ET and ETC guns require large amounts of electricity—but how does one deliver it to a thousand tanks maneuvering on a battlefield?

Between four and eight megajoules of electrical energy are needed to turn water into plasma. A single megajoule of energy is the equivalent of a one-ton pickup traveling at one hundred miles per hour.[55] So, to fire a single shot, somewhere between four and eight times that much energy must be delivered in a quick pulse each time the gun is to fire. This means that a portable generation system of remarkable quality and prodigious size must be installed in each tank. Alternatively, there must be a breakthrough in battery and capacitor design that will allow a tank to repeatedly produce bursts of this sort.[56]

It is likely that the power problem will be solved. However, there remains a basic physical problem with all plasma weapons: the maximum speed of wave expansion—a problem similar to the maximum rate of gas expansion in dry-powder weapons. Plasma waves travel at about four thousand meters per second. Their speed apparently cannot be increased beyond this. This means that at maximum efficiency, an electrothermal gun of any design will be able to fire a projectile about three or four times faster than a conventional gun. This is certainly an improvement, but not a revolution. But a revolution is exactly what is needed, otherwise, tanks will simply find a way to improve their armor using advanced materials, and everyone will have run very hard to remain in the same place. In the end, as tanks up-armor to cope with the threat of higher-velocity rounds, the electrothermal projectile will run up against the same limits that powder technology did.

What is needed is a gun that leaves wave propagation aside altogether, that translates energy directly into propulsion without using a medium to pass it through: in other words, the electromagnetic gun. The electromagnetic rail-gun system uses a pulse of electrical current traveling down one rail inside the weapon's barrel, across an armature at the base of the projectile, and up another rail on the other side. This creates a magnetic field that accelerates a metal-clad projectile to incredible speeds—depending on the amount of energy in the pulse. Again, the energy in the pulse is the crucial factor. Firing a hyperkinetic round requires more than 3 million amps of electricity, or enough electricity to light up a hundred-watt lightbulb for every person in the United States.[57]

The problem is weight. The rail gun currently converts about 30 percent of electrical energy into kinetic energy—an incredibly efficient conversion rate. However, the 9-megajoule gun currently planned requires about 30 megajoules of electrical input, which requires a capacitor bank weighing about eleven tons. A tank with a 15-megajoule rail gun would need a 50-megajoule capacitor bank small enough to fit in a tank weighing about 50 to 70 tons, powered by a 1,500-horsepower engine.[58] This is asking for a lot—and it will cost even more.

But if the weight problem were overcome, the tank would be able to fire a kinetic round at a speed of 300–6,000 mps. In February 1993, the British Defense Research Agency's test center at Kirkcudbright, Scotland, was able to fire a 90-mm round at 7,500 feet (about 2,500 meters) per second—with the expectation that it would be able to double this without major design changes. As a first try, this is extremely promising. But the problem, again, was that the batteries and capacitors used were enormous. Another problem emerged as well: at the extraordinary accelerations generated by the gun, the immediate problem is simply developing projectiles that will stay in one piece when fired.[59]

The U.S. Army has decided that electromagnetic technology is more promising for tank armament than electrothermal technology and is planning to test-fire a 90-mm electromagnetic gun. While cuts have been made in the electromagnetic project, it is a sign of the urgency felt by tank designers that this relatively exotic project is still alive—and fairly well funded. Among the very real advantages of the EM gun are velocity, silence, no muzzle flash, and possibly usefulness against nonarmored targets, such as aircraft. The very real problem is energy—batteries and capacitors to store and deliver huge amounts of energy.[60]

It is clearly possible to increase the velocity of guns mounted on tanks using electrothermal and electromagnetic technology. But would it represent anything new or be merely another phase in the growing senility of the gun—its reductio ad absurdum? Line-of-sight encounters between main battle tanks would continue to depend on the ability of one tank to see the other first, aim its weapon, and fire first, except that the cost and complexity of doing this will have soared automatically. The cost of the gun, both in terms of research and development and production, would be much greater than the older dry-powder variety. The cost of the armor would be immeasurably greater than the old RHA. And finally, the need to fire the first shot would raise the costs of target acquisition and fire control until they began to approximate those of aircraft. Most important, **the tank would still retain roughly the same probability of lethality and vulnerability as it had before.**

Moreover, given the dramatic increases in kinetic energy, the sheer quantity of armor that would be necessary to make the front hull invulnerable to rounds packing dozens of megajoules would raise the weight of tanks to between eighty and one hundred tons, from the current fifty to sixty-five tons.[61] This would pose tremendous problems for the industrial plant and logistical system, which would have to supply greatly increased amounts of petroleum merely to keep the tanks operating at previous levels. The costs of protecting tanks from attack are much greater than the cost of designing new threats—one of the most ominous signs of senility.

Consider the difference between the last generation's M-60A3, fielded in 1979, and the current M-1A1 Abrams tank. The total per mile cost of operating the M-60A3 was $50.39; the cost per mile for the Abrams is $159.74, more than triple. The cost of fuel jumped three and a half times, and spare parts jumped five times. The M-60A3's HEAT round cost $127; a single Abrams HEAT round costs $1,033. The kinetic-energy round went from $148 to $711. The strains modern armor places on the supply system have become so great that, during the Gulf War, in spite of a multinational logistical force, the Allies had to launch their attack with only 5.6 days of petroleum, oil, and lubricants on hand.[62] Had the famous left hook failed, disaster would have been a real possibility.

Consider that the United States is planning, as a successor to the TOW antitank missile, a line-of-sight, guided, kinetic-energy missile called LOSAT (*line of sight, antitank*). In five tests, the missile struck each time with a force of *sixty megajoules*, in one instance knocking the turret entirely off the target tank. Traveling at speeds in excess of two kilometers per second, it can hit an enemy tank at a distance of three kilometers in less than 1.5 seconds, giving the tank no chance for maneuver. It is guided by a forward-looking, infrared sensor located on the launcher, which transmits course corrections by laser. It adjusts its course through the use of small radial thrusters behind its nose.[63] Most important, the missile does not have to be fired from a tank. Smaller vehicles, helicopters, and fixed-winged aircraft can all be used.[64] LOSAT's per unit cost will be around $30,000.

There is little doubt that this weapon and its kin will make life difficult if not impossible for tanks on the battlefield of the future—even if they are armed with the most exotic guns.

The issue, therefore, is how to break the technological and financial spiral that armored warfare has found itself in—how to simplify warfare again. The solution involves breaking the tyranny of line-of-sight weapons, just as the tyranny of ballistic weapons has been broken.

Whittaker assumed that the only serious threat to tanks came from line-of-sight weapons. This meant that an area generally hidden from the line of sight, such as the top, did not need much armoring. In reality, the fundamental threat to the tank comes from the possibility that the old assumptions about vulnerability are no longer true. The biggest challenge will come from weapons that are not based on line of sight, weapons that can not only attack the top of the tank (or anywhere else) but can find the tank if its precise location is unknown or has changed. With a sensing system and a terminal guidance system, non-line-of-sight weapons will create a nightmare for tanks and compel a fundamental change in the way they are designed. They will also redefine how we wage war. In the end, they will drive tanks to senility.

6

Sensing Senility

The limits of human vision and the limits of ballistic projectiles imposed the logic of line of sight on tanks. But once this connection is broken, the tank makes little sense. Improvements in range and accuracy have made a short-range line-of-sight tank gun unreasonable. Above all, the revolution in sensors that took place in the 1980s under the aegis of the Star Wars initiative has left it too visible, too vulnerable, and too expensive to survive on the battlefield.

No Place to Hide: Sensors and the Tank

Saying that it has become extremely difficult to hide a tank is about as obvious as saying that it's hard to hide an elephant. No matter how it is camouflaged, something that big will be seen by some sensor in some spectrum. The sensors used in tracking tanks are not unfamiliar to us—we have already encountered them in our previous discussion.[1] These include:

- Visible light—utilized through electro-optical systems (such as television) that pass information to a human controller or to an onboard computer system; also the low-light-enhancement sights and goggles that multiply available light thousands of times, making it possible to see in the dark.
- Laser—a narrow band of visible light (red, blue, green, etc.) that can be used to designate a point on which a projectile can home in or a gun can aim, providing information on angle and distance.
- Imaging infrared—turns data from the infrared position of the spectrum into a visible image, in the same way as electro-optical systems.
- Infrared homing—locates heat sources using cryogenically cooled sensors and then homes in on the source.
- Synthetic aperture radar—used by satellites and reconnaissance aircraft to locate tanks that are hidden or dug in.
- Millimeter wave radar—radar in the shortest radar spectrum (the L band), which can create images of objects not visible to the naked eye—particularly useful on munitions that are guiding themselves to the target.
- Acoustical—homes in on the sound emitted by tank engines.

An extraordinary amount of data, from a wide range of spectra and gathered by a wide variety of sensors and platforms, is now available to the tank hunter. The issue is how to use this data to kill the tank.

The traditional sequence for firing a gun at any target consisted of three steps: seeing an object that may be hostile; identifying an object as hostile; aiming and firing a projectile at the target. In the past, each of these sequential functions was carried out from the same place, under the direct control of a human being. The human being saw and identified the target; human eyes, brains, and hands controlled the weapon. The introduction of sensors, sensor platforms, computers, and complex communications systems has both automated and decentralized the firing sequence. The linear path has been replaced by a complex sequence with many branches and options built in:

- Target acquisition—This can take place from any number of platforms, ranging from a special operations soldier, to a high-flying reconnaissance aircraft, to the warhead of the projectile itself. Moreover, initial acquisition of the target can take place in any spectrum.

- Data transmission and distribution—Data about the target is transmitted to the appropriate station, or the warhead of an antitank missile.
- Target recognition—Data from the target acquisition is turned into a confirmed target. This can be done by humans, intelligent warheads, or a tactical intelligent platform, such as a UAV (unmanned aerial vehicle).
- Fire control—This concept includes everything from sighting and shooting a gun to guiding a projectile to the target.

Traditionally, the projectile has been a hunk of iron—sometimes with an explosive inside—trapped by its trajectory. The new projectile both senses its surroundings and can make sense of it. Thanks to the development of lightweight, relatively low-cost sensors and the ubiquitous microchip, the controls and intelligence that had previously been concentrated in the tank gunner are now being reconcentrated in the projectile itself, with men pretty much excluded from the loop. The entire range of weapons—from cruise missiles right down to the hoary antitank round fired by a gun—has been affected by this shift toward the centralization of function into the projectile.[2]

For example, the M-1 Abrams tank has now been given a depleted-uranium kinetic-energy round, the X-ROD, which has in its nose a millimeter wave radar sensor and a computer chip. The tank fires the round toward an enemy tank, then the X-ROD searches the area for a possible target—while in flight! It finds a possible target and compares the image of the target tank to an image programmed into its onboard computer. If they match, X-ROD fires a rocket motor at about a kilometer out—accelerating the round into the tank's hull.

Another munition available to the M-1 is the XM943 smart, target-activated, fire-and-forget (STAFF) round, which can be fired at tanks that cannot even be seen by the Abrams's gunner. The round is fired high, in indirect-fire mode, then the millimeter wave radar finds the enemy tank, identifies it, and uses a rocket to attack its thin top armor. Even tanks have abandoned the simple, line-of-sight, dumb round, in favor of rocket-assisted, intelligent, and maneuverable rounds.

The revolution in tank ammunition is even more dramatic in artillery shells, which have historically been impotent against tanks. The Swedes, for example, have developed the lowly mortar into a deadly antitank weapon. The Strix is a 120-mm shaped charge, which is launched in typical mortar fashion, in a high arc. In its nose is an imaging infrared system that feeds data to onboard computers. The computers, in turn, control side thrusters that guide the mortar round to the top of any tank that has an engine running—and that therefore has an infrared signature.[3] British Aerospace has developed an 81-mm, man-portable mortar that performs the same job, using a millimeter wave radar for terminal guidance.

A wide ranger of indirect-fire weapons can be used to bring firepower to bear on the top of tanks. The Copperhead program was an early American attempt to devise a top-attack, 155-mm round, but it can be used against any target. For example, the United States and Germany are designing a multiple-launch rocket system (MLRS), in which a rack of launchers, each containing a rocket, would be fired in rapid succession, saturating an area. The basic round would have a maximum range of thirty kilometers and would guide submunitions armed with millimeter wave seekers to the target. Imagine a twelve-tube launcher firing twelve rounds simultaneously, each dividing into three or more rounds, each with an extremely high unlikelihood of hitting its target. The effect would be devastating.

The MLRS brings us to another new distinction—between projectiles and munitions. In an odd way, this is a return to the earlier days of gun warfare when the different parts of a round were divided and joined together in the barrel. The modern formulation divides the round between the projectile, responsible for the long-range transport of the system, and the munition, responsible for destroying the target. Until recently, the projectile *was* the munition—it never separated from its warhead, striking the target together. Now the projectile becomes, once again, just the delivery system, transporting the destructive munition to the general area of the target.

The advent of highly maneuverable and intelligent delivery systems, from the tube-fired rounds able to travel several thousand feet to cruise missiles capable of traveling several thousand miles, provides an opportunity to attack armor at its most vulnerable point. The situation is dramatically improved by the ability of the main projectile to dispense intelligent submunitions—armed with their own sensors, computers, and means of maneuvering. In this way, each projectile is able to attack anywhere from several to a hundred or more targets rapidly and from great distances.

During Desert Storm we saw the Tomahawk cruise missile at work. Fired hundreds of miles away from the target, it can travel at subsonic speed to a designated target, using either a terrain-following radar or ground positioning system to guide it. Once it reaches the target, the Tomahawk slams into it, detonating with about a thousand pounds of explosives aboard—the projectile and the munition are one. This is effective, but not early as efficient as it could be. Imagine a successor to the Tomahawk reaching the target area and then dispensing a load of submunitions, each of which locates a target and attacks it individually. Imagine further that the cruise missile would then turn around and return to its base, to be used again later.[4] The effectiveness of each attack would soar, and the cost of each attack would plummet.

Developing such a system is a high priority for the U.S. Army. The key is a submunition called the brilliant antitank munition, or BAT. Designed by Northrop, BAT was originally designed to be carried on the triservice standoff attack missile (TSSAM), a short-range cruise missile. Twenty-two BAT submunitions can be carried on board, each weighing about forty-four pounds and about three feet long.[5] It has folding wings and tail fins that deploy when it is released—giving it a sinister appearance. In a recent modification, Northrop has equipped the BAT with what it calls a *finute*—a finned, balloon-parachute combination, which is stored in a one-inch case in the rear of the munition and released when the munition begins free fall.[6] This gives the BAT more time to survey the scene below and spot a target.

BAT is one of the first true multisensor weapons, combining both an infrared sensor and an acoustic sensor, and will identify any tank whose engine is running.[7] Indeed, the Army has decided to upgrade the BAT with millimeter wave radar, which will identify targets whose engines are turned off.[8] It is unlikely that space constraints will permit all BATs to carry all sensors in the immediate future—although progress in miniaturization and improvement in onboard microprocessing does not make this at all inconceivable. In the immediate future, BATs with a mix of sensors will be released.

Perhaps the most extraordinary aspect of the weapon is that it processes data *on board*. Two things are striking about this. One is that megabytes of data are processed in a matter of seconds—quite remarkable given the size of the files that are generated by sensors of any sort, let alone sensors refined enough to provide data on which to base the kind of judgments that the BAT will have to make.

This is the second extraordinary part—that the BAT can in fact interpret the data at all, let alone in a few seconds, so that it can locate the target and distinguish it from buildings, brush fires, car engines, and so on. It requires a level of programming that, if not yet approaching artificial intelligence, certainly demands a level of mathematical sophistication to create algorithms that discriminate so finely and quickly among masses of data. Once the target is located, the BAT turns on a rocket motor and, using its wings, guides itself to an attack, using a conventional-shaped high-explosive antitank round to penetrate the relatively thin top hull of the tank.

There are a number of ways to deliver the submunition package. Northrop currently plans to package the BAT so that it could be launched by the multiple-launch rocket system (MLRS). The MLRS has twelve rocket tubes, and Northrop intends to put two BATs in each rocket. So, on a single salvo, twenty-four BATs could be fired—albeit at the fairly short ranges of a tactical rocket system. But the MLRS would also be able to launch two TSSAMs, which, carrying twenty-two BATs each, could launch forty-four BATs over one hundred miles.[9]

Ultimately, the BAT, or successor munitions that are even more brilliant and deadly, could be launched from any platform. They could be attached to spare platforms or dropped from manned aircraft. They could be carried to the target by artillery shells from two miles away, or by cruise missiles from thousands of miles away. BATs could also be attached to unmanned aerial vehicles (UAVs) hovering tens of thousands of feet above the battlefield, scanning the battlefield for targets. When a cluster of tanks is spotted, the BATs would be dropped toward the general target area. At a few thousand feet, they would release their parachute-balloons and activate sensors. Spotting their prey, they would turn on their rockets and attack.

Obviously, the BAT—and weapons like it—is a deadly complication for the tank. The traditional threat against a tank came from other tanks or men or planes, platforms that were vulnerable to tank countermeasures. Now the tank will no longer be able to strike at its tormentors. It will be seen by sensors thousands of miles away; deadly submunitions, fired from virtually any distance, will strike at its most vulnerable parts with stunning accuracy. Against this, the slow, heavy tank will offer a line-of-sight gun with a range of a few kilometers. **The twentieth-century tank will not survive into the twenty-first century.**

Consider the following scenario. A low-orbit satellite with synthetic aperture radar sweeps over an area and locates and maps a concentration of tanks. The information is passed via relay satellite and ground station to a theater commander, who orders a UAV hovering above the battlefield at seventy thousand feet with a multisensor package on board—including optical, infrared, and radar—to focus on the area. The UAV scans the surface below and transmits data to a communications satellite overhead, which retransmits the data via several relay satellites to an earth station within the battle theater but several hundred miles from the battle zone. Analysts and computers identify a battalion of tanks moving toward the battle area. The theater commander, having an array of delivery systems to choose from, selects a submarine-launched cruise missile, which, when launched, locates itself via a ground positioning system, is fed area coordinates, and moves toward the target. On reaching the area, the missile releases its submunitions, then turns to land at an airfield several hundred miles away. The submunitions scan the area, locate the tanks, and swoop down, each having a hit/kill probability in excess of 50 percent.

Not once, until the very end, is the tank in a position to threaten the attackers, and even then its only option is to try to swat down the incoming submunitions. This is what is dramatically new. The tank can be attacked; but it cannot, in turn, counterattack. In effect, the weapons system attacking the tank is thousands of miles wide and long, hundreds of miles high, and linked by data flows. Satellites, UAVs, aircraft, missiles, BAT—all of these are part of a single system against which the tank is quite helpless.

Platforms and Computers: Gathering and Using Data

What further complicates life for the tank is that the sensors used to detect it need not be in the projectile or the munition. Three of these platforms are already operational—satellites, reconnaissance aircraft, UAVs. Each currently provides detailed pictures of the combat environment but suffers, in varying degrees, in its ability to transmit data to tactical combat units in a timely fashion. Each has already brought about tremendous changes in warfare.

- **Satellite reconnaissance**—The virtue of satellite reconnaissance is that it can provide information about large swaths of terrain. Its weakness is that, as an orbiting platform, it cannot remain on station for an extended time—it cannot loiter in a controlled fashion, unless it is in geostationary orbit, and then it cannot map terrain very well. This means that orbiting satellites provide *snapshots* of enemy activities, rather than ongoing monitoring. The only way to provide continuous monitoring is to orbit clusters of satellites—an expensive business. So satellite data is useful for higher levels of command, such as theater commanders, but of little value to tactical commanders, who need minute-by-minute updates. Satellites can be used to locate enemy armor in concentration, but munitions will not be able to home in on data they provide.

▪ **Airborne reconnaissance**—There are several virtues to airborne reconnaissance: with refueling, the aircraft can remain on station indefinitely; its areas of coverage can be controlled and modified; it can carry a wide variety of sensing equipment, and a large enough staff that processing of data into information, and even real-time distribution of that information, can be managed from the aircraft.

During Desert Storm, the United States deployed an experimental aircraft with these characteristics—joint surveillance and target-attack radar system (JSTARS)—which was a resounding success. JSTARS is similar to AWACS (airborne warning and control systems), except that where AWACS is designed to manage the air battle, JSTARS is designed to manage the ground battle. The aircraft is a Boeing 707 with a twenty-six-foot-long radar housing underneath the fuselage.[10] JSTARS currently carries a multimode radar system. One of these modes, the wide-area surveillance/moving-target indicator (WAS/MTI), is a Doppler radar that detects the motion of objects. In the WAS mode, the movement of larger armored formations can be detected, while in the MTI mode, individual vehicles can be seen and tracked vehicles distinguished from wheeled vehicles. An interesting sidelight is that the main contractor for the JSTARS is United Technologies Norden System, the corporate descendant of the manufacturer of the old Norden bombsight.

The second mode on board is synthetic aperture radar (SAR). SAR permits JSTARS to produce maps of target areas, or photograph-like images of particular installations. In addition to mapping any stationary target—bridge, aircraft on the ground, tanks standing still—it can also map metallic structures buried beneath the surface. The depth that it can reach is classified.[11] As munitions buried beneath the surface become more important—and mine warfare was certainly a critical part of Desert Storm—this ability will be crucial. JSTARS was able to map the Iraqi minefields in southern Kuwait, allowing U.S. Marines to deal with them with minimal casualties.

In its WAS/MTI mode, JSTARS can cover a fifty-thousand-square-kilometer region—a square about 130 miles on a side.

This meant that during Desert Storm, a single JSTARS aircraft (there were two) was able to provide complete coverage of Kuwait. It can also focus down to a twelve-by-twenty-mile segment, where it can distinguish individual vehicles. A fleet of JSTARS can provide coverage of any size battle area.[12]

JSTARS has several problems. One is managing accumulated data; the other is the general problem of the relationship between accumulated information and battle management— JSTARS is a data-production system that needs to grow into a battle-management platform. Another problem is more critical—JSTARS is a plane, and a large, obtrusive, nonmaneuverable one at that. In an age of brilliant weapons, JSTARS is a prime candidate for being shot down.

■ **Unmanned aerial vehicles**—The smallest, least expensive, and potentially most effective sensor platforms are unmanned aircraft. Not dissimilar in principle to the inexpensive radio-guided aircraft used by hobbyists, the UAV has several virtues. Consider the Pioneer, developed by the Israelis and used by all U.S. services during Desert Storm. Appearing very much like a radio-controlled model, it has a range of more than a hundred miles and a cruising altitude of 12,500 feet, traveling at speeds of up to sixty miles per hour for five hours. Best of all, it is virtually undetectable at altitudes over three thousand feet, due to its size and the quietness of its engine.[13]

The Pioneer is controlled by radio, with a controller monitoring its movements and the terrain using a TV camera that transmits data. In this role, the Pioneer was used in a wide range of tactical and operational roles.[14] The success of the Pioneer foreshadows important developments among UAVs. The United States has committed about $6.6 billion to developing its own fleet of UAVs.[15]

Just as aircraft started as purely reconnaissance platforms and evolved into a weapons system in their own right, so too the UAV began life as a data gatherer and transmitter. However, a series of planned UAVs code-named RAPTOR TALON is being designed at Lawrence Livermore Labs as a platform for intercepting missiles. Flying at extremely high altitudes, and powered by solar engines, or by batteries recharged by microwave beams from ground stations, the RAPTOR loiters quietly and invisibly, possibly for days.[16] On board would be several TALONS, brilliant kinetic-energy missiles that would intercept incoming missiles.[17]

Another innovation that had been considered by the Air Force in conjunction with the Applied Physics Lab at Johns Hopkins University—which is working on the National Aerospace Plane—is a reconnaissance drone with intercontinental range and speeds up to Mach 20. While unlikely to be built in this cycle of weapons development, it gives us a chance to consider the consequences to reconnaissance and battle management in the next generation.[18]

There are, then, three classes of sensor platforms already in existence. The most immediately useful is the aircraft-based system, which probably has the least likely future simply because of the vulnerability of a large, manned aircraft. Satellite reconnaissance will continue to be useful strategically. The operational and tactical need for satellites able to loiter over the battlefield will require dramatic technical breakthroughs. Either sensor systems must improve so that they will function from geosynchronous orbit, or a prodigious power source must be found so that satellites can loiter without falling out of orbit. The former is far more likely than the latter. Obviously, the UAV holds out the greatest promise over the next twenty to thirty years. It is relatively inexpensive so its loss is not a disaster, it is small and unobtrusive, it can shift position, and in due course it will be able to carry multiple sensors and even weapons.

The success of these systems creates a paradox—the sensors are becoming too successful, they are generating too much data that cannot be translated into information useful to the warrior. There is a crucial difference between data and information. Data is the material gathered by sensors. Interpreting that data, transforming it into useful, real-time information is a more difficult task.

There are five parts to the data-information problem:

■ Gathering raw data—Sensors accumulate information about the battlefield in the various spectra, as a mass of digital material or mathematical formulas.[19]
■ Transmitting the raw data from the sensor to an interpretation node—computer or human—which transforms it into information.
■ Fusing the data—Data and/or information from different sources is joined into a single comprehensible package on enemy deployments and capabilities.
■ Interpretation of information—The information is analyzed and its meaning and uses are determined.

- Transmitting the information—Information is delivered to the combat command and distributed to all appropriate levels, while maintaining security and data integrity.

The key to all of this is *time*. In tactical combat, minutes, even seconds, matter. Knowing that a company of enemy tanks is approaching from beyond a hill is vital information, but only if it is delivered in time to do something about it. Information arriving even minutes late is useless. Information arriving in a format requiring detailed analysis beyond the ability of commanders operating in a stressful battlefield environment is similarly worthless. The closer the contact with the enemy, the more urgent the need for this information. A theater commander is interested in threats that may take days or weeks to show themselves. A divisional commander concerns himself with hours and days; a battalion commander deals in minutes and hours. For a company and platoon commander, seconds count. And wars are fought by squads and platoons. Conventional tactical warfare requires the rapid processing of intelligence with customer lists including every squad in the Army.

Data management has thus become the pivot of tactical land combat. The problem is no longer gathering sufficient intelligence, it has become screening and distributing what has become an unmanageable amount of data. The U.S. Army has focused on these issues in a program called the Army Tactical Command and Control System (ATCCS), and in a key subprogram, the All-Source Analysis System (ASAS). The program is described by CECOM (Communications and Electronics Command):

> ASAS is the central nervous system guiding field commanders to successfully execute the air/land battle. . . . ASAS automates command and control of IEW [Intelligence and Electronic Warfare] operations and intelligence fusion processing. It generates a near real-time picture of the enemy situation to guide employment of maneuver forces and systems and provides coordination to systems within the ATCCS arena. Many sophisticated sensor systems provide targeting information; however, the capability to process and respond to that information is limited, today, by manual and partially automated methods. ASAS uses state-of-the-art computers to speed the process and improve its accuracy.[20]

ASAS is intended to tie together all forms of intelligence—human, electronic, signals, imagery—and fuse them into a single information system, then distribute them to combat commanders at all levels via the new Joint Tactical Information Distribution System (JTIDS).[21]

ASAS has come under attack by critics. Undoubtedly, as the first attempt at a fully automated battle information system for all levels of combat, ASAS has many shortcomings and will be abandoned in favor of other programs. Nevertheless, ASAS shows a great deal of what combat is beginning to look like. It also shows the extent to which American military leaders are aware of the problems and solutions of sensor-based warfare. The fusion of data from all sources coupled with real-time distribution of just the right amount of material to the appropriate command level has become a combat priority. Too little information, and the commander will be caught by surprise. Too much information, and he will be overwhelmed and unable to understand let alone act on it. Indeed, in Desert Storm, Gen. Norman Schwarzkopf designated the lack of timely intelligence as one of the biggest failures in the war.[22]

Computers, screening and fusing data, will transmit information not only to commanders but also to their subordinates. In fact, it will be transmitted to weapons themselves. Long-range and short-range missiles will be guided to targets via system information. We already saw a low-level example of this in Desert Storm, when ground positioning system (GPS) satellites permitted everyone and everything, from lone infantryman to B-52 bombers and Tomahawk cruise missiles, to locate themselves on the battlefield. Satellite imagery was transmitted to commanders via graphics terminals. The minefields were mapped by synthetic aperture radar located on satellites and aircraft. The next step will be to fuse this data into a single information system, then to fuse the information system and weapons systems into a single operational entity, in which distance means less than time and data flow.

If Price Is No Object: The Tank Struggles to Survive

The immediate problem that tanks face is that they must protect themselves against a set of projectiles designed to attack the weakest portion of their hulls. The wide array of sensors that might be used in the attack increases the difficulty. Protecting against any single sensor type—optical, infrared, radar, ladar (laser radar), acoustical, and so on—might well be possible and even cost-effective. Protecting against all of them is a much more complex and costly task. When we add to the problem that the tank must mask itself not only from the projectile's sensors but from "third party" sensors as well, we see the tank facing potentially insurmountable odds. And, it must be added, these are merely the *current* problems confronting the tank; it says nothing about the emergence of new sensor technologies.

One solution for designers might be to worry less about the tank's being seen and more about its being so well armored that it cannot be hurt. This, after all, was the principle that made the tank so effective at its inception, during World War I. This would mean that the entire tank must be made invulnerable. Top, bottom, sides, and rear would have to resemble the front hull. Unfortunately, it has been estimated that up-armoring the top of the tank alone with contemporary armor would raise the weight to 150 to 200 tons—making it three to four times heavier than today's behemoths.[23] Advanced armors such as laminates may reduce weights by 20 percent for equivalent protection, but this is a marginal gain, as the pressure to up-armor will quickly wipe out the advantage of using composites.[24]

The problem was acknowledged by Col. William Miller, program manager for the U.S. Army's tank-survivability program: "The [tank-killing] bullets are getting bigger. . . . We've come to the point now where we can't make the armor as substantial as we like because we have mobility problems."[25] Tanks must cross bridges, operate on roads, be airlifted in planes that can only take so much weight. Increasing their tonnage will destroy both strategic and tactical mobility and, perhaps most serious, will increase fuel consumption astronomically—straining an already stressed logistical system. If tanks are to survive, some way must be found to increase their protection—without resorting to armor. We must fundamentally rethink armor.

All contemporary armor is passive. It works by resisting the energy after a force has struck it. Since armor works by passive resistance alone, it must be substantial enough to overmatch the energy of the active force—it must weigh a lot. Assume, however, that instead of being purely passive, armor became active, resisting the attack by active rather than passive means. Then energy must be substituted for mass, and armor could weigh less.

Under a scheme named the Smart Armor Program, the Advanced Research Project Agency (ARPA) of the Department of Defense is looking for a way to generate energy against energy. As we have discussed, this has already been achieved, to some extent, with reactive armor, used as an appliqué over the armor. When a HEAT (high explosive antitank) round strikes the appliqué, it explodes. The HEAT round's extrusion is disrupted *before* it reaches the armor itself, and the round fails to penetrate. However, reactive armor is useless against kinetic rounds, which, having struck the appliqué, still have sufficient energy to penetrate in spite of the explosion.

What ARPA proposes to do is have the reactive armor—or better still, a less volatile and bulky successor—explode a split second *before* the round strikes, with sufficient energy to deflect it. As with angling of armor, placing an angle on the rod-armor collision enormously increases the amount of armor that has to be penetrated. In addition, the angle places lateral stress on the depleted-uranium extrusion, possibly shattering it when it contacts armor. How can the armor know when to explode? A mesh containing a large number of microsensors would cover the tank all over the hull, about a foot out from the actual armor. Small explosive tiles would cover the hull proper. When a round hits the mesh, sensors would instantaneously measure its size, speed, and point of impact, instructing an appropriate number of tiles to explode, deflecting the kinetic round or disrupting the HEAT round's extrusion.[26]

Obviously, this would require computerizing the hull of the tank. But even this is, in many ways, too passive. Another approach might be to make tanks able to defeat incoming projectiles by destroying them before they strike—a mini–Star Wars or mini-Aegis system. In this scheme, a sensor system called the top-attack threat-detection (TATD) system, currently being developed by the U.S. Army's tank command and Delco Corp., uses Doppler tracking radar to identify incoming projectiles.[27] An onboard computer would select the proper response to the threat, ranging from hypersonic projectiles to high-energy beams to reactive armor, and order action. While an interesting thought, the idea is conceptually flawed. The tank, an essentially immobile object, would face an indeterminate, but theoretically unlimited, number of incoming projectiles traveling at many times the speed of sound and, possibly, approaching on an erratic and unpredictable course. The tank's survival would depend on its ability to shoot down every one of those missiles, without ever failing. And of course it would have to do this while still carrying out its basic mission. Such a system would obviously become *very* expensive.

Another possibility is to make the outer layer of the tank smarter, able to heal itself. The Air Force has been looking for a way to do this with aircraft—to layer the outer two inches with sensors that would have the ability to make repairs when they sensed damage. For example, if the skin was torn, signals and power would be rerouted around the affected areas, and the aircraft could continue to fly. In addition, microscopic structures embedded in the skin and advanced materials able to return to their original shape regardless of insult would increase the ability of the aircraft, and the tank, to absorb punishment. The cost, of course, would be astronomical, and in the end the attacker would be sure to inflict more damage than could be repaired.[28]

In the long run, the best solution might be one that was at first rejected—trying to make the tank less visible. This is a daunting task, since visibility is no longer confined to only the visible spectrum. So the first step might be to see if the tank could be made smaller.

In a way, the tank exists to protect the crew—the larger the crew, the larger the tank. Most contemporary tanks have four-man crews—commander, driver, gunner, loader. Soviet and Japanese tanks have already eliminated the loader, by replacing him with a mechanical loader. As the tank rounds get larger, this will become necessary anyway. Reducing the tank crew further has been difficult; however, with developments in internal controls—called vetronics—the role of the commander might be combined with either that of driver or gunner.[29] A separate commander was needed to observe what was going on around the tank, either by standing in the turret or by using the tank's periscope. There are now ways around this. In 1993, the Army Research Laboratory undertook a project designed to make the tank "transparent." Bob Miller, head of the lab, said:

> We'll put head-mounted displays in front of the operator that will give him the impression that he's driving out in an open vehicle. As he turns his head, [the system] will sense that he wants to look in a different direction and will access different blocks of memory from sensors to give him what he would see if he were looking in that direction. We certainly expect to have multiple sensors—IR, maybe millimeter wave radar, and optical, which could also be combined with sensor fusion.[30]

In other words, the commander will no longer have to stand in the hatch of the turret to see what is happening around him. The tank crew would then function more like a two-man fighter crew—a front-seat driver/commander and a rear-seat weapons officer.

The reduction in crew will reduce the size of the compartment that will have to be enclosed. A further development might be to eliminate altogether the second-largest part of the tank—the turret. The armored-gun system, until recently under development by the U.S. Army, addressed this possibility. It envisioned leaving the gun itself *outside* the armor, mounted on a pivot, with the ammunition and loader inside. All of this would reduce the weight of the tank somewhat—from seventy tons to perhaps thirty tons—and its dimensions might shrink as well. But the tank will still be a large mass of engine, treads, and armor. Other means will have to be taken to make it unobtrusive.

As with the B-2 bomber, the real solution has to be a technology that will cloak the tank in all spectra. The U.S. Army's Tank-Automotive Command (TACOM), in a study released in 1992, gives us a sense of the complex hurdles facing tanks trying to be invisible on the modern battlefield:

> The high technology sensor systems of the future battlefield will necessitate that commanders rapidly block or screen their vehicles from these threats. Infrared decoys and smoke can assist field commanders in managing the signature of their vehicles to minimize detection. The technology focuses on smoke and obscurants that are effective in aiding deception on the battlefield. Technology exploration includes infrared screening compositions, atmospheric effects, multispectral screening compositions, infrared emissive smokes, aerosols to protect against high energy lasers and high power microwave weapons, advanced dissemination concepts, smoke elimination concepts, real time obscurant characterization and non-toxic smoke/obscurant. Rapid dispersal devices necessary for the multispectral screening smoke/obscurant will provide countermeasures to react quickly to homing munitions employed by the enemy.[31]

The analysis concentrates on ways of obscuring tanks in the infrared spectrum by means of heat shielding and diffusing smoke and spray. This is crucial since tanks generate a tremendous amount of heat, which leaves by the way of the exhaust and defuses from the engine generally. Passed over rather quickly is the need to generate multispectral screens that will protect against ladar and millimeter wave radar as well as the more conventional radar.

Also left undiscussed is the spectrum in which the tank is perhaps most vulnerable and about which it can do least: acoustical sensing. Fifty-ton machines moving at thirty miles per hour are necessarily noisy. It should be noted, however, that retired Army general Phillip L. Bolte mentions that there are projects under way to counter acoustic and even seismic sensing—which implies that we must be doing research in seismic sensing to hunt enemy tanks.[32]

There is an ironic twist in Bolte's discussion: "Just as signature reduction would be used to reduce its probability of detection by an adversary, signatures could be used to detect an adversary vehicle."[33] It is the dilemma of all masking devices. Turn the B-2 question around. If American air defense experts discovered that a foreign power was developing a stealth bomber, they would immediately attempt to devise a technical counter, and in all likelihood they would achieve it. Along the same lines, having explained what would have to be done to make their own tank less visible on the battlefield, the folks at TACOM then turned around and explained how their study of signature reduction on our own vehicles would, in turn, make it easier to detect adversary vehicles. There is no sense of irony in this, no sense that the enemy will also be doing the same as well. Indeed, the Defense Department has created a secret Joint Counter–Low Observable Office, which deals with defeating the extremely expensive stealth technologies the rest of the department is creating.[34]

Using the complex of stealth technology, signature masking and electronic warfare capability will certainly make it more difficult to see tanks and to engage them, but it will also make tanks much more expensive and therefore more valuable. Continuing the analogy with the B-2, just as it is difficult to define a conventional mission in which it would make sense to risk a $600 million bomber in order to deliver twenty-five tons of munitions, so too it becomes difficult to define the circumstances under which a $10 or $20 or $50 million vehicle can be exposed to risk so that its single tank gun might fire twenty hundred-pound projectiles a distance of five thousand meters. The cost of the tank will undoubtedly soar, but its lethality will still be limited to the single direct-fire gun, with perhaps a missile launcher. **We will have a weapon too costly to risk and not deadly enough to accomplish its mission.**

An M-1 currently costs about $3 million. A SADARM—search and destroy missile—using both millimeter wave and infrared sensors costs about $7,000; a laser-guided Hellfire can be had for about $50,000; a millimeter wave–guided Hellfire about $200,000.[35] If SADARM comes even close to living up to expectations, the cost of the Abrams will soar, as more and more measures will have to be taken to fend off weapons systems that can attack it. These include projectiles that are not merely smart but "brilliant," such as the BAT—not only able to sense the presence of enemy armor but to recognize it as such, guide itself to it, fend off countermeasures, and detonate in a precise manner.

Unquestionably, means can be found to allow the tank to continue to function on the battlefield. Sensors detecting incoming projectiles, assuming they were not too stealthy, could automatically trigger masking agents to make the missile's homing impossible. Alternatively, the sensors could trigger the launch of an antiprojectile projectile, or swarm of projectiles, that would destroy the incoming round.[36] The Soviets have already developed such a system under the code name Drozd.[37]

On Senility: The End of Armored Warfare

We now reach the heart of the issue. The purpose of the tank was to move a large gun around the battlefield, destroying fortifications, killing infantrymen, disrupting lines of supply and communication. As invariably happens, the cost of defending the tank from threats is beginning to dwarf the benefits of the initial mission. Put differently, if the initial purpose of the tank was to place explosives with precision at key points on the battlefield, then it is no longer clear that this cannot be done more cheaply and efficiently with long-range projectiles. You simply do not need a tank to destroy a tank, nor do you need tanks to envelop and isolate enemy formations.

There have been two stages in the growing senility of tanks. The first was when the tank abandoned the primary antipersonnel/ antifortification role it had been given at its inception for an increasing antitank role. Yet at that point, it was still in a position to influence the battle, indeed, to control the battle's outcome, as could be seen in tank-to-tank encounters such as Kursk, during World War II, or the Chinese Farm, during the 1973 Arab-Israeli War. Defeating enemy armor laid open the enemy's war-making infrastructure.

The second phase of the senility came when the tank was no longer in a position to strike at threatening weapons platforms. So long as the basic threat was line-of-sight weapons or at least weapons that had to be launched from platforms within the reach of tanks, tanks still retained an offensive capability. Thus, while wire-guided or fiber optically guided missiles increased the probability of a given round hitting a tank, the missiles' platforms, even to some degree the man-portable antitank missile, remained vulnerable to the tank.

During the 1980s, however, indirect fire placed tanks in a new and untenable position: they could no longer strike at what menaced them most; all they could do was defend themselves. With the major threats beyond their range, they became like a turtle on its back—able to survive but unable to function. The tank could do nothing to defend itself beyond swatting down missiles as they approached, or hunkering down beneath an armored shell. Had the Cold War not ended, the United States was prepared to pay *$60 billion* to upgrade its armored force.

Modern warfare has revolved around delivering a quantity of explosive to a particular point on the globe and firing them at the enemy without, in turn, being destroyed. To do this, a weapons platform has to move within range of the enemy—which means being very close. The tank was invented so that this move could occur without the crew being killed. Because of a revolution not only in range but also in speed and accuracy, a projectile does not have to be driven to the target's neighborhood—it can fly there.

Eventually it will no longer be necessary to combine the weapons platform and the fire-control system. Quite the contrary, the National Research Council, in a study for the U.S. Army, has made it clear that in the future it will be necessary "to separate weapons physically from the system that performs the targeting."[38] The ability to see is conceptually separate from the ability to shoot and practically different as well, since vision should be located at the highest, clearest point, while the weapon should ideally be shrouded in the ground clutter. As new technologies mature, it will make little sense to put these functions together **in one slow vehicle and then drive within a few miles of the enemy.**

7

The Rise and Fall of the Gunboat

Naval warfare is experiencing the endgame of the gun-based warfare, part of same process that is transforming land-based warfare. And the surface ship will be the most vulnerable to the changes that are taking place. The replacement of guns by missiles as the primary weapon of the surface warship occurred gradually after World War II and was in place by the 1970s. It had been dictated by aircraft carriers, whose ability to project firepower over vast distances meant that most battles would be fought beyond gun range. Ironically, when the United States decided to take action against Iran by destroying an oil platform in the Persian Gulf in 1989, it found to its shock that its vessels, heavily armed with expensive missiles, lacked the firepower to destroy the platform. Ships had only small-caliber mounted guns on deck, primarily used for ceremonial functions.

But carriers, like all other surface warships, were the victim as well as the beneficiary of this shift. Because of missiles, they were at risk at far greater ranges than before, and as missile accuracy and intelligence increased, they had a greater likelihood of being struck. As important, missiles could not carry out several functions nearly as well or cheaply as guns. For example, missiles were a wrenchingly costly way to carry out shore bombardment, and during the 1980s, the United States took battleships out of mothballs to increase its capabilities.

In essence, we are witnessing the closure of a process that began in the fourteenth century, matured in the nineteenth, and grew senile in the twentieth century. In 1340, at about the same time that guns were making their appearance on land, the English used them at sea, at the Battle of Sluys.[1] Their appearance dramatically changed naval warfare. Prior to the advent of guns, warfare at sea was not dissimilar to warfare on land. Ships, primarily powered by oars, sometimes by sail, maneuvered in groups, to be able to approach enemy ships on the most advantageous terms. Fleets would close with each other, grappling lines would be tossed onto enemy ships, lashing attackers and defenders together, sometimes dozens of ships at a time, and then the real weapons of the fleet would show themselves. Sailors, armed with swords, knives, spears, clubs, and even rocks, engaged in hand-to-hand combat, just as infantrymen did in land warfare.

At times, ships used proto-artillery—mechanical catapults—to hurl heavy objects, flammable chemicals, or burning cotton at other ships. The side that lacked artillery solved the problem by closing rapidly with the enemy and grappling, thus negating the effectiveness of artillery. The inefficiency of this early artillery, both in rate of fire and targeting, meant that closure was a practical solution. This is what made cannon so attractive.

Traditional naval warfare placed a premium on a large number of smaller vessels that were simple to operate so that crews could carry out multiple duties. Projecting power at long range was obviously difficult, as the offensive force inevitably had fewer men and ships than the defenders. Cannon made it possible to damage enemy ships while keeping them at a distance—eliminating the manpower advantage and giving the upper hand to the technically advanced aggressor.

The introduction of cannon separated ships, much as cannon separated armies on land. The Mediterranean powers, particularly Venice, were conditioned by geography. Because the naval skirmishes in the Mediterranean usually took place in coastal waters in which oar-driven boats could manage more efficiently than sailing ships, galleys remained dominant. Galleys operated by using the controlled forward momentum imparted by oarsmen to ram other ships, then using naval infantry armed with bows, swords, and small arms to board enemy vessels. The galley, although perfectly suited for the inland waters of the Mediterranean and for nations able to concentrate relatively large numbers of ships and men because they were fighting close to home, fell easy prey to the larger sailing ships that were armed with cannon.

The Origins of the Gunboat

It is important to consider why European seagoing powers created the long-range sailing ship. The answer lies in the continent's geography and global politics in the fourteenth and fifteenth centuries. Europe, particularly Western Europe, had access to the Atlantic Ocean basin. Christian Europe's archenemy, Islam, controlled the prosperous centers of the Eastern Hemisphere, the Indian Ocean, and the Mediterranean Sea. The geographies of Islam's waters—the Mediterranean and the Indian— differs dramatically from the Atlantic's geography.

The Indian Ocean and the Mediterranean are both *enclosed* bodies of water. The Mediterranean has only one exit, Gibraltar, and that is easily closed. The Indian Ocean has several exits—the Cape of Good Hope around Africa, the long voyage around Australia, and the narrow passageways through the various Indonesian straits—although each is difficult to navigate and is subject to closure by hostile military action. Because the entire body of water is enclosed, and because of the configuration of land forming the enclosure, it was possible to navigate it without leaving coastal waters—at most moving across large bays or straits. More important, nothing outside these basins was as attractive as what was inside. Neither the Atlantic nor Pacific offered appealing alternatives to the rich commerce of the Mediterranean Sea or the Indian Ocean. The far side of the Indonesian archipelago was tempting, perhaps, but Chinese power was daunting. Islam, therefore, had every reason to remain within its wealthy basins, and little political or economic incentive to risk leaving them.

Long ocean voyages held no attraction for the Chinese either. Nothing to the east, in the Pacific, was worth the arduous voyage, while passage into the Indian Ocean basin meant naval conflict that could not be supported by ground troops. The various American Indian empires were similarly unmotivated to attempt a trip into the vast and empty expanses of the Atlantic and Pacific.

Nations didn't pour their national treasures into great explorations without compelling need and geographical opportunity. Exploration is fiendishly expensive. From Queen Isabella's financing of Columbus's trips to NASA's nation-breaking budgets, exploration, far from being a marginal adventure, has required the commitment of substantial portions of the national treasure. It must, therefore, arise from urgent necessity and not an adventurer's whimsical sense of excitement.[2]

The Moslems and the Chinese had no pressing reason to explore outside their regions. The European case was dramatically different. The part of Europe that fronted on the Atlantic Ocean was poor compared to better-endowed, warmer regions on other coasts. Given climate and geography, this situation was unlikely to improve. Thus, there was ample incentive to seek wealth elsewhere. However, the sea to the south, the Mediterranean, was closed by the Moslems and their alliance with the Venetians. To the east there was the even greater poverty of Central and Eastern Europe and the vast distances to Asia—hardly likely to attract traders and adventurers. To the north, these was even less to seek, in the bare existence of near-arctic Scandinavia. But if one could get around the Moslems to the south, reaching the west coast of Africa or the Indian Ocean itself, the Europeans, and particularly the Spaniards and Portuguese, knew that there was wealth to be found. They certainly had no idea what else they would encounter.

Ironically, Atlantic Europeans—technically proficient in both naval and military science, but poorer than citizens of most inland countries—out of necessity roamed through the maritime rimland of Eurasia and invaded the sequestered terrain of the great civilizations and imposed the first genuine global culture in human history, turning the rest of the world into the periphery of the Eurasian system.[3]

The seafaring skills necessary for navigating the Mediterranean or the northern Indian Ocean were far more modest than those necessary for the limitless bounds of the Atlantic, let alone the Pacific. Thus, the Atlantic Europeans were forced to develop navigational skills appropriate for open oceans.

Geography forced Atlantic Europe to create large, long-range vessels able to survive combat against an enemy that greatly outnumbered them in circumstances where there was no hope for reinforcement. Where Mediterranean vessels, galleys, were of short range but could be precisely controlled to concentrate overwhelming force close at hand, Atlantic vessels had to keep their distance and keep their options open. Global navigation required larger ships that were powered by some method other than oarsmen alone as well as a means of asserting power once the destination was reached. Sailing around the world with a fleet of galleys and a large infantry force was not feasible. Atlantic Europeans had to arrive in a few large ships, none designed for grappling or ramming, and force the natives to do business on the Europeans' terms. They needed a means for keeping natives at a distance, sinking their smaller ships, and pounding their ports. Quite obviously, guns were the answer.

If anything, guns at sea were even less accurate than cannon on land. A ship was an unstable platform, and firing had to be precisely timed to accommodate rolling decks. In the early sixteenth century, a crucial breakthrough occurred: cutting firing holes belowdecks. By doing this, the number of guns that could be brought to bear in a broadside soared, thus increasing the probability of a hit. By angling the guns right and left, the warship had nearly a 360-degree range of fire. Very quickly, an arms race was set off; various European powers—including the English, French, and Swedes—sought to mount the maximum number of cannons on board their ships. The Portuguese probably won the race, building the São João in 1554, carrying no less than 366 guns.[4]

Around the same time, the galleon, the first optimized sailing warship, was introduced. The galleon shared its lines with the galley. It was sleeker in design than traditional ships, lacking the ornate castles; it was also more stable and more maneuverable. Yet because it was longer relative to its beam, it could carry a full complement of cannon without detracting from agility.

The very size of these vessels contributed tremendously to their global capability. Large ships needed to carry large numbers of men, but they could also carry a great deal of supplies. A small group of ships could carry enough provisions for a multiyear journey. European vessels were able to carry out extended military operations without returning to base. These large ships, powered by sails, were not fast, but they had range and striking power.

Apart from changing the nature of ship design, the appearance of cannon aboard ship had another, world-historical consequence. To this point, the individual ship had always been extremely vulnerable, particularly in distant waters. Entering the coastal waters of foreign powers was especially deadly, as there was a great danger of being approached, boarded, and captured, either by the forces of foreign governments or mere pirates. Galleys manned by oarsmen could never travel far under any circumstances, as the stamina of the men limited speed and distance and, therefore, limited the chances for prosperous trade or plunder. But even the sailing ship, whose ability to travel was greater than the galley's, would have been severely limited had it only recourse to conventional arms.

A ship carrying cannon had an enormous advantage. Even traveling alone, it could move into areas where armed ships without cannon were the rule and prevent these ships from approaching. Indeed, the closer these ships came—which they had to do to allow boarding parties to storm the intruder—the more vulnerable they became to cannon fire. Thus, a small handful of cannon-armed ships—or even a single ship— could engage and either destroy or drive off enemy vessels, *even in coastal waters*. The closing in short for grappling gave the advantage to cannons. An additional and significant advantage was that the ship was able to project fire ashore, supplementing landing parties or in the place of such parties.

The cannon permitted the Europeans to go where they wanted and do what they wanted. Portuguese sailors went into the heart of the Indian Ocean basin, sweeping aside opposition. Vasco da Gama entered and dominated the Indian Ocean basin as early as 1502, through the sheer brutality made possible by his guns:

> We took a Mecca ship on board of which were 380 men and many women and children, and we took from it fully 12,000 ducats, and goods worth at least another 10,000. And we burned the ship and all the people on board with gunpowder, on the first day of October.[5]

By the end of this voyage, da Gama was able to force the Sumari of Calcutta to surrender and left behind a fleet of ships—with cannon—to enforce the peace.

The cannon completely tilted the balance of force at sea. Ships armed with cannon could survive in hostile environments. They could project force ashore, both by supporting landing parties with cannon fire and by removing cannon from the ship to the shore. Without equivalent armament, local leaders were forced to capitulate or reach some sort of accommodation. Thus, the Portuguese and Spaniards, who first moved out from European coastal waters, were able to impose imperial relationships on a mammoth scale through the use of the new artillery. By the seventeenth century, guns became the conventional weaponry on ships.

The English Commission on Reform in 1618 reported:

> Experience teacheth how sea-fightes in these days come seldome to boarding or to great execution of bows, arrows, small shot and the swords, but are chiefly performed by the great artillery breaking down masts, yards, tearing, raking and bilging in the ships wherein the great advantage of his

Majesty's navy must carefully be maintained by appointing
such a proportion of ordnance to each ship as the vessel will
bear.[6]

The Battleship as Reductio ad Absurdum

The sailing ship was indeed the foundation of European power, but it
could not sustain that power. Time was against it. Europe's empire was,
after all, global in scope. The fastest means of communications was no
faster than the fastest ship, and the movement of ships was subject to the
vagaries of tide, current, wind, and weather. Uncertainty in communications meant uncertainty in intelligence. Because news from the colonial
frontiers lagged by weeks or even months, European capitals did not
have any real sense of what was going on in their empires. By the time
word of a crisis reached the home country, it would already have been
resolved, usually unfavorably for the occupying power.

As a result, both political power and military force had to be dispersed throughout the globe. In violation of sound military principle,
forces could not be concentrated and held in reserve until decisive
action was necessary. They had to be located wherever trouble *might*
occur. Moreover, since crises could arise without warning, each distributed force had to be large enough to cope with problems without the
certainty of reinforcements. At the very least, troops were expected to be
able to maintain an acceptable stalemate pending the arrival of a larger
force.

Few European powers were able to deploy a force large enough
to protect their coastlines, their lines of supply to their colonies, and
at the same time provide a sufficient naval presence to control political and military developments within a colony and protect it from
encroachments by other imperial powers. When we add in the need
to maintain large land forces to protect against land neighbors in
Europe, we can readily see that Britain, without any land borders, had
an advantage over its European competitors. We can also see why lesser
powers—such as Belgium and the Netherlands—unable to protect
their borders at all, were able to develop enormous imperial interests.
Being in a geographically vulnerable situation allowed them the luxury
of concentrating their forces in their colonies, while relying on Britain
to control the sea-lanes and maintain the balance of power on the
Continent.

The dispersal of force not only required military power that was on the spot, it required political leadership as well—from viceroys to bureaucrats—dispatched by the colonial office with current intelligence, able to make command decisions as needed. The formal links of empire frequently hid the informal reality, which was that the colonies were effectively independent of the metropole in any practical sense. Indeed, at times colonial officials pursued policies that were quite at odds with the wishes of the capital.

The fate of the European empire could have followed the Roman model, as local warlords pulled away from the authority of the capital, barricading their own regions, making war on neighboring warlords and even on the capital itself in an effort to gain imperial power. Certainly, the underlying centrifugal forces were present within the European imperial system, and there are many examples of tension. Disintegration was headed off not because of the inherent superiority of European administrative techniques over Roman, but because technical innovations increased the speed with which politico-military intelligence moved from the imperial periphery to the center and decreased the time it took for forces to move from the center to the periphery. This resulted in more direct political control being exercised on regional governors, who now had to call on the capital to dispatch forces that would continue to be under central, not regional, control. At the same time, because it was not necessary to predeploy a large number of troops, the cost of maintaining an imperial force was dramatically decreased, to the advantage of the central government.

The ability to increase the speed of intelligence and the projection of force was of obvious interest to all the European powers—but particularly to those, like the British, who were passing into a conservative stance after acquiring vast imperial holdings. For them, the problem was holding on to what they had. Thus, the British were particularly eager and able to take advantage of technical innovations. Steam-supplemented sailing ships and sail-less coal-powered ships allowed them and the other European powers to reach their colonies faster, as well as to overcome transitory factors such as weather and current. One might say that, for the first time, communications and movement became predictable and could, therefore, be planned. The introduction of wireless communications improved intelligence and, thus, increased the need for speedier transportation.

Certainly, the main threat to getting troops to troubled regions quickly and at will did not come from the occupied territories, but from other European powers. The inability of Europe to form a united entity during the nineteenth century meant that each nation had an interest in undermining the other's imperial hegemony, and each was motivated to increase its own imperial power. As France's territorial holdings grew, its ability to mobilize the resources of North Africa and Indochina grew as well, increasing its power in Europe. As Germany united in 1871, it found that its power was limited by its lack of colonies that could be exploited, the most attractive ones having long been usurped by its rivals. It badly needed to pry at least some of these colonies out of the hands of the other imperial powers as well as to seize what little was left untaken.

Thus, the nineteenth century became a period of naval competition, as powers sought to preserve their national security at home by securing colonial resources. The emergence of a great Central European power in the late nineteenth century—Germany—which was both politically insecure because of its geopolitical position and capable of tremendous industrial efforts, challenged the global dominance of Atlantic Europe for the first time. The stable relationship established after 1815 between Britain and France was undermined by the German thrust for geopolitical parity. Add to this brew the emergence of the U.S. Navy as a global power at the turn of the century plus the simultaneous rise of Japanese naval power, and Atlantic Europe suddenly saw itself exposed on all fronts.

The long line of supply that ran from the northwest corner of Europe through the Suez Canal or around the Cape of Good Hope into the Indian Ocean basin to China and the Pacific was vulnerable to interdiction. European power depended on the ability of each nation to control the sea-lanes to its colonies and protect those colonies from encroachment. A weapon was needed that could protect coastlines, control sea-lanes, and project power ashore into coastal areas, all without requiring reinforcement.

The battleship was born out of this geopolitical problem. The distances involved required fast ships. The mission demanded powerful guns, both to enable engagement at a distance beyond the range of other ships and to penetrate the armored hulls of the enemies. To carry these guns, large vessels were necessary, and to survive the impact of shells, heavily armored hulls were required. The result was a constant race between speed, armor, and armaments while the traditional surface warship matured into the battleship and the battleship grew and grew.

In 1850, the average battleship displaced about 7,000 tons. By 1900, this amount had grown to over 15,000 tons; by 1920, to over 30,000 tons; and by World War II, to nearly 60,000 tons. Thus, in one hundred years the battleship had grown nearly 900 percent.[7] The weight of guns surged as well, from 4.1 tons and a 110-pound shell on HMS *Warrior* in 1861, to 57.7 tons with 850-pound shells on HMS *Dreadnought* in 1906.[8] The geopolitical effectiveness of the vessels did not, however, increase. The earlier ships had been just as effective against merchantmen and contemporary warships as their behemoth descendants. The forced growth was simply parasitic, reflecting the increased burden of carrying out the same basic mission.

The battleship was deadly to anything that came within its range. The problem was that its range was severely limited. Even as its aim became more and more accurate and the size of its projectiles grew exponentially, the distance it could fire its shells increased but a matter of miles. At best, a battleship could command about a thirteen-mile radius—a little over five hundred square miles of ocean. The Mediterranean, for example, is 971,000 square miles in size, and it would take 1,942 battleships to provide complete coverage. The alternative to complete coverage would be to concentrate forces at strategic points or to maneuver enemy formations into the range of a group of battleships.

The obvious problem with this strategy is that, in bringing its guns to bear, the fleet must expose itself to enemy gunfire. Marginal advantages in range and speed are important, of course, but to take advantage of that benefit, perfect information about the enemy's deployment and capabilities is necessary, as well as perfect command and control of one's own forces. More likely, both fleets will come into the other's range, and the result will be a brawl, with victory to the quickest and luckiest.

The usual tactical solution to combat between sides whose weapons are closely matched is surprise. Surprise could create temporary advantage, but it would be difficult to achieve with a weapon that had to close to within thirteen miles to fire. Particularly after the introduction of seaborne radar, the chances of surprising a battleship fleet became minuscule. But even prior to this, the limitations of shipboard gunnery meant that the enemy fleet would be seen either by the battleship or by destroyers screening the fleet's movements.

The battleship had moved to stalemate. One of the sidelights of the senility of the battleship was that the differentiation of the technical capabilities was marginal at best. Indeed, the last great battleship fleet action, Jutland, during World War I, ended in a tactical draw, and victory to the British only by default. As with Goliath's and Saul's armor, the solution was not to be found in upgrading the offense and defense incrementally, but in cutting the Gordian knot with a radically new, simpler technology.

Like Goliath, the battleship could defeat anything it could reach. It could move around within a very small area at only about twenty knots. Thus, the trick to defeating the battleship was to engage in a variation on David's theme—find a weapon that was beyond the battleship's reach and hurl inexpensive projectiles at it.

Though the force driving the development of the battleship was the threat of other battleships, prevailing against enemy vessels was not an end in itself, merely a means toward an end. The primary goal was control of sea-lanes: having the ability to sail merchantmen from home to the colonies, to interdict the flow of merchantmen belonging to other nations, to protect against amphibious assaults, and to bombard enemy coastal positions. The problem was, all of this could be achieved by vessels other than the battleship. Indeed, a larger number of faster, smaller, and more lightly armed vessels would have been preferable for the task, since the targets were unarmed or lightly armed merchantmen. The battleship was necessary for the secondary mission of surviving until the primary mission could be carried out.

What began as an efficient, indeed elegant, solution to the problem of sea-lane control became encumbered with extraordinarily costly defensive measures that aided the primary offensive mission not one bit.[9] One cannot underestimate the expense. From 1850 to 1945, the expense of a single battleship increased more than two hundred times.[10] The *Maine,* ordered in 1896, cost the U.S. Navy $4.7 million. The *North Carolina,* ordered in 1937, cost $77 million, while the *Iowa,* ordered in 1940, cost over $100 million.[11] Improvements occurred over time, but not on the order of increased costs.

The pressure that battleship construction placed on national budgets was dramatic. In 1899, the German naval budget was about 133 million gold marks. The *Wittensbach,* whose keel was laid in that year, cost 22.7 million gold marks—nearly 20 percent of Germany's naval budget was being spent on a single weapons platform. By 1906, the naval budget had risen to 233 million gold marks, an increase of about 75 percent. The *Helgoland,* ordered about then, cost over 46 million gold marks, more than double the cost of the *Wittensbach.*[12] During those years, German GDP increased by only 27 percent.[13] Defense spending in general, and the cost of battleships in particular, were clearly outstripping German economic growth—as it was outstripping growth among all of the great powers.

This meant that the battleship began to cut into the strategic and operational strength of a nation. Resources that could be spent on other purposes—expanding land armies, constructing defensive fortifications, artillery, and, later on, aircraft—were set aside for a decreasing number

of more expensive battleships. And while it is true that their performance improved, their effectiveness did not. The *Maine,* for example,
had a sailing radius of about five thousand miles at ten knots, and a maximum speed of eighteen knots. It had four twelve-inch guns.[14] While the
North Carolina had twice the cruising range of the *Maine,* the effective
range of its much larger guns was only marginally better.[15] Thus, while
the *North Carolina* was certainly a more capable ship, it was not necessarily a more effective ship.

Though the balance of power remained stable throughout the first
part of the twentieth century, the effect of maintaining that stability
shifted—until the burden became something that only great powers
could bear, and then only with difficulty. It was no surprise, therefore,
that at the end of World War I, the United States and Britain sought to
stabilize the relative strength of their navies by imposing a treaty limiting
the number and tonnage of battleships. They were trying to maintain
strategic superiority—and the economic benefits deriving from it—by
arresting the arms race. They failed because they sought to preserve a
status quo that was inherently unacceptable to the third great naval
power—Japan. Thus, the arms race began again.

The battleship's primary mission, sea-lane control, was no more
effectively carried out in 1913 or 1938 than it had been in the 1870s
when it was first introduced. The sea-lanes of the great imperial powers
were no more secure in 1900 than they were in 1880. Improved performance did not expand effectiveness, it merely preserved it. Firepower
against unarmed merchantmen stayed steady, as did the ability to prevail
against enemy ships of war. Speed increased marginally at best—but so
did the logistical complexity of fueling the hungry beasts. To improve
the battleship's capabilities, greater and greater resources had to be
diverted away from other missions.

The battleship was the classical case of senility in a class of weapons:

- It served a critical function throughout its history, maintaining
  sea-lane control for Europe's maritime empires and, later, for
  American imperial interests. It operated autonomously at long
  distances, projecting great firepower against lesser naval vessels
  and shore targets without requiring reinforcements. It could
  move fast, hit hard, and survive.
- Under pressure from other battleships, it maintained its relative
  survivability and lethality without substantially reducing its performance—albeit at great cost.
- The cost of maintaining its effectiveness undermined a nation's

ability to field other forces and maintain its economic well-being.

- The battleship's technology—cutting edge in the late 1890s, fueling development in related civilian fields—became obsolete in the 1930s, draining resources from newly emerging technologies.

- Lacking an effective rival, the battleship continued to dominate, growing larger and more expensive with each new keel.

- Finally, the senile battleship was blasted out of existence by a simple technology able to take advantage of its weaknesses, take over its mission, and contribute to new revolutions in civilian technology.

The ultimate problem of the battleship was that, when all was said and done, a ship costing a large portion of a nation's treasure could hurl a few hundred pounds of explosives a little over a dozen miles. The problem was not that it did not do this well—given what it cost, it would have to. The problem was that this was all it did. It could and did destroy any vessel clumsy enough to blunder within its tiny kill zone, but with increases in naval intelligence from sensors—for example, maritime reconnaissance in World War I and radar in World War II—the probability of such blunders declined. In the twentieth century, the battleship grew into an enormous, deadly Goliath. Like the original Goliath, the battleship appeared fearsome, but it had its weaknesses. One came from beneath the sea. But the David that actually slew Goliath came from above—the carrier-borne plane.

After the Fall: Surface Warfare after the Battleship

It is generally agreed that Pearl Harbor marked the rise of the aircraft carrier and the fall of the battleship. But this argument is not altogether correct. Pearl Harbor was not a battle of fleets but an assault on immobilized, virtually undefended ships. The battleship was not killed by the Japanese at Pearl Harbor but by the Americans at the Battle of Midway, some six months later.

Midway was a genuine naval battle, and a true test of the aircraft carrier's capabilities. The Japanese realized that they had committed a potentially fatal error during the attack on Pearl Harbor. While the first two waves of attackers had destroyed the ships and planes, a third wave, designed to obliterate oil tanks and dry docks, had not been launched, and Pearl Harbor not only remained usable, it was becoming the core of a vast American buildup that the Japanese couldn't hope to match.

By taking Midway the Japanese could base long-range bombers on its airfield. This meant they could harass U.S. ships at Pearl Harbor and, by increasing the dangers, potentially force the still fragile fleet to withdraw to San Diego or San Francisco. This would forestall the feared American counteroffensive and make any buildup in Australia difficult if not impossible.

This mission was so important that the entire Imperial Japanese fleet sailed, under the command of Admiral Yamamoto himself, along with the entire Imperial naval staff. The fleet was divided into three parts. The fast carriers were to pound Midway from the air, while screening the landing group from attack by American carriers. Supporting these two groups was the bulk of the Japanese fleet, including Japan's huge battleships. Facing them were three American carriers, including the crippled *Yorktown* and a handful of cruisers and destroyers as screens and escorts.

By a stroke of luck—and by having penetrated the Japanese naval code—the Americans were able to strike first, sinking Japanese carriers, while losing only the *Yorktown*. Japan's landing force was intact, and the surface battle group under Yamamoto's direct command was untouched. Yet the battle was over. Yamamoto broke off contact and abandoned the attack, and Japan went over to the strategic defensive, never again to mount a serious offensive against the United States. Yet the bulk of its fleet was unscathed. Only a handful of carriers had been lost. Why did Yamamoto give up the battle?

The two American carriers would have been helpless had they encountered Yamamoto's battleships within range of their huge guns. Yet Yamamoto did not attempt to pursue them, because he understood that to do so, his battleships would have to pass through the hundreds of miles of deep kill zone surrounding the carrier *before* they could bring their guns to bear. The time a battleship would have been at risk negated, in Yamamoto's mind, the prize that could have been won.

At Midway, an intact, experienced, well-led surface fleet built around superb battleships declined to do combat with two aircraft carriers. Range had redefined the correlation of forces. The lethality of the guns, and even their accuracy, meant nothing if they could not reach the enemy. A carrier with a complement of torpedo planes and dive bombers was, therefore, safe from attack by battleships. The battleship was at risk from the moment it came into range of the planes.

The obvious lesson, learned by the United States after World

War II, was that the battleship represented a drain on resources without adding substantially to the combat power of the Navy. The descendants of da Gama and Drake, Nelson and Farragut, began to pay homage to the mighty aircraft carrier, which could patrol so wide an area that two or three of them were enough to provide complete coverage of the North Atlantic. The famed battle lines of the past were reorganized into carrier battle groups. Few roles remain on the high seas for the warship except to provide support.

In the carrier battle group, surface weapons served two functions. Small, agile destroyers guarded against enemy submarines that might try to slip in and torpedo the carrier. Cruisers and some destroyers were assigned the task of supplementing the carrier's own combat air patrol, which screened enemy aircraft hundreds of miles out. These ships, carrying the antiaircraft systems of the day—from old pom-pom guns to today's Aegis antiair system—guard the fleet against aircraft that might slip through or against missiles fired from outside the combat air patrol's range.

The great expense of building carriers froze virtually all powers other than the United States out of the game. Some nations continued to build carriers that could launch vertical takeoff and landing (VTOL) aircraft, but the combat performance of these was so poor when compared with the conventional aircraft launched off American carriers that there was no point in pursuing it.[16]

In effect, most Eurasian nations have abandoned the blue waters or traveled there at the sufferance of the United States. The only real requirement they made of their navies was to protect coastal waters from intrusion. Even the Soviets, when the structure of their navy is carefully examined, used their surface ships primarily to patrol their northern shores against intrusions by NATO. To the extent that they planned to cut supply lines between the United States and Europe, their plan was to use submarines and long-range, land-based aircraft such as the Backfire bomber. They gave no serious thought to sorties by surface vessels against carrier battle groups.

But along the coasts of Eurasia, surface vessels flourished. One reason was that land-based aircraft could support surface operations along the coast, much as carriers supported surface ships. Another, and perhaps more important, reason was that coastal vessels abandoned their traditional guns and replaced them with surface-to-surface and surface-to-air missiles. Coastal boats thus became gunboats.[17]

The Soviets had pioneered this evolution with the development of the Styx missile. When fired from a small surface ship such as one of the

Komar class, it destroyed the Israeli destroyer *Eilat* in 1967. Having learned their lesson well, the Israelis acquired from France, in part by theft, a fleet of Saar-class fast-attack craft. Instead of mounting machine guns or small-caliber weapons on these agile boats, they installed their own surface-to-surface missile—the Gabriel.

The marriage of fast-attack craft with Gabriel missiles proved a brilliant stroke during the 1973 war, when the Israeli Navy used them to devastate the Syrian Navy, then turned these missiles on coastal targets. Even more important, Israeli Saars attacked the northern flank of the Egyptian Army in the Sinai, smashing their concentration of surface-to-air missiles and opening a hole through which Israeli aircraft could fly to strike at the Egyptian rear. Within the limits of its range, the fast-attack-craft/Gabriel-missile combination proved brilliant and deadly.

There were two key factors to this:

- The Gabriel wasn't a ballistic missile by any means; it was a guided missile—guided by human and its own programmed intelligence. It was steered to its target first by a shipboard radar and radio guidance system and then by a sensor. The probability of one Gabriel hitting its target was in excess of 50 percent; fire two of them, and your target was probably killed.
- The Gabriel was designed to be fired from patrol ships as small as fifty tons, which were designed to evade enemy counterfire. Small size makes a ship hard to see, fast, and maneuverable.

Fast-attack craft (FAC) and patrol boats are between 90 and 150 feet long, weigh between 150 and 300 tons, and can accelerate quickly to speeds in excess of 35 knots.[18] Their weakness is their virtue—their size. They cannot be used for the projection of global power. And they have limited range—about 4,000 miles at 17.5 knots. FACs are well suited for the role chosen for them by the Israelis and other secondary powers with naval interests that do not take them out on the high seas, but the intercontinental, multimonth tours of duty required by American geopolitical interests are beyond them.

The United States has emulated part of the Komar/Saar strategy by arming larger warships with antiship missiles such as the Harpoon. Though the missiles are deadly, large American warships are, by virtue of their size, extremely vulnerable to countermeasures. The ability of an Iraqi plane to fire an Exocet missile at the USS *Stark,* a well-armed guided-missile cruiser, and hit it dead on at night is an indication of just how vulnerable. The situation is this: surface vessels capable of extended global travel must be large enough to sustain life with a degree of comfort, remain on station indefinitely, and handle well on the high seas. They must also be hard to see, highly maneuverable, and very fast. As can easily be seen, the first and the second sets of criteria are incompatible.

One option is to permit surface vessels to operate only in the company of aircraft carriers. But, as we shall see, this may well prove to be no solution at all, as aircraft carriers may be the most vulnerable ships of all. The entire future of the carrier battle group is at stake and, with it, U.S. naval strategy. A second option is to design a surface ship that is both global in reach and survivable. This is a task to which the U.S. Navy has devoted a great deal of thought—as well as money. In 1983 the Navy began a black project utilizing the same concept, the same secrecy and facilities, that the Air Force had selected for its bombers: stealth. On the surface, the stealth ship, dubbed the *Sea Shadow,* resembles the stealth aircraft and is intended to achieve the same end—to present as small a cross section as possible on enemy radar, by absorbing what radiation it can and deflecting what it cannot.[19]

The attempt to make a 560-ton, 160-foot-long, 70-foot-wide ship invisible to radar was, as one would expect, extraordinarily expensive.[20] Some things were learned from it that have been applied to the design of new ships. Other things were learned as well—the limits of stealth.[21] It has been reported, for example, that some new designs avoid radar so well that they give away their position because of the blank area on a radar-reflecting sea. But while it is certainly possible to avoid radar detection in the visible spectrum, avoiding detection in all spectra at the same time is difficult if not impossible. Ships give off heat, noise, reflect light, have wakes, and on and on. It is extremely difficult to hide them in all spectra.

The *Sea Shadow* was honorably retired in the summer of 1995. What could be learned had been learned.[22] Unfortunately, the lessons were learned by everyone. Even the Swedes were building stealth ships.[23]

One aspect of the *Sea Shadow*'s design needs to be noted—SWATH (small waterplane, area-twin hull). By creating a hull that lifts out of the water, or lowers itself down, and which planes across the water as well, designers of the *Sea Shadow* found a way to make smaller vessels that operate in shallow littoral waters manageable on the high seas. By doing this they opened the door to new designs that hold the key to all forms of stealth—small size. One suspects that the *Sea Shadow* will be remembered for this characteristic more than for the putative virtues of stealth.

Other types of surface vessels, sea-effects ships (SES) and air-cushion vehicles (ACV), promised to provide high speeds, extreme maneuverability, and the ability to operate in a wide variety of depths.[24] Even if they solve the high-seas handling problem, their small size limits their ability to endure extended deployment. Their pared-down crews and limited living accommodations make them more suitable for coastal patrol and amphibious warfare than for sea-lane control.[25]

The U.S. Navy—with its continuous concerns about blue-water, sea-lane control issues—continues to think of surface warships in fairly conventional ways. The Navy is particularly proud of the Arleigh Burke–class (DDG-51) guided-missile destroyer, regarded as a genuine innovation in surface warfare. In a fascinating article arguing for the Burke-class destroyer, Navy commander Alan G. Maiorano focused entirely on the survivability of the $850 million warship rather than on any new offensive capability it would have.[26] The most innovative feature of the Burke is its ability to integrate Aegis technology with next-generation, supersonic, highly maneuverable, sea-skimming antiship cruise missiles. But in the end, the *offensive* capability of the Burke will be antiship missiles and antisubmarine helicopters, not unlike those flown by predecessors. As with bombers and tanks, considerable money is being spent to keep the surface warship alive—not to increase its effectiveness.

The same can be said for the warships being planned for the twenty-first century. This new generation of surface warship, dubbed the SC-21, incorporates interesting technical innovations, such as new electric-drive propulsion systems, and concepts, for example, the idea that the core of the ship should be interchangeable with destroyers, frigates, and other types of ship. But even when everything learned from stealth technology and SWATH is included, it remains a large, unmaneuverable sore thumb sticking out of the ocean.

What is striking about the SC-21 proposal is how hard it strains to be original, and how little originality is possible. The frigate version of the SC-21—dubbed FF-21—has as planned performance goals:

Cruise speed sustained	18+ knots
High speed	32+ knots
Endurance	5,000 nautical miles at 16 knots per hour
Stores and provisions	30 days[27]

As with fighter aircraft, the warship's ability to perform in dangerous circumstances has increased. But its speed, range, and endurance remain the same. It cannot go faster or farther than its predecessors. Again, it is a case of running hard to stay in place.

According to a draft issued by the chief of naval operations, the SC-21 is supposed to be able to "launch and support precision strike weapons . . . provide fire support for amphibious operations, and protect friendly forces from enemy attack through the establishment and maintenance of battlespace dominance against theater missile, air, surface and subsurface threats."[28]

The SC-21, like the DDG-51, will integrate the Aegis antiair, antimissile system, albeit in a more advanced and, therefore, more expensive form.

Both the DDG-51 and the SC-21 are Aegis systems to the core. As such, they are fundamentally defensive platforms. They are designed to survive. Now, survival is an essential prerequisite for carrying out the offensive mission, but it is not, by itself, a mission. Neither the Burke nor the SC-21 will carry increased sea-control power. The ability of both these vessels to engage enemy warships and destroy enemy merchantmen rests on the same suite of weapons—albeit upgraded—that preceding destroyers deployed.

The Harpoon antiship missile and its variants are, of course, deadly systems. The issue, as with tanks and bombers, is whether delivering them requires that an $850 million vessel go in harm's way. As it becomes possible for surveillance and targeting provided by satellites and unarmed aerial vehicles to cover the bulk of the oceans from land bases, the point of sending the SC-21, with all of its fiendishly expensive defensive systems, into combat will become more and more dubious.

The primary purpose of the SC-21, FF-21, and DDG-51, like the other destroyers, frigates, and cruisers before them, is not to exercise sea-lane control but to protect the aircraft carriers that are intended to be the real instrument of sea-lane control. This fact alone should trouble us. The idea that the aircraft carrier can only carry out its mission with a cluster of near-billion-dollar air-defense and antisubmarine platforms at its side raises the real question: Does the strategic effectiveness of the aircraft carrier justify the expense of keeping it alive, or is it too becoming senile?

Thus, this discussion of the vulnerability of the surface warship is merely preliminary to the key issue for the U.S. Navy—the future of the aircraft carrier. If the aircraft carrier can survive, then so can any other surface warship. If the aircraft carrier is destroyed, everything else will go down with it. Of course, the question is not simply whether the carrier can stay alive, but what price will have to be paid for literally and figuratively keeping it afloat. Should it survive at a cost that undermines the national security as a whole? How does the cost and effectiveness of Aegis compare to the power of the carrier battle group? Can the surface warship survive?

8

The Aircraft Carrier as Midwife

The modern aircraft carrier—along with the attending warships that together constitute the carrier battle group—is the military foundation of today's Pax Americana. It serves two functions. The first is to guarantee U.S. domination of the sea-lanes. American carrier battle groups can establish control virtually anywhere on the blue ocean or in the littoral surrounding the continents. Through wars in Korea, Vietnam, the Persian Gulf, and the former Yugoslavia, they have never been challenged at sea. Rather, they have been impregnable fortresses that could hurl airpower hundreds of miles inland, without themselves being at risk. Indeed, this is the second mission of carrier battle groups—to project airpower ashore.

In countless operations since World War II, the United States has been able to deliver to a region an air force that dwarfed the indigenous one. These aircraft have served to support ground operations, cut lines of supply and communications, attack command and control centers, and strike at the industrial infrastructure of the enemy. So efficiently does the aircraft carrier perform its role that in countless near-war circumstances around the world, the arrival of one or more battle groups offshore has had a sobering effect on political adventurers.

Aircraft carriers, like most weapons, began their history as inexpensive alternatives to a much older and more costly technology—battleships. Initially, they seemed to be much cheaper to produce. But as usually happens, the need to counter emerging threats made the carrier's price tag soar, even as it retained its effectiveness. Between 1920 and 1940, carriers and their aircraft increased two hundred times in unit cost. From 1920 to 1980, it increased two thousand times.[1]

To a great extent this growth was caused by the greater price of aircraft and, as such, is part of the cycle of senility we have discussed earlier. But the price of the carrier itself also rose dramatically. In addition, this estimate does not take into account the cost of the ships that attend the carrier—including cruisers and destroyers that protect it from air threats that its own combat air patrol can't handle; destroyers and frigates as well as hunter-killer submarines for antisubmarine warfare duty; and numerous support vessels.

One of the reasons American carriers have been so effective is because the aircraft that have flown off them have been as capable as land-based aircraft. Land-based aircraft—even single-engine fighters without heavy bomb loads—are heavy, which means that they need long runways for their engines to reach takeoff speed. These long runways are not available on carriers. The American innovation, simple in conception, complex in design, and revolutionary in political significance, was the steam-powered catapult. What the catapult did was accelerate the aircraft dramatically, so that, with its engine at full throttle, it would be hurled off the end of the carrier deck at a speed sufficient to make it airborne. The stresses involved in takeoff and landing meant that the aircraft had to be particularly robust, designed to withstand high g's. This drove up the price of carriers and their planes dramatically. But because of the catapult, attack aircraft could carry substantial tonnage of munitions to the target. The catapult was the key to the sustained naval air campaigns in Korea, Vietnam, and the Persian Gulf. It also was what prevented other navies from challenging American naval hegemony.

After World War II, each of the prewar carrier powers fell behind the United States. Japan, obviously, had had its carrier force destroyed. The British pursued a catapult-carrier policy for a while, then abandoned it in favor of other, lesser platforms. The French, dreaming of glory, did build some catapult carriers, but they never had enough to do more than assert a regional influence, nor, until very recently, were the aircraft used on board superior to first-rate land-based aircraft. A few other nations, such as Argentina and India, purchased outmoded carriers, but these were ineffective. Today, however, a more serious attempt to sail carriers is under way in India and possibly China.

The most important challenge to American hegemony, of course, came from the Soviet Union. The Soviets, tied down with land disputes during the Cold War, were not in a position to divert resources to a massive naval buildup. Until the mid-1960s, they were forced to be content with a coastal navy designed to deter intrusions and amphibious operations. The Cuban missile crisis drove home the fact that, without naval power, Soviet global interests could not be protected. The ability of the United States to seal Cuba off from Soviet supplies, and the inability of the Soviet Union to do anything in response beyond deploying submarines that were themselves vulnerable to U.S. ASW, caused a massive expansion under the command of Marshal Gorshakov, a brilliant naval strategist and father of the Soviet naval expansion.

The Soviets understood that the complex technical skills required to build catapult carriers and aircraft able to survive catapult takeoffs and carrier landings and to train the pilots and crews for such operations would not be acquired anytime soon. Indeed, development of the first Soviet catapult carrier was just getting somewhere when the Soviet Union collapsed. Rather than pursuing catapult-carrier operations, the Soviets chose other routes.

One route was to find an aircraft that could take off from a carrier deck without needing to be accelerated to ordinary takeoff speeds—this was the very short takeoff and landing (VSTOL) plane developed by both the Soviets and the British. The British Harrier (also used by the U.S. Marines) and the Soviet Yak-38 took off and landed with minimal roll or a shorter deck, tilted upward in ski-jump style. Powerful engines deflected thrust downward, thereby lifting the aircraft off the deck by brute power, then accelerating it to achieve aerodynamic lift.

The problem with VSTOL aircraft was that a structure designed to endure these stresses and an engine designed to generate these thrusts tended to perform less effectively in air-to-air combat than either land-based or advanced U.S.-carrier-based aircraft. In addition, neither the British nor the Soviet VSTOL carriers were large enough to transport sufficient aircraft for sustained operations against a relatively sophisticated enemy. Nevertheless, the development of VSTOL raised the possibility that nonconventional aircraft could equal conventional aircraft in performance. Were this to be achieved, it might be feasible to abandon the carrier altogether and distribute aircraft throughout the fleet—even aboard merchantmen.

The Soviets knew they could not challenge U.S. carriers, guarding the supply lines to Europe, with surface vessels and that their VSTOL aircraft were not up to the job. They also knew that their long-range aircraft, such as the Backfire bomber, did not stand a chance against F-14s standing guard hundreds of miles from the combat group.

Their solution was long-range missiles, which could be fired before the Backfires were in range of the F-14s and which could use their own homing devices to find and destroy the American carriers. The possibility that these bombers and their missiles might be deployed set off an earthquake among U.S. naval strategists. The purpose of the carrier's air patrol was to intercept enemy aircraft before they came into bombing and torpedoing range. Soviet antiship missiles were threatening because they could be released *before* the aircraft carrying them came within range of the U.S. planes. Unless the size of the combat air patrol could be expanded dramatically, it would be necessary to intercept the missile itself. This was difficult, if not impossible, considering fuel needs and the limited number of aircraft available when the threat began to emerge in the 1950s.

What was needed was a defensive technology that could deal with missiles. The system that was eventually developed consisted of sensitive, high-speed radars able to intercept missiles at various distances. Close-in weapon systems (CIWS), radar-guided, rapid-fire guns that looked like and were called R2D2 from the movie *Star Wars,* would be the final line of defense.

The name of all of this was Aegis, and it was so complex and cumbersome that it needed its own ship—a cruiser in fact, although some destroyers could sometimes be used—to house it. The Aegis cruiser, or sometimes two when the direction of the air threat could not be known with certainty, now had to accompany the aircraft carrier, adding billions to the cost of the carrier battle group's construction and billions more to its operation.

One problem with Aegis was that it added nothing to the carrier's offensive mission. It was merely a means toward that end. Another problem was that the goals of Aegis were beyond attainment. No matter how many missiles the Soviets fired, every last one of them had to be destroyed or decoyed. If even one leaked through, the carrier might be rendered inoperable for a time—if not destroyed outright. The effectiveness of the carrier therefore depended on a system that had to be foolproof against a potentially limitless number of incoming missiles.

Indeed, as missiles developed—their speeds and maneuverability increasing, the angle of attack shifting and becoming less predictable—the Aegis system could only respond with more and more expensive countermeasures.

Unless aircraft can carry greater loads—requiring a new breakthrough in jet engines—or carriers can grow much larger, making them even more of a target, the tonnage of munitions they can throw at the enemy cannot expand much. In the end, the cost of defending them outstrips the damage they can inflict.

The definitive question is this: Is the carrier the only way to place a certain tonnage of explosives on a certain point of the globe at a certain time? When it becomes clear that a new alternative is available, then the carrier will go from senility to extinction. Its career—short, brilliant, and doomed—will prove to be a bridge from one epoch to another.

Between Epochs: The Aircraft Carrier's Rise and Fall

In one sense, the aircraft carrier is a classic, European-epoch warship. Powered by hydrocarbon engines—except for a few recent American specimens powered by nuclear reactors—it is a surface warship that moves at about the same speed as other surface warships can move. It differs from the traditional vessel only in its armament: a hybrid between the old and the new. If we think of aircraft as projectiles filed by the carriers, then we see the fundamental difference between these projectiles and those fired by a battleship: these projectiles are intelligent. The pilot is embedded in the carrier aircraft, and with intelligence he guides the projectile to its target. Paradoxically, throughout most of the history of naval warfare, the intelligence of the pilot was obviated by his reversion to ballistic warfare at the very moment he released his dumb bomb or torpedo. Fortunately, this aberration is being rectified today.

The airplane revolutionized not only the architecture of the warship but the structure of naval warfare as well. It increased the effective range of naval gunnery from a few miles to hundreds of miles. Assume that a carrier-launched aircraft had an operational range of 250 miles. A carrier would have an area of coverage just under two hundred thousand square miles, about four hundred times greater than a battleship. Five aircraft carriers could cover the Mediterranean, where nearly two thousand battleships would have been required. And they could do this for .05 percent of the cost.

Obviously, as range increased, the quantity of munitions available per square mile for a given period of time decreased. During World War II, a North Carolina–class battleship carried nine sixteen-inch guns firing 2,700-pound shells about 41,600 yards or 23.6 miles. Each gun was able to fire about two rounds a minute.[2] The ship carried a theoretical maximum load of seventy-five rounds per gun, or 675 rounds for a total of 1,822,500 pounds of total explosive tonnage—911 tons.[3] The North Carolina class, therefore, could cover an area of 1,750 square miles with 911 tons of munitions in about 37.5 minutes, firing its guns at maximum rate for the maximum period of time. This gives a saturation rate of about a half ton of explosives per square mile, or 0.0139 tons per square mile per minute. Alternatively, we could say that the North Carolina–class battleship could fire 24.3 tons of munitions a minute for up to 37.5 minutes. Within the range of the battleship, therefore, the battleship was a devastating weapon system.

The North Carolina–class battleship was a slightly prewar design. A reasonable comparison could be made with the Essex-class aircraft carrier, which made its appearance during the war. An Essex-class carrier, operating during 1943, carried only about 424.8 tons of munitions, less than half that carried aboard a battleship. Unlike a battleship's main armament, this was broken down into a variety of specialized munitions:

MUNITIONS ABOARD ESSEX-CLASS CARRIER IN 1943

Pounds/	100	500	1000	1000	1000	1600	2000	325	100	2000	Total
Type	GP	GP	GP	SAP	AP	AP	GP	DB	INC	Torp	
Quantity	504	296	146	129	110	19	19	296	296	36	1851
Total Tonnage	25.2	74	73	64.5	55	15.2	19	48.1	14.8	36	424.8

GP = General Purpose   SAP = Special Armor Piercing   AP = Armor Piercing   DB = Depth Bomb   INC = Incendiary   Torp = Torpedo

Assuming an effective combat radius for the Vought F4U of three hundred miles, an Essex-class carrier could cover 282,600 square miles of ocean. On the other side of the ledger, its 424.8 tons of munitions, allowing only 0.0015 tons per square mile, compared to 0.5 tons per square mile for a North Carolina–class battleship, gave a coverage only 0.3 percent as dense. Moreover, where a battleship could fire its entire tonnage in a little over a half hour, a sortie by all ninety aircraft aboard an Essex-class carrier, even assuming that each of them was assigned a strike role and that all were able to take off with maximum ordnance load, would carry about ninety tons of munitions, requiring about five sorties of all aircraft to deliver the full load, a matter of days rather than hours. In reality, where thirty-six of the planes on a carrier were dedicated air-superiority fighters, eighteen were scouts capable of bombing, eighteen were dive bombers, and eighteen were torpedo planes, a maximum surge would involve only fifty-four tons, perhaps two sorties per day, and a whole four days to deliver the full load. Where a battleship could deliver 24.3 tons a minute, a carrier could only deliver 0.07 tons per minute, about 0.2 percent of the battleship's rate.

Within the effective range of their guns, however, battleships were far more deadly than were aircraft carriers. The quantity and quality of munitions available per square mile, per minute of combat, were sub-stantially greater. They were more likely to detect a threat within their range and more likely to destroy it. Yet none of this mattered. Range of weapons overwhelmed every other consideration. The aircraft carrier could engage the battleship—or any gunned ship—miles and hours before the battleship could return fire.

It is important to remember that the first use of aircraft at sea was as scouts, as intelligence-gathering platforms. In this sense, the develop-ment of naval airpower at sea closely paralleled the development of air-power on land. And, in the same way that airpower on land quickly began locating targets beyond the range of artillery, so too naval aviators were able to extend far beyond the range of naval gunfire. In both cases, it was necessary to increase the effective range of weaponry to coincide with the increased range of aircraft. Guns were incapable of achieving this. Therefore, logically, explosives had to be delivered by aircraft. Scouting gave way to bombing.

On land, airpower theory quickly focused on the question of whether airplanes should be used to support existing ground forces in combat—that is, serve as airborne artillery—or whether they should strike more deeply, at the very heart of society, at its industrial plant and its workforce. The latter argument held sway during the interwar period. At sea, a variant of the argument took place, one that limited the theo-retical role of airpower in peacetime, while dramatically increasing it during the war.

Mass saturation bombing efforts of enemy territory did not apply to naval warfare. First, the primary mission of the navy was sea control. Second, the geography of sea-lane control placed the heartland of Eurasia beyond the range of naval aviators. Finally, and most important, the sheer numerical limits imposed by the ships bearing aircraft—limita-tions in numbers of planes and total quantity of munitions as well as munitions per plane—made a sustained assault on the core of enemy society impossible.

Naval warfare differed from ground warfare in two fundamental respects. Compared to the number of tanks and artillery pieces on land, the number of warships was quite small. At the beginning of World War II, the United States had 17 battleships, 36 cruisers, about 150 destroyers, and 216 surface combatants, including 7 carriers and not counting 112 submarines.[4] The entire Pacific was covered by only 102 surface ships. Losing a battleship meant the loss of a substantial amount of firepower and sea-lane control. There was no equivalent platform on land.

This led to the second difference: the smaller number of platforms meant wider dispersion. Saturation bombing, which worked well on land, depends on a target-rich environment to make up for the inherent inaccuracy of bombers. Massed formations were flying against targets whose locations were fixed and known and relatively close together. It was possible to concentrate against these targets. At sea, the target was unknown save for scouting reports, and even then it remained eminently mobile between the report and the arrival of the strike.

Therefore, at sea, targets were sparse, mobile, and required prior intelligence. The intelligence requirement demanded dispersal of forces, to saturate the potential strike zone with intelligence-gathering platforms, and to avoid unanticipated attack by the enemy. At Midway, the U.S. Navy defeated the Japanese Imperial fleet because it had broken the enemy code and could, therefore, concentrate its forces strategically. Even so, the lack of tactical intelligence—not knowing the precise location of the Japanese carrier strike force—meant that the air attack on the Japanese was not nearly as effective as it might have been. The Japanese, on the other hand, lacked strategic intelligence—they did not even know if American carriers were near Midway—as well as accurate tactical data.

Because the information provided by scouting aircraft was frequently wrong or inadequate, air strikes were frequently being dissipated. Unable to compensate for inaccuracy with masses of planes and bombs, the Navy needed to increase the likelihood that each mission would result in the destruction of the targeted vessel. To do this, it had to maximize the accuracy of each projectile launched.

The manned bombers of the Army Air Corps released their bombs to drop vertically. The bombs were aimed at the target, but the real expectation was that because of the dispersal of bombs from multiple aircraft, at least some of the bombs would hit the target. Once released, the bomb's path was subject to the laws of physics. The bombardier with his bombsight was tasked with measuring the operant variables and releasing the bomb in such a way that those forces would guide the bomb to its target. Usually, the bombardier lacked sufficient information or skill to guide the bomb to its target. Ultimately, the Army Air Corps relied on luck and lots of chances for success.

The Navy had fewer planes than the Army Air Corps with many fewer projectiles. Thus, aimed at small and dispersed targets, vertical bombardment was guaranteed to fail. Indeed, the sorry record of B-17s in the Pacific against Japanese warships indicates the weakness of uncontrolled vertical bombardment at sea. During the battle of Midway, B-17s flew forty-nine sorties against the Japanese fleet without a single hit. During Guadalcanal, B-17s sank a single destroyer.[5]

To increase the probability of a hit with a single projectile, the Navy concentrated on two kinds of attack aircraft: the torpedo plane and the dive bomber. Each in its own way carried an intelligently guided projectile. The aircraft, under the control of the pilot, would point itself toward the target. A torpedo or bomb attached to the fuselage of the plane would be released on an angle of attack determined by the pilot. The projectiles would proceed on this preset motion—the movement of the torpedo augmented by an onboard motor turning a propeller. The plane would continue on course for a few moments to assure that the projectile's momentum was not disrupted, then would take another course to evade enemy fire. If the pilot set the course properly and survived to release the bomb, the probability of his hitting the target was surprisingly high. In some ways, the torpedo most closely resembled contemporary missiles. Rather than relying on gravity, it moved toward its target under its own power. Although aimed at its target by the plane and impelled by its gravity, the final approach was with a motor turning a propeller. The standard air-launched torpedo, the Mark XIII, weighed over a ton and carried a six-hundred-pound warhead. Dropped at extremely low altitudes—as low as fifty feet—and at a shallow angle to prevent a deep dive, it traveled at 110 knots. In its most important use, during the battle of Midway, the Mark XIII proved an utter failure. All but four attacking torpedo planes were destroyed by the Japanese, without a single hit. Following this disaster, the Mark XIII was refitted so that it could be dropped from as high as 2,400 feet, with its speed increased to over 400 knots. For the rest of the war, the results were nearly amazing. Out of 1,287 torpedo attacks, 40 percent resulted in hits.[6] This comes extremely close to the probabilities achieved with contemporary precision-guided munitions.

The torpedo and the dive bomber together revolutionized naval warfare by increasing the hit-kill probability along with the range of naval platforms. The carrier was able to locate enemy surface warships before they came into gun range, then attack them effectively, even in the face of ship-based antiair. The combination of range and precision meant that while the destruction of enemy battleships, cruisers, and other surface vessels was not a certainty, it was a certainty that these ships could not strike at the aircraft carrier. In any battle in which one system is invulnerable to attack by another, the long-term outcome of the series of engagements is preordained. The carrier had merely to catch the battleship; it would eventually destroy it.

Recall the example of David and Goliath. The battleship, like Goliath, was powerful and deadly—within its range. The aircraft carrier, not nearly so powerful, substituted range and precision. By staying outside the range of the battleship, the carrier, like David, could not lose. All it needed to do was strike home once with its aircraft. A single blow could cripple the battleship, giving the carrier the chance to finish it off at its leisure, as David did with Goliath.

The key to the carrier's initial success was a David-like simplicity. Like David, it overcame the previous generation's increasingly complex defensive methods with more elegant technologies. That redistilled the essence of the offensive. It also opened the door, just a crack, to the future: precision and range, the intelligent projectile. Yet, as in all things, senility inevitably reinstated itself. The chief threat to the carrier came, of course, from enemy aircraft, whether based on land or on other carriers.

Aegis: Precision-Guided Missiles vs. the Carrier Battle Group

The third battle of the North Atlantic was never fought because the third battle of Europe was never fought. During the First and Second World Wars, the fate of Europe hinged on the ability of Britain to keep the North Atlantic clear of German submarines and warships that were seeking to cut Britain's line of supply with its colonies and with North America. Had the Soviets tried to seize control of Western Europe by cutting the supply line running from the United States to Europe, the world war would have been fought over the same terrain.

The U.S. Navy's job was to keep those sea-lanes open, by making sure the Soviets stayed away from the convoys. To make the trip in the shortest possible time, the convoys had to travel as northerly a course as possible. Thus, the Soviets had two goals. The first was to sink enough ships so that NATO operations in Europe would be disrupted because of a lack of supplies and reinforcements, or failing this, to force the ships as far south as possible and, thereby, to impose severe delays on the delivery of supplies.

As we have seen, the vagaries of geography made this a particularly difficult task for the Soviets. Certainly, no Soviet surface warship could expect to enter the Greenland–Ireland–United Kingdom (GIUK) gap and exit alive. NATO controlled the surface completely. Rather, the Soviets had to bypass the surface routes in ways that could not be intercepted. Submarines were one solution, but this was made difficult by superb American underwater sensing capabilities, called the SOSUS (Sound Surveillance System) line.

The second solution was to use aircraft. But to reach the convoys, the Soviets would first have to pass over the carriers dispersed in the North Atlantic, along the GIUK gap or just south of it. With each carrier carrying at least twenty air-superiority fighters, it was virtually inconceivable that the Soviet aircraft would engage the carriers directly. Given the ranges involved, their aircraft would not be able to carry sufficient fuel to dogfight with NATO forces. The only possibility was to neutralize the land-based fighters in Scotland, Iceland, and Greenland, as well as in Norway, knock out at least one carrier, and approach close enough to the convoys to sink a substantial number.

To attack both land and sea bases, the Soviets would have to launch weapons from beyond the range of fighters, or at the very least, at the extreme point of the range, thereby limiting the fighters' ability to dogfight.

The key was to develop a missile that had the following characteristics:

- A range measured in hundreds of miles.
- A guidance system able to detect enemy vessels and home in on them.
- Sufficient speed and agility to survive whatever defense systems were in place.

The Soviets' first attempt at an antiship missile was the AS-1, code-named Kennel. Kennel, introduced in 1958 as a coastal defense weapon, was later mounted on larger bombers. Thus it was essentially a MiG-15 fighter loaded with a 600- or 1,000-kg warhead. Because its course was controlled by the bomber that launched it, and that bomber had to stay in sight of the target until impact, Kennel lacked both range and speed.

More important, it did not, at least initially, have its own guidance system. These were essential if it was to attack ships from outside their defensive envelope and survive air defenses.

The AS-2 Kipper, introduced in 1961, increased speed to the supersonic range—Mach 1.4 to 1.6—using a turbojet engine. Unlike the AS-1, it did not depend on the bomber to guide it to its target but before it was launched locked its own radar on the target, although the bomber could track its flight and override its internal controls if necessary. This meant that it had to be launched from a fairly high altitude, making the launch aircraft visible to ship-based radar. The missile would then undertake a shallow dive (about three degrees) toward the target.

The AS-4 Kitchen represented the real breakthrough in Soviet anti-ship technology. Extremely long (over thirty feet) and thin (about three feet in diameter), it weighed over three tons and could carry either a nuclear warhead or a one-ton warhead. AS-4 was a true rocket, carrying liquid propellant, which added to its weight and to the difficulty of maintaining and mounting it. Nevertheless, it had a range of over two hundred miles, which meant that it could be fired against an American carrier at the extreme range of the fighter cover. It also had a speed up to Mach 3.5. Unlike its cruise-missile predecessors, AS-4 could dive sharply at its target, closing at supersonic speeds, making it extremely difficult to intercept. Its weakness was that, in its earlier versions, it carried nothing beyond terminal guidance—which meant that if the general location of the carriers was unknown, the missile was useless. And if only the general, not the precise, location was known, the missile would be useful only when armed with a nuclear warhead. This left the Soviets with the option of going nuclear or forgoing engagement with the carriers. It should be added that since it was so difficult to shoot down, the U.S. Navy found it necessary to deploy a nuclear-armed interceptor missile against it.[7]

The introduction of initially passive and later active radar to the AS-4 represented the Soviet's first step to use PGMs to solve their problem. Later versions offered greater improvements. The AS-6 Kingfish increased range to about 350 nautical miles while maintaining supersonic speed. Most important, it was reportedly able to lock on after launch, which meant that the precise location of the carrier battle group would not have to be known at the moment of launch.

The Soviet emphasis on developing heavy, long-range, supersonic missiles posed a challenge to U.S. strategy. If Soviet bombers were able to exit the Kola Peninsula and fly through the GIUK gap, it became apparent during the 1960s that U.S. strategy required a defense system able to protect both the screening carriers and the convoys' previous supplies. In fact, the goal became even more ambitious—to create a single integrated weapons control system to protect carrier battle groups from air, surface, and submarine threats. The first attempt, canceled in 1963, was the Typhoon. The next attempt was named after the shield that Zeus gave to Athena—Aegis.

We have spoken of strategic weapons. If the aircraft carrier is one of the foundations of the Pax Americana, then Aegis is what makes it possible for the carrier to survive in an increasingly dangerous environment—particularly from the air. It is, of course, a danger sign when a purely defensive weapon begins to play a fundamental strategic role. It indicates that the threats to the offensive mission have become so great that core resources are being expended merely to protect it. This is certainly the case. Aegis, as we shall see, is costly in terms of both money and expertise. And like all such defensive weapons, it can only become more costly over time—as more and cheaper threats to the carrier emerge. That time has not yet come, and therefore we can marvel at the complexity and brilliance of Aegis, even as we see its very beauty as a mark of on-rushing senility.

The aircraft carrier has always been a vulnerable ship. Necessarily large, its primary purpose caused it to be a vast, flat target. Beneath its deck were magazines filled with explosives and tanks full of aviation fuel. It could never expect to engage a surface vessel of any substance and survive; even though it was fast, it was hardly maneuverable and was certainly poorly armed and armored. Its safety consisted in its aircraft— scouts that could spot enemy ships hundreds of miles away, bombers that could destroy them before their guns could be brought into range, and fighter planes that could destroy enemy bombers on their way to attack the carrier. The one threat that the carrier's planes could not defend against were enemy submarines. Thus, throughout World War II, carriers were accompanied by destroyers whose task was to screen the carrier from lurking subs as well as carry out other assorted missions such as pilot rescue, shore bombardment, and so on.

It was always understood, however, that the carrier was helpless should planes, ships, or submarines actually penetrate the defensive screen. During the Coral Sea, Midway, Great Mariana Turkey Shoot, and the other carrier engagements of World War II, these vessels proved themselves extremely vulnerable to direct attack. Protecting the carrier against Soviet submarines and missiles was, therefore, a well understood operation for which conventional sonar and radar were sufficient.

From the first perception of the threat in the early 1960s, through to the 1980s, the only protection afforded the carrier battle groups was geography—the long, tortuous path Soviet maritime bombers would have to fly to strike the carriers. The path gave the U.S. fleet time to detect them as they flew from the Kola Peninsula, down the Norwegian coast, time to intercept them as they passed the GIUK gap, and time to engage and destroy them before they fired their missiles. During the 1970s, the U.S. fleet entered a period of increasing vulnerability. The AS-6 Kingfish, fired by the sophisticated, supersonic Backfire bomber, opened the door to attacks on the fleet from beyond the range of its combat air patrol. Given the missile's speed, angle of attack, and range— several hundred miles—it was possible that Soviet aircraft could strike at the carrier with a thousand pounds and more of high explosives before being engaged by U.S. fighters.

To offset the Soviet threat, a comprehensive solution was required. Special radars would be necessary to track objects capable of moving in excess of Mach 3 and descending at extremely steep angles. As important, countermeasures would have to be devised, ranging from systems to confuse the missiles' sensors to weapons designed to destroy those that could not be decoyed. Finally, and most important of all, the sensors and weapons would have to be tied together so that information from the radar could be directly fed to both a combat information center where the battle could be managed and to the weapons themselves.

The key to Aegis was system integration, which was made possible by the advent of computers. Human beings remained in the loop—manning the consoles in the combat information center (CIC)—but computers made everything possible. Computers managed the data produced by the radar, fed it to computers in the CIC and to the weapons, guided the weapons, and allowed the weapons and CIC personnel to query sensors for more information.

It should be remembered that computers able to process the necessary amount of information in time to intercept a supersonic missile about to strike a carrier did not exist in the 1960s. Indeed, Aegis was first made operational in 1983, when the first dedicated Aegis cruiser, the USS *Ticonderoga*, was commissioned.

The emergence of sophisticated computers posed a military problem. How should an integrated air defense—and an antisubmarine warfare (ASW)—system be deployed? The ship to be protected was the carrier, and it would appear logical that the carrier would be the appropriate site for such a system—as well as the most cost-effective base. However, as far back as the 1960s, during the Typhoon development, the Navy found that it had to develop a dedicated platform, a ship that did nothing but shield the fleet from air attack. This was an expensive solution, but a necessary one. A carrier required both a clear deck and limited electronic emissions. A carrier could not be cluttered with weapons and radars. It needed all of its space for its primary mission, launching and landing aircraft. Moreover, if the purpose was to screen the carrier, a separate ship, able to interpose itself between the carrier and the threat axis, would be more efficient than the carrier's own onboard systems.

The result was the Ticonderoga-class carrier, designed and built for one purpose—to house Aegis. Displacing over 9,500 tons, 567 feet long, carrying a crew of 358, each Tico class costs over $1.2 billion—and as of July 1994, there are twenty-seven of them, as well as twenty-eight Arleigh Burke–class destroyers, which are nearly as big, with a crew of 303. At least one accompanies every carrier battle group into combat, while others are used to provide area coverage during amphibious operations, to protect convoys, and so on. Each is a marvel of engineering. But in the end, each is a vessel designed to carry out a mission in which a single error can be catastrophic. In the long run, such a mission cannot succeed.

Aegis's paradox is that at the same time it is attempting to imbue projectiles with intelligence, it casts those projectiles in an operationally defensive mode. It uses extraordinary new technology to defend increasingly senile systems—rather than as a way to transcend those systems. Nevertheless, there is much to be learned from Aegis, particularly as it evolves from being a system wedded to a single platform such as a cruiser or a destroyer to a diffused system.

Aegis can logically be divided into three parts: sensors, management systems, and weapons.[8] At the core of the Aegis system of sensors is the AN/SPY-1 radar. SPY-1, as it's called, is a phased-array system. That consists of four antennae able to give 360-degree coverage over a radius of eighty nautical miles. In addition to providing a warning of incoming aircraft and projectiles, the SPY-1 can also provide precise tracking of multiple targets, while managing midcourse guidance for missiles being fired by the ship.

- The AN-SPS-49 provides coverage of a much larger area—out to about 250 NM—but is primarily a search and warning radar, without any of the battle-management capabilities of the AN-1. It is also useful in detecting low-radar cross-section objects, from small missiles to stealthy aircraft.
- The SPS-55 is a surface search radar used for tracking small surface objects, such as submarine periscopes, as well as low-flying aircraft and missiles. SPS-55 also serves as a navigational aid.

- SPS-67 is an additional surface search radar.
- AN/SLQ-32 is a system designed to identify and track targets in an environment filled with electronic countermeasures and also serves as an ECM system on its own.

In addition to radar, Ticonderoga-class cruisers carry a wide array of sonar. Sonar, a descendant of the British antisubmarine detection equipment (shortened to ASDIC by the British) used during World War II, is an acoustical sensing system, whereas radar is an electrical sensing system. Since water conducts sound with great efficiency, it is possible to hear other ships moving through the water at fairly long distances. This is particularly useful in tracking submarines, or for submarines tracking surface ships, since visual and electronic means of detection are not available. Sonar comes in two modes. The passive mode simply listens for the sounds of a ship's engines and movement through water. As with an ear, listening does not give any hint of the listener's presence. In the active mode, sonar radiates sounds that echo off the hulls of ships. It is an effective means of location since the time it takes sounds to return and the direction of the return permits a ship to be located with some precision. Active sonar, however, gives away the position of the listener, or at least his presence, which can obviously be dangerous.

Ticonderoga-class cruisers carry a full suite of sonar equipment:

- The SQS-53 is mounted on the hull and used for both search and attack.
- SQR-19 is a passive, towed (to keep it away from the sound of the cruiser's own engines and hull) system.
- Later Ticos use follow-on systems for the same purpose.

Thus, the Tico operates within a ball of sensors—it can see out as far as 250 nautical miles in the air, and dozens of miles undersea, depending on water conditions. Aegis can also integrate data from other sources—such as reconnaissance satellites, unmanned aerial vehicles, or the LAMPS (light airborne multipurpose system) helicopters, which fly off the cruiser's decks and use sonobuoys to detect enemy submarines.

Aegis's ability to see the enemy comes at a high price—Aegis can also be seen. The multiple radars and active sonars that are necessary for Aegis to detect threats are visible hundreds of miles away. Since every radar transmitter has a peculiar signature, it is possible for an enemy to pick up the signals of an Aegis cruiser or destroyer and know not only that a particular ship is in the area but also that other valuable assets are in the area or that an important mission is under way. Aegis cruisers and destroyers are not deployed casually—and when they are deployed, they light up the electronic warfare screens for hundreds of miles, giving warning and providing targeting information.

This is one of the reasons for separating Aegis from the carrier. By placing the system on another ship, it is possible to offset the transmitters from what is, after all, the prime target—the carrier. Missiles launched will have to take their bearings on Aegis, permitting the system to destroy it before they can locate and lock onto a carrier.

All of these sensors, radar and sonar, feed into the command decision system (CDS), which helps assess threats, assign priorities, and task weapons controlled by the weapon control system (WCS). At the core of both of these systems are a series of computers that operate the system in one of four modes:

- Automatic, in which data is fed directly to CDS, then to WCS, then to individual weapons.
- Automatic special, in which controllers can preset priorities for targets, designating Backfire bombers as greater threats than other air threats.
- Semiautomatic, in which human controllers interface with the automated systems and exercise judgment.
- Casualty, which operates as systems are knocked out by enemy action, rerouting data and retasking weapons.

The human, physical core of this system is the combat information center. The CIC is an actual, physical place on board a Ticonderoga-class cruiser—an enclosed room filled with eighteen computer terminals and men manning them. Each console has a specific function:

- navigation radar
- radar system supervisor
- radar track manager
- electronic support measures, such as communications interception
- antiair warfare coordinator
- missile engagement controller
- air intercept controllers (2)
- fire control system supervisor
- antisubmarine warfare consoles (4)

- surface-warfare manager
- Tomahawk fire control
- Harpoon fire control
- gunfire control
- AN/SLQ-32 electronic warfare system

The final element in the system is the actual countermeasures that can be taken. A Ticonderoga-class cruiser is also equipped with

- 122 Standard Missiles (SM-2) with a range of about forty to seventy miles (depending on the model) at Mach 2.5, which can be guided by fire-control radar on the vessel or by its own terminal-guidance radar. The SM-2, launched from a Mk 41 vertical launcher, can reload in seconds after firing, critical for engaging multiple targets.
- 2 Vulcan Phalanx close-in weapon systems (CIWS) designed as the final defense against approaching missiles. Resembling R2D2, the robot in *Star Wars,* Phalanx is a rapid-fire gun system able to fire five hundred to nine hundred 20-mm rounds of ammunition a second. Able to depress to −25 degrees and elevate to +85 degrees, Phalanx can shift from full depression to full elevation in a bit over a second and traverse 126 degrees a second. (Note its weakness in dealing with objects descending from directly overhead.)[9] Guided by its own internal radar, which tracks both the object and its own bullets, Phalanx is designed to destroy rapidly closing missiles.

In addition to guns, torpedo tubes, and a large array of flares and chaff designed to decoy both radar-guided and infrared missiles, a Ticonderoga carries a Tomahawk cruise missile, a Harpoon antiship missile, and Antisubmarine Rocket (ASROC) antisubmarine missiles.

Note that of all the weapons on board the Ticonderoga class, only two are dedicated to its primary air-defense mission—the Standard surface-to-air missile and the Phalanx CIWS. Both are effective weapons, but the issue has always been the extent to which these defenses can be saturated by multiple targets, and the extent to which the system managing these weapons can respond to threats.

The key weakness in Aegis can be found in the interface between the humans in the CIC and the automated systems in the CDS and WCS. In 1987, for example, the USS *Vincennes* shot down an Iranian civilian aircraft that was taking off. The SPY-1 had the correct data, but it was not properly accessed. Battle managers misinterpreted the data and shot down the Airbus.[10]

The Aegis sensors therefore have two problems, one peculiar to Aegis, the other to all battlefield-management systems processing large amounts of information. First, Aegis sensors are not suited for operations close to shore. Since contemporary U.S. naval doctrine is focused on littoral warfare as opposed to blue-water operations, this is a severe drawback, requiring substantial redesign. The other problem is that the number of data sources, coupled with the amount of information being gathered by each, can overwhelm both the human and automated battle-management systems. Turning data into information and then acting on that information in real time are the critical problems in all of modern warfare. Given the centralization of naval warfare, due to the small number of weapons platforms available, this problem is inevitably magnified. A single error can have strategic consequences.

The speed at which information is received is determined by the tempo of events. As multiple targets threaten the carrier battle group, data has to be absorbed, transformed into information, fed into the weapon control systems, and acted upon. The speed at which targets appear and close determines the speed at which information has to be transformed. In the Persian Gulf, a target detected 75 NM from the carrier, traveling at Mach 1.5, is traveling at about one-third of a mile per second. From the moment of detection to impact, it would take about 225 seconds to strike. In those 225 seconds, detection, evaluation, and countermeasures must be taken. As speed increases, time contracts. Even more, when several objects are hurtling toward the target, the threats need to be prioritized and dealt with simultaneously.

Technically, Aegis and the SPY-1 radar system are designed to do this, and numerous weapons exist with which to engage targets. But the more threats exist, the greater the time pressure, and the less able the system is to respond. It is certainly not known to those without the highest security clearances in the Navy what the actual saturation point is for the Aegis. It has to be remembered that Aegis has never been tested under intense combat conditions—where a sophisticated enemy has launched more projectiles than its systems are designed to cope with.

Aegis stands as an extraordinarily sophisticated attempt to protect an extraordinarily effective weapon system. But Aegis is not an infallible system. Designed to operate in the unimpeded waters of the North Atlantic, Aegis did not do particularly well in the Persian Gulf. There, clutter from coastlines, pollution from oil fields, electronic emissions from the shore nearby, reduced its capabilities dramatically. Fighters, for example, could not be controlled more than seventy-five miles away from radar sources.

And almost by definition, it must fail. Aegis is a shield. It can protect a limited territory—defined by the strength of its radar and the range of its weapons. It cannot engage an enemy projectile before it reaches the outer limits of the shield. The faster the projectile is traveling, the less time there is to intercept it; at a certain speed, there is no time at all. Under certain environmental circumstances—such as coming in too close to the shore—the shield contracts. Moreover, the closer in to the shore you are, the more inexpensive launchers will be available to fire projectiles in response. Therefore, the absolute limit of the shield, defined by the number of counterprojectiles available, is compounded by a relative limit—the number of projectiles that can be handled during any one time period.

Certainly Aegis can evolve, and obviously it will have to do so. For example, the envelope of engagement is expanding. New missiles, such as late-generation Exocets, are both sea skimmers and extremely maneuverable, making it difficult for them to be destroyed with Phalanx guns, and impossible to do so with missiles. There are reports that the Soviets have developed a new antiship missile described as "aerodynamic"—it approaches the ship at high altitude, then attacks straight down at supersonic speed. The Phalanx, which has a maximum elevation of 85 degrees, would have difficulty engaging a target coming straight down, while a standard missile, attacking head-on, would require far better guidance and proximity fuses than are presumably currently available.

A great deal of development has focused on the Aegis's Theater High Altitude Air Defensive (THADD) program—the theater interceptor descendant from Star Wars. In a sense, this is a more promising approach for Aegis than the tactical role it has been created to serve. Because of the ranges involved, theater-level missiles provide greater warning time. For example, by linking Aegis to the DSP (Defensive Support) satellite program, as was the case during Desert Storm's SCUD hunts, projectiles coming from relatively greater distances—and passing out of the atmosphere—will allow defenders superior time frames and angles of interception than in the complex tactical situation being encountered.

One promising outgrowth of the THADD program is the cooperative engagement concept. Here, rather than depending on single platforms such as cruisers or destroyers to provide integrated air defense for carrier battle groups, the entire battle group would be tied together in a wide-area network, which would function as a single defensive system, locating targets, tasking weapons, and so on. A powerful concept, it suffers from the obvious defect of all wide-area networks—electronic countermeasures, or just unanticipated glitches bringing down the entire system.

Certainly, this is a criticism that can be made of any modern weapons system that is dependent on sensors. But there is a fundamental difference between offensive and defensive weapons systems. The failure of an offensive weapon system represents the failure of the attack but not necessarily of the weapons platform that launched it. The failure of a defensive weapon system represents the destruction of the weapons platform itself. A failure of a Tomahawk missile launched from a Ticonderoga means that an enemy target survives to fight another day; the failure of Aegis could mean that a carrier battle group is destroyed.

Aegis and Naval Warfare after the Carrier Age

The defensive side of combat is always more desperate than the offensive. This is why rational commanders seek to take the offensive whenever possible. The aircraft carrier began as a purely offensive weapon; the growing threat of enemy counteraction resulted in Aegis—the attempt to maintain the carrier's offensive capability. But at root, the very complexity of Aegis, as opposed to the relative simplicity of Russian antiship missiles, repeats the problem. Aegies can only be justified as a fail-safe system, and there are no fail-safe systems.

The purpose of Aegis is to permit the carrier to continue air operations. We do not know how many thousand-pound explosive packages it would take to sink a carrier, nor how many it would take to render it inoperative for a given period of time, but we do know that a mine could disable the USS *Princeton,* while two Exocets rendered the *Stark* inoperative. Thus, there is little doubt that multiple strikes with large antiship missiles would at least temporarily place a carrier at risk and limit its mission capability.

Here's another way of looking at the question: Given the enormous cost of a carrier battle group and its strategic significance to the United States, what sort of mission would we be willing to risk one on? For example, during Desert Storm, had it been merely possible—let alone likely—that a carrier could be struck by Iraqi missiles or attacked by Iranian Silkworms at the Strait of Hormuz, would we have been prepared to allow multiple carrier battle groups to move into the Persian Gulf? Given the Navy's own defined mission of littoral warfare, in which carriers will be operating within range of both land-based aircraft and land-based missiles, and given what we know about the unpredictability of explosives, can commanders accept the risk or likelihood of even single strikes on their carriers?

The carrier is a highly visible target in the midst of lesser targets. Traveling at Mach 2.5, a missile can travel one hundred miles in about three minutes. A carrier traveling at thirty knots cannot move quickly enough to evade being hit. The only option is to shoot down or decoy each and every missile fired—or to destroy the launchers ashore.

Now, there is no question but that Aegis is and will be able to destroy the vast majority of missiles fired at the carrier battle group. We might even concede that in most cases it will be able to destroy all of the missiles that are launched. But it will not be able to destroy them all; one or more will leak through, some striking at the carrier's waterline, others smashing into the superstructure, others hitting the deck. How many will have to hit before the carrier is destroyed? We will know that the minute after it happens. Our entire grand strategy rests on only twelve battle groups. What can we risk them on?

We are seeing an explosion of high-speed, agile antiship missiles. For example, the Soviet SS-N-22 Moskit, developed during the 1970s, is a surface-launched antiship missile that achieved speeds up to Mach 3. A follow-on missile mounted on aircraft has been reported to reach speeds of Mach 4.5, with further development planned to increase the speed to Mach 6. More important, since Moskit is ramjet rather than rocket technology, it is expected to have substantial standoff range as well as speed.[11] Alternatively, there is the Russian 3M-80 Sunburn. A combination rocket and ramjet, it cruises at 1,700 miles an hour out to a maximum range of fifty-five miles. What is most treacherous about the Sunburn is that it travels at an altitude of sixty feet and then attacks the ship at twenty feet, reaching its target in under two minutes from maximum launch point while skimming the sea. Sunburn is so impressive that the U.S. Navy has been trying to buy one for itself.[12]

Even more disturbing is the diffusion of the super/hypersonic antiship technology. The Chinese, for example, have reported through the China Precision Machinery Import and Export Company that they are now producing a missile with a 1,100-pound warhead called the C301, able to fly at Mach 2 at an altitude of three hundred feet, then dropping down to attack at thirty feet, with a range of eighty miles. The French are also developing a follow-on missile to their subsonic Exocets that is intended to be extremely maneuverable at very low altitudes and to strike at the waterline.[13]

Now, when we reach the point that not only the Russians but the Chinese and French as well are developing ramjet technology and anti-ship missiles, it is reasonable to assume that this technology will diffuse even more rapidly over the coming years. The Chinese and the French have both been noted for their tendency to sell advanced technologies to secondary and tertiary powers, for example, the North Koreans, Iranians, and Iraqis, to gain political leverage and reduce the cost of their own procurement projects. And given German and Israeli interest in supersonic missiles, further diffusion is inevitable.

Thus, U.S. defense planners need to evaluate the risk facing the fleet on the highly probable assumption that the speed, agility, and intelligence of antiship missiles are about to take a quantum leap. The United States does not have an intense supersonic antiship-missile program in place, and this has been taken as an argument against the need for such missiles.[14] However, this is not much of an argument. While the United States does not need to cope with Aegis-type defenses, other nations have focused on such defense measures, which are inherently more complex than the offensive. Israel may not foresee an operational need in this direction, but many of the customers it serves may well see such a need.

The growth of antiship missiles and their diffusion to secondary and tertiary countries necessarily imposes a conservative strategy on the use of the carriers. Given America's emerging strategy of littoral warfare against secondary powers rather than blue-water confrontation with greater powers, there will be few conflicts whose outcomes are so important as to justify risking one or more carrier battle groups. At the same time, we will find few conflicts where the carrier battle group will not be at risk—antiship missiles being both cheap and commonplace. During World War II, carriers were precious although not nearly as expensive and irreplaceable as they are now. But the stakes were enormous and we were prepared to risk losing them. During Korea and Vietnam, the issue did not really come up—nothing could threaten American carriers. During the Persian Gulf, after careful consideration, it was determined that Aegis could easily manage the threat to U.S. carriers—should the threat ever materialize, which it did not.

In the event of a crisis, the evolution of the Aegis system would argue against promiscuous deployment. Losing a single carrier—let alone two or three—could undermine our fundamental strategic interests, such as sea-lane control and the ability to intervene at points on the littoral where such intervention would be really critical.

Even if Aegis were able to absolutely protect the carrier—which is uncertain—it would be at an enormous cost. In the final analysis, a carrier's true value can only be judged in terms of the damage it is able to do. That, in turn, is measured by the aircraft on board. Consider the air wing carried on board a Nimitz-class carrier—the largest nuclear-powered carrier in the fleet:

Type	F-14	F/A-18	A-6E	EA-6B	S-3A/B	E-2C	SH-3/6
Number	20	20	16	4	6	4	8
Mission	Fighter	Attack	Attack	EW	ASW	AWACS	'Copter

A Nimitz-class carrier has seventy-eight aircraft. Only thirty-six, or 46 percent, are used for the main offensive mission. Others are essentially part of the ship's defensive system; that is, they are designed to get the aircraft through to its target.

In the normal configuration in which there is a threat to the carrier, at most forty attack aircraft are available for combat, slightly over half the complement. Each can carry between seven and eight tons of ordnance. Assuming that all of these are used for air-ground combat, and assuming that all aircraft can sortie, the total capability of the air wing is about three hundred tons of ordnance per mission. If three sorties were flown per day—which is an extraordinarily high number—a Nimitz-class carrier can deliver nine hundred tons of ordnance to targets in each twenty-four-hour cycle.

The reality is much grimmer. During Desert Storm, the Navy flew over one-third the number of sorties the Air Force did and dropped one-fourth the amount of ordnance, about the same as the Marines. On one Navy raid, sixty aircraft from two carriers flew a mission in which thirty-two tons of laser-guided bombs were dropped by only eight A-6s.[15]

Now, it must be understood that once it is transported into the theater, the Air Force has a much easier time operating than the Navy. Carrier operations are more complex, the distances flown are frequently more extreme, and the aircraft are obviously more limited in size—the Air Force is counting B-52s and F-111s in their sorties, while the Marine Corps flies shorter distances for close air support and is therefore able to ferry more bombs. That said, the fact is that the cost of putting six carrier battle groups on station during Desert Storm dwarfed the amount of damage they did.

On a conceptual level, the carrier has twin problems. First, its ability to survive as a sea-lane control platform in the face of advanced missiles from enemy aircraft is necessarily dubious—which is why more of the support ships are carrying antiship missiles such as Harpoons and Tomahawks. But it is simply no longer clear that the best way to suppress enemy shipping or to combat enemy warships is by deploying carriers. Israel, for example, has made excellent use of coastal fast-attack craft and missiles. It is not clear that destroyers and cruisers carrying longer-range missiles could not do the job as well.

The second problem is with power projection. Though the carrier can move an air wing within range of an enemy, it cannot do this in coastal regions where the enemy has sophisticated surface-to-surface or air-to-surface antiship missiles. In other words, where the carrier is most needed, it is most vulnerable. Most important, the quantity of ordnance that the carrier battle group can bring to bear is frankly insufficient for most missions. During the attack on Libya in 1986, two carrier battle groups had to be supplemented by F-111s flying in from England. During Desert Storm, had every carrier battle group in the inventory been deployed in the region, they would still not have been capable of supplying more than one-third of the ordnance that we dropped.

The carrier is both increasingly vulnerable and increasingly ineffective. Its vulnerability derives from the shift in the offense-defense relationship at sea. Where it was once a primarily offensive system, it is increasingly put on the defensive by the antiship missile. Where it was once able to project power offensively and influence events on land, it now has only a minimal opportunity to do so because its resources must be devoted to self-protection. The carrier suffers from all the defects of the manned aircraft and is on the wrong side of the life-cycle curve, increasingly inefficient because of increasingly efficient threats.

Yet, while the carrier may be approaching senility, beyond the help of the complex inefficiency of Aegis, its mission remains absolutely essential to the United States. The United States must control the world's oceans and must be able to project power ashore in Eurasia. We must be able to approach the coastlines of Bosnia and Korea, deliver fighting forces where needed, and not be overwhelmed by the enemy's inexpensive and multitudinous missiles. We must have ships that we can afford to risk in order to fight where our interests take us. On land and in the air, we have turned to our advantage the same technology that has undermined our capabilities. So the question is: How can we do this at sea?

9

First Thoughts on Airpower

In 1848, the city of Venice rose up against the Austro-Hungarian Empire. The Austrian fleet, unable to approach the city because of shallow waters, could not bombard Venice. The swamps surrounding Venice prevented land-based artillery from moving into range. The commander of the Austrian forces, Field Marshal Joseph Radetzky, was frustrated and turned to an invention by an Austrian artillery officer, Franz Uchatius. The invention was a hot-air balloon that carried a cast-iron bomb filled with gunpowder, which would explode on impact. The bomb was connected to the balloon by ropes, which were attached to a fuse. The bombardier would calculate the amount of time needed for the wind to carry the bomb over the target and light the fuse. Theoretically, the fuse would sever the ropes at the right moment, dropping the bomb on the target.[1]

As one might expect, the contraption did not work. Yet, as with many technological failures, it carried within it the promise of a wholly new way of understanding reality. From the first moment that men went airborne on November

21, 1783, when a physician and an army officer rose to a height of three thousand feet over Paris in a hot-air balloon, the stunning possibilities of airborne reconnaissance were inescapable.[2] On a clear day, you can see for nearly seventy miles from three thousand feet.[3] This meant that from a single aerial platform it would be possible to clearly observe an entire nineteenth-century battlefield, as well as the approach of reinforcements. Aerial reconnaissance could—and eventually would—transform command, control, communications, and intelligence systems, but the real achievement of the airborne platform was to become a weapon in itself from which explosives could be dropped on enemy forces beyond the reach of guns and artillery. Exponents of airpower argued that the airplane was so revolutionary it would make all other weapons obsolete. This was not an unreasonable assumption. Neither was it an accurate one.

The Airplane as Warplane

On October 23, 1911, Lt. Giullio Gavotti carried out the first recorded bombing mission from a heavier-than-air platform, striking at Turkish troops in Libya. He dropped one bomb on one target and flew on to a second target, an oasis, where he dropped three others. The bombs, little more than hand grenades, about the size of grapefruits, weighed about four pounds, and he armed them by removing a pin with his teeth. The bomb attack did little beyond boosting Italian morale—the Turks barely noticed it.[4] Still it proved an important point, that aerial bombardment was possible. The side doing the bombing frequently operated under the illusion that its bombardment was doing more than it was—an illusion that badly distorted judgment and frequently led to ruin. The Germans drew great comfort from bombing London, the Allies from bombing Germany. Most of all, the Americans comforted themselves with the bombing of Vietnam, imagining that it would bring victory in weeks or months.

In World War I, aircraft accounted for little operationally, except as reconnaissance platforms. But the ability to see from great heights made up for their other shortcomings—especially when it was combined with the wireless radio, an event that presaged a long, intimate relationship between airpower and electronics. Introduced by the British, the wireless permitted the plane to range far over the battlefield, providing up-to-the-minute intelligence for the first time. In a critical instance during the Battle of the Marne, French reconnaissance aircraft spotted a growing gap between the German First and Second Armies. This information permitted the French counterstroke that stabilized the front and helped determine the fate of France.

In spite of this, the airplane's most important effects were promises and theories. The promises were that aerial bombardment would make conventional war obsolete. Not only could airplanes move without restriction over terrain or even enemy ground forces, they had no theoretical range limitation nor limit on bomb load. This meant that, again theoretically, aircraft could rain explosives wherever they wanted to. It followed, therefore, that airplanes were the ultimate weapon. They could not be stopped, and if enough were built, they could destroy anything. The only question remaining was, what should their target be?

The foundation of the modern army is the factories that produce the weapons and supplies necessary for waging war. A regiment, for example, might well be destroyed from the air—but it could be replaced in hours. But by destroying the enemy factories that produce rifles, all regiments could be paralyzed, if not immediately, then soon thereafter and for a long time. Thus, if the heart of modern warfare is the industrial system that supports it, it follows that the target of airpower ought to be that industrial system. Not only the factory but the factory's workers and the social structure that supported those workers ought to be targets.

Giulio Douhet, an Italian general and the father of modern airpower theory, wrote in his *Command of the Air:*

> The guiding principle of bombing action should be this: **the objective must be destroyed completely in one attack, making further attack on the same target unnecessary.** Reaching an objective is an aerial operation which always involves a certain amount of risk and should be undertaken once only. The complete destruction of the objective has moral and material effects, the repercussions of which may be tremendous. To give us some idea of the extent of these repercussions, we need only envision what would go on among the civilian population of congested cities once the enemy announced that he would bomb such centers relentlessly, making no distinction between military and nonmilitary objectives.[5]

Billy Mitchell, the father of American air doctrine, wrote:

> Heretofore, to reach the heart of a country and gain victory in war, the land armies had to be defeated in the field and a long process of successive military advances made against it.

> Broken railroad lines, blown-up bridges, and destroyed roads necessitated months of hardships, the loss of thousands of lives, and untold wealth to accomplish. Now an attack from an air force using explosive bombs and gas may cause the complete evacuation of and cessation of industry in these places. This would deprive armies, air forces and navies even, of their means of maintenance.[6]

Douhet and Mitchell, as did Hermann Göring, Arthur Harris, Carl Spaatz, Curtis LeMay, and other great aerial warlords, believed that aerial warfare was an assault on the urban centers of modern society. Partly the purpose was operational and partly psychological.

Airpower would destroy the infrastructure of modern cities, forcing massive social dislocations and making production impossible. It was also assumed that the inability of the state to defend civilian populations against air attacks would delegitimize the regime, driving a wedge between ruler and ruled, possibly leading to rebellion, and certainly to social and political distrust. It should be noted that, to a great extent, this was also the theory behind nuclear war—that destroying the mechanical infrastructure of a city, in addition to inflicting huge causalties on the population, would make the city uninhabitable—something no rational nation would risk. Behind this theory was an assumption about the fragility of urban life. Unlike agricultural societies, which live together with their food stores, urban society requires constant replenishment. For example, it requires the delivery of water through complex engineering systems and the removal of wastes through equally complex systems. It seems reasonable to assume that raining explosives down on the city would easily disrupt the support systems, destroying the life of the inhabitants and the functioning of the city's factories.

Conventional bombardment, of course, never lived up to this promise, at least in the case of the sprawling world-class cities such as London, Berlin, and Tokyo. Battered though they were, they continued to function without complete infrastructure collapse or social chaos. Even lesser cities such as Hanoi or Frankfurt continued to function. Experience has shown that conventional bombardment imposed great anguish with little overall effect when it came to bringing an end to the war. Nuclear weapons were created to impose even greater anguish with much greater effect.

What seemed most impressive about the aircraft was its invincibility. At first sight there was no defense against the bomber. This was certainly Douhet's impression:

> Viewed in its true light, aerial warfare admits of no defense, only offense. **We must, therefore, resign ourselves to the offensives the enemy inflicts upon us, while striving to put all our resources to work to inflict even heavier ones upon him.** This is the basic principle which must govern the development of aerial warfare.[7]

Douhet was expressing the first pure vision of a new weapon. He understood the bomber to be simple and unstoppable. **The history of airpower has been the history of defensive pressure deforming its early pure form, until the bomber becomes an extraordinarily complex, delicate, and inefficient weapon, devoting most of its resources to simply surviving in an impossibly hostile environment. The path from Douhet's vision to the B-2 is the path from the revolutionary debut of a weapon to its senility.**

Initially, the greatest threat to aircraft came from surface weapons. Realizing this, aircraft kept their distance and altitude. It followed that, if the planes wouldn't come to the gun, then perhaps the gun could be brought to the plane, but the only way to do this would be to mount weapons on pursuit aircraft. The pursuit plane had to be quick and agile in order to close with the enemy plane. This meant that it could not carry bombs—only weapons. The first of these weapons were pistols and rifles, single-shot weapons of little consequence. Then, on October 5, 1914, a French mechanic named Louis Quienalt, in a Voisin piloted by Joseph Frantz, shot down a German plane with a Hotchkiss machine gun.[8]

The first bombers had a single requirement: they had to be stable enough for the bombardier to sight the target with some hope of hitting it. As the environment became more threatening, bombers became more than stable transport vehicles. They had to carry armament of their own. Engines became more power, and speeds rapidly surged. In 1914, for example, aircraft speeds ranged from 115 kmh to 165 kmh; and by 1917, from 165 kmh to 225 kmh (the Bristol F2.b).[9] All of this was driven by defensive considerations. To evade any enemy aircraft, planes had to become faster. **Dropping bombs was simple. Staying alive was hard.**

The primary mission of airpower was to drop bombs. A blimp could do that, but it could not do that and survive. Survival drove aircraft development. Air superiority, or control of airspace so that bombers could carry out their mission, was merely the preface to aerial warfare, but sometimes achieving air superiority became so daunting that it consumed all of the energy of an air force.

Over time, the number of engines required to deliver a load of bombs soared as bombers sprouted additional engines and were accompanied by fighter escorts and, later, flak-suppression aircraft. Paralleling the increase in the number of engines was the increase in the quantity of fuel that had to be burned to deliver a given load of bombs. The number of pilots and mechanics needed to drop a load of bombs also increased. Like Goliath and his armor, the attack began to choke on defense—a harsh reality that grew more apparent during World War II.

First Stabs: The Battle of Britain and the British Response

Douhet was nothing if not persuasive. The U.S. Army's Air Corps Tactical School handbook, *Air Warfare,* stated that "air warfare may be waged against hostile land forces, sea forces, and air forces, or it may be waged directly against the enemy nation. The possibility for the application of military force against the vital structure of a nation directly and immediately upon the outbreak of hostilities is the most important and far reaching development of modern times."[10] It was not a uniquely American idea that attacks against the enemy's economic and social infrastructure would determine the outcome of the war.[11] The Germans and the British also shared this view.[12] In fact, it was a general consensus that would shape World War II.

Nonetheless, it should be noted that from the beginning of the war the Germans had a more complex view of the uses of airpower, determined in large part by Germany's geography. Unlike the United States and Britain, Germany faced powerful land enemies on its borders. German ground forces would be in harm's way at the outbreak of war and would need immediate air support. As a result, air resources had to be diverted from strategic missions to more mundane tasks such as close air support of troops in combat and interdiction of roads and bridges immediately behind the front. The Luftwaffe was committed to strategic bombardment; German reality imposed more pressing demands.

The same was true for the Soviets, whose ground forces were also immediately engaged at the outbreak of war. There was never a point in which their ground forces did not require the primary commitment of the air forces.

The British and Americans paid relatively little attention to close air support until later in the war and, instead, concentrated on strategic bombardment. The Japanese, for whom the 1941–45 war was primarily a naval conflict, treated airpower and naval power as part of the same system, adopting the U.S. Navy's perspective as a national commitment. Thus, where the classic American planes would be the B-17 and B-29, the classic Japanese planes would be the carrier-based Kate bomber and Zero fighter.

The classic German plane would be the JU-87—the famed Stuka dive bomber—which could carry only a little more than a thousand pounds of explosives out to a range of barely two hundred miles.[13] But the Stuka was extraordinarily precise. By diving toward the target, the plane served as a projectile in its own right—a manned and guided projectile. It would release the bomb a few thousand feet above the target and then continue on the dive path before pulling out. The bomb would continue on the path, striking the target at which the pilot had aimed his entire aircraft.

The Germans did not have the luxury of waiting for strategic bombardment to take effect—they were not protected by the English Channel or the Atlantic Ocean. The ability to take the long view requires a tremendous sense of security, provided either by being overwhelmingly powerful (as was the United States in Vietnam) or by possessing impregnable geographic borders. Then, and only then, can aerial resources be diverted from the raging battle. For Germany, such moments were rare. One came in the summer of 1940. Having overrun France, the Germans found themselves in a completely unexpected circumstance. Invulnerable to attack, they were also unable to strike at the British with their ground forces. This was a moment for which advocates of strategic airpower had waited—the moment for a strategic bombing campaign.

The German attempt to destroy Britain from the air was doomed from the start. The BF-109, their primary fighter, could only reach southeastern England; the primary bomber, the HE-111, could reach the Midlands industrial area, but unless it had a fighter escort, it would be slaughtered. This left Britain's industrial heartland outside the range of the Luftwaffe. The German focus on continental operations had bred aircraft without the range or bomb capacity needed to carry out a strategic bombing campaign against Britain. In addition, the Germans lacked VHF radios, having only HF, which meant that they had only short-range communications with their planes. Unlike British planes, German aircraft could not be controlled from the ground once airborne, which mattered little since they also lacked radar, and ground control had no idea what was going on in the air over England anyway.

The British, on the other hand, were fighting over their own airfields. Not having to travel to a target, they had plenty of fuel with which to maneuver and fight. Having used up fuel on the trip over, the Germans had to get to their target, release their bombs, and get home. This made them extremely vulnerable to RAF fighters, and the result was heavy casualties. Moreover, having radar meant that the RAF command knew both where the Germans were attacking and when. British aircraft did not have to orbit in wasteful patrols, waiting and watching for an offensive maneuver, but could take off only when needed, preserving both fuel and machinery. Given the limited range of German aircraft, the RAF had a built-in sanctuary for retreat.

All of these factors contributed to the failure of the German air offensive, the first strategic bombing campaign in history. But the primary reason the Luftwaffe did not succeed was because aerial bombardment was both extraordinarily inaccurate and inefficient. It required tremendous expenditures of resources with extremely uncertain rewards. Indeed, the Germans had a word for the tactics they were forced to engage in—*Knickebeinen,* "blind bombing." The Germans tried to make it less blind by using radio beams to guide their bombers, but this gave them minimal guidance and was easily jammed by the British.[14]

Douhet and the early theorists of strategic bombardment were not altogether clear on whether they meant that the bomber would be used as a precision instrument, dropping bombs on carefully selected targets, or whether the bomber would be used as a blunt instrument, saturating an area with bombs. For the Germans, London was within reach, and best of all, it was large enough to hit, even at night. But the Luftwaffe could not carry sufficient tonnage across the Channel to obliterate London or destroy London's infrastructure. Neither could they drive a wedge between the people and their government. Strategic bombardment turned out to be neither a blunt instrument nor a scalpel. It was spectacular suicide, costing Germany its air force.

The British counter-Blitz did not fair much better. British economic intelligence was so poor that it simply had no idea which German factories had to be hit. When they knew the location, their bombers could not find it; when they could find it, they could not hit the factories; when they hit them, they could not destroy them.

It is absolutely extraordinary that during the first two years of the war, **the Germans frequently had no idea what targets the British bombers were trying to hit.** From their point of view, the British air attack appeared to be a random strewing of bombs around the countryside. D. M. Butt, a member of the British war cabinet, analyzed one hundred raids. He found that only one in five planes came within five miles of the intended target, with many bombs falling off target by dozens of miles.[15] A cabinet study undertaken by Capt. Harold Archibald in January 1942 claimed that in the previous two years, only 10 percent of the bombs that had been dropped fell in the *target area*—doing indeterminate damage.[16] Solly Zuckerman, the operations researcher who had done much to improve the efficiency of the British military, estimated that under the best of circumstances, even massed bombardment would reduce German wartime production by only 7 percent.[17] A study done in 1941 by P. M. S. Blackett, Zuckerman's colleague in operations research, showed:

> The average number of bomber sorties per month, then mainly by Wellingtons, was 1,000 and of these some 40 were lost with their crews of five men, giving a loss of airmen, all highly skilled men, at the rate of 200 per month. Comparing this with the estimated killed, that is, 4,000 men, women and children . . . it was concluded that in the matter of personnel casualties the 1941 bombing offensive had already been nearly a dead loss.[18]

As a matter of fact, Blackett's estimate of civilian casualties was twice the actual amount.[19]

British bombers had to confront two problems. First, there was the simple problem of navigation. Finding a target in the dark is not an easy matter. At the distances and speeds that aircraft traveled, minor errors in fixing the flight path, from miscalculation of wind to misreading of stars, could lead to errors of many miles. From altitudes in excess of ten thousand feet, the human eye could not assist in locating positions even during daytime. Help was sought in artificially augmented vision. British inventions with code names such as GEE, OBOE, and H2S (for "home sweet home") sought to exploit the electromagnetic spectrum to guide bombers. Radio beams from bases in Britain provided flight paths to the target, while intersecting beams permitted aircraft, through triangulation, to locate themselves in relation to the target and, even more important for the crews, in relation to home.[20] Trained crews, flying in Pathfinder craft, would use electronic means to locate the area, drop incendiary bombs or flares, and allow the rest of the force to key on the flames.

If they found their targets, bombers then encountered the second and more intractable problem—hitting them. Hitting a factory from two or three miles up proved to be an almost impossible task. The British solution to the inaccuracy of bombs was much the same as the solution to the inaccuracy of rifles: saturate the area. Having lots of aircraft bomb the area around the factory at the same time increased the probability that the factory would be hit. Since factories were frequently located in populated areas, this tactic meant that large numbers of civilians would be killed as well. Thus, the assault on industrial targets became an assault on the German people.

The Holy Grail: The American Search for Precision

The Americans did not accept the British and German view that precision bombardment was impossible. They entered the war convinced that daylight precision bombardment was both possible and necessary. Indeed, the American Army Air Corp's high command believed that the war could be won this way. As early as 1918, Lt. Col. Edgar S. Gorrell had argued that the only way to stop the Germans was to destroy the factories that produced German weapons.[21]

In a way, the Americans were Douhet's best students. They not only believed in the decisiveness of airpower, they also believed in the invincibility of the bomber. Americans assumed that multiengined bombers could always outrun single-engine fighters, and that, as a result, bombers were unstoppable—unless destroyed before they were built.[22] This added urgency to the air war.

There was another side to the American commitment to precision bombardment. The Americans had observed the effect of Japanese air raids on Chinese population centers in the 1930s and noted something interesting. Rather than breaking public morale and driving a wedge between the Chinese people and their government, the raids solidified anti-Japanese sentiment and actually strengthened morale.[23] The Americans became convinced that German terror raids on London would thus be counterproductive. This, added to the logistical problem posed by saturation bombardment—wasting ordnance on nonmilitary targets strained production and transportation capabilities—led the Americans to an obsession that continues to this day: minimizing collateral damage, avoiding hitting populated areas, and maximizing the accuracy of bombardment by all technical means available. Precision became the American holy grail, an obsession that distinguished American air doctrine from other countries' and shaped air doctrine from World War II right through to Desert Storm.

The obsession with precision was not only a moral concern—although it was that as well. The Americans were also concerned with minimizing the waste and exertion of war. Precision permitted an economy of force that Americans required, given their logistical realities. Thus, Americans cared for precision far more than did other nations. This is not to say that they achieved precision. The limitations imposed on aerial bombardment meant that, with the best intention possible, bombing turned into slaughter. By the mass bombing campaigns over Japan, precision had, for all practical purposes, been abandoned.

The Americans set themselves an inherently difficult task, given the fact that gravity bombs dropped from several miles high were wildly inaccurate. There was a mismatch between doctrine and capability, a mismatch that was to continue to mark American airpower for several generations. We did not know how to do what we were committed to doing.

The American concentration on strategic bombardment was driven, in part, by the type of aircraft the United States was forced to design and build in the years between the First and Second World Wars. The Americans had no nearby enemies—Mexico, Canada, Cuba, were hardly threatening. Potential conflicts were thousands of miles away, with vast oceans separating American power from combat zones. To protect American shores and project American power in Eurasia, the United States had to first control the oceans. In the Atlantic, the North Atlantic convoy routes were of primary importance, as well as the approaches to North America in the western Atlantic. In the Pacific, the United States had to control the line of supply to Hawaii and the approaches to the West Coast. Thus, the U.S. strategic bomber fleet was designed for long-range, land-based maritime bombing as well as ground attack. The bombers had to have sufficient precision to strike enemy warships from great heights—which would put them beyond enemy firepower. This strategy required precision, endurance, and above all a strong logistical support base. It required long-range, technically advanced bombers that were expensive to build and expensive to maintain.

Throughout the 1930s the Americans tried to design and build a long-range, high-altitude, precision bomber with a large payload, capable of defending itself against enemy fighters. In 1936, for example, the Air Corps Matériel Division called for the design of a bomber with an eight-thousand-mile range and a speed of 230 mph—able to bomb Europe from the east coast of the United States or Japan from Hawaii. The Douglas Aircraft Company completed the XB-19 in 1941. The plane suffered from several flaws—the engines of the day could not support the concept—and it was never produced in quantity.[24] The dream of a true intercontinental bomber would have to wait until the B-52.

The failure of the XB-19 did not deter American air planners. A more modest attempt at constructing a long-range bomber was undertaken by Boeing. The XB-17, to become the B-17 of World War II fame, had the speed of the XB-19 but not the range. It could fly only 2,100 miles, stripped. The B-17 Flying Fortress did have sufficient range to be able to patrol the oceans out to a distance of about a thousand miles. This meant that when operating from bases in the continental United States, it could fly out as far as a line running from the eastern tip of Newfoundland south to Bermuda and then Puerto Rico. From bases in Puerto Rico and Newfoundland, the B-17 could provide coverage almost to Iceland, throughout the Caribbean, and as far east as forty-five degrees longitude. This would cover all the Atlantic approaches to North America. From bases in Hawaii, Guam, and Manila, the B-17 could provide complete coverage of the Pacific as far south as Singapore, as far north as Okinawa, and throughout the Japanese mandate. Thus, the B-17 was, in theory, the perfect defensive sea-control weapon for the United States.

While the B-17's lack of intercontinental range would have been a real problem in a classic strategic bombing campaign against Japan, the European theater was different. With bases in England and North Africa, the B-17 could strike at both Italy and Germany.

The mature B-17 carried a bomb load of 6,000 lbs. (maximum load, about 7,000 lbs.), at a maximum altitude of 35,800 feet and a maximum speed of 287 mph (cruising speed of 182 mph). It carried thirteen .50-caliber machine guns in its nose, tail, belly, top, and waist, so it could project deadly fire in all directions.[25] Because planners wanted the bombers to defend themselves—and not require protection from fighters—the aircraft went out alone against flak and fighter. Until the summer of 1944, monthly bomber losses ran as high as 6.5 percent per sortie and never lower than 1.9 percent.[26]

British air marshal Harris had also pursued the pure air offensive, ignoring both fighter escorts and heavily armed bombers—but he suffered an attrition rate of over 65 percent for the campaign, with 47.5 percent of the men in the aircrews killed.[27] The casualty rate, unparalleled in any contemporary air or ground campaign, was the result of the British saturation strategy. Concentrating the bombers over the territory created a target-rich environment in which even at night the probability of flak's hitting something soared. Horrendous casualty rates did not only affect the British. During the American raid on Schweinfurt, Germany, sixty out of three hundred bombers crashed.[28]

It should be remembered that the antiaircraft gun suffered from the same defect as the bomber; both were ballistic weapons, both were inherently inaccurate, and both compensated for inaccuracy with masses of weapons. It was estimated that it took between four thousand and eight thousand shells to shoot down one plane. A 75-mm shell took twelve seconds to rise to 16,500 feet. If a plane was flying at a height of 8,500 feet at 100 mph, 162,000 shells would have to be fired simultaneously to guarantee bringing it down.[29]

Air defenses were not sophisticated enough to stop the bombing campaign. What they did accomplish was the diversion of resources from bombing to defensive measures—flak suppression and missions to confuse radar. Just as important, air defenses forced bombers to attack targets using inefficient approaches and bombing patterns. This decreased the likelihood of destroying the target but increased the probability that aircraft would survive the mission.

The Americans were much more sensitive to taking casualties than were the British and began compensating for enemy defensive measures. At first they attempted to structure bombing formations so that the bombers' machine guns could support each other and provide complete coverage against enemy fighters—creating a fortress in the sky. Under this plan, the bombers would fly in tight formation, and instead of each plane's conducting a bomb run on the target, the group would bomb together. The bombardier on the lead aircraft would conduct the bombing run, with all other bombardiers releasing on his command. The problem with this was that it reduced the accuracy of bombardment— already below mediocre levels—by about one-third of what it would have been had the planes operated on individual runs.[30]

A second direction was the creation of a class of aircraft whose sole purpose was to protect the bombers—the fighter plane. Already present in World War I, the dedicated antiair fighter craft was further perfected during World War II. The fighter plane was designed to be fast and maneuverable in combat. Which meant that it had to minimize its weight relative to engine size. Not only could it not carry bombs, it could not carry much fuel, and little fuel meant limited range. Thus, planes like the Spitfire, which had won the Battle of Britain, could never reach Germany.

What was needed was a fighter that could fly vast distances that was also quick and agile. The development of longer-range fighters, such as the P-51 Mustang and P-47 Thunderbolt, was aided by the invention of the drop tank, an auxiliary fuel tank that could be jettisoned when emptied or if the aircraft found itself in a dogfight. These fighters could escort bombers all the way to Berlin and even beyond.

When the bomber could reach the target area, it still faced a tremendous problem—hitting the target. For the Americans, who were attempting high-altitude, daylight precision bombardment, the answer rested with technology. Nothing could be done about the bombs, except for using fins to stabilize their fall. Nothing could be done about the aircraft, except to fly it straight and level. If the plane was stable and the bomb fell true, what remained to be done was to perfect the bombardier's aim—to make certain that the bomb, when released, would be drawn by gravity to the target. The human eye couldn't work this miracle. It could barely see the target, and it could not compensate for the speed of the plane, crosswinds, or any of the other variables that influenced the path of the bomb. The American solution was to invent a machine that would compensate for the inaccuracy of the bombardier's eye—the Norden bombsight.[31]

The Norden bombsight was, first of all, an automatic pilot. During a bomb run, the pilot would relinquish control of the aircraft to the bombardier, who in turn depended on the autopilot—officially the "automatic flight control equipment"—to stabilize the aircraft. The stabilizer consisted of electric gyroscopes stabilizing the craft for direction and attitude, and a complex mechanism consisting of over two thousand mirrors, lenses, gears, and other parts, which was the sight proper. The most dangerous part of the bomb run, the part requiring straight and stable flight, was reduced to only fifteen to twenty seconds on the Mark XV model.[32] The bombardier, situated in the nose of the bomber, input data on altitude, wind, and so on into the Norden bombsight, peered through a low-power telescope with crosshairs, and when the crosshairs were on target, released the bombs, returning control to the pilot.

The Norden bombsight was extremely important in ways having nothing to do with World War II. Between the Norden bombsight and the Manhattan Project, a culture was created that saw technology as the foundation of American power. It also understood that technology was portable, and that it had no national allegiance. Therefore, hiding technological innovation became an obsession.

The obsession with secrecy about the Norden bombsight began reasonably enough. If daylight precision bombardment could win the war, it followed that the technology that made this achievement possible was a matter of the highest national security—that if it fell into the hands of the enemy, it could spell U.S. defeat. Americans saw the technology they developed as a threat to them as well as to the enemy. Where other nations guarded their national strategies with their lives, the United States guarded its national technology. For the Germans and the Russians, the precise deployment of forces was the deep secret. In the United States, it was always relatively easy to find out who was deployed where. What was guarded were the technical secrets behind our weapons. From the Norden bombsight through the Manhattan Project and the B-2, the United States was obsessed with the fear that its technical expertise would be turned against it.

The bombsight was produced and shipped under tremendous security. The bombsight's inventor, Carl Lucas Norden, a Dutchman, was constantly accompanied by bodyguards during the war lest he be kidnapped and reveal the secret of the sight. It was removed from the aircraft after each mission and reinstalled before takeoff by armed guards. Crews were ordered to destroy the bombsight if they had reason to believe that they would be forced down over enemy territory.[33]

The paradox was obvious: it took thousands of workers to manufacturer the sight and thousands of bomber crews would have access to it. Most important, many of the planes using it would be shot down over enemy territory. There was no possible way to keep the Norden bombsight secret. At the very least, it had to be *assumed* that the secret of the bombsight would be penetrated. There was something zany in the security precautions—but also something very comforting. The bombsight's value derived in part from the security that guarded it. Anything guarded that carefully had to be important—and effective. This culture of security would elevate the importance of many weapons systems of dubious worth.

However, security would not be the Norden bombsight's Achilles' heel. Its chief defect was that it did not work very well—even though it worked better than anything else around. In 1943, the average error of daylight bombardment was 450 yards—about one-quarter mile. When there was poor visibility, a commonplace event in Northern Europe, this error increased to 1,200 yards. At night, the error was about three miles.[34] Increased altitude resulted in greater inaccuracy, far beyond the ability of the Norden bombsight to compensate. Though precision bombardment was still beyond the B-17/Norden combination, the inflated American claims had a serious political consequence. During the pre-invasion bombing of France in 1944, American bombing raids killed thousands of French and destroyed vast swatches of property. The French, who had believed U.S. propaganda about the effectiveness and accuracy of its bombing, were convinced that it could have chosen to avoid the casualties and damage.[35] American callousness seemed more believable than American boasting. The idea that Americans could not hit the proverbial side of a barn with their bombs simply did not occur to the French. It bred a great deal of anti-Americanism that would poison Franco-American relations during the Cold War.

The more immediate problem would be the purpose of the American strategic bombing campaign. Gen. Haywood S. Hansell, who played a critical role in both the European and Pacific air campaigns, described the effort needed to destroy the German power grid. Hansell calculated that forty-five power stations would have to be destroyed along with eleven other transformer sites. Each target measured about four acres. Hansell estimated that two hits with 1,100-pound bombs would knock each out for several months, and that three hits would do the job from six to eighteen months. The average radial error of a bomb was 875 feet. A bomb pattern delivered by a single combat box of 18 bombers had a 75 percent probability of one hit out of 108 bombs. A combat wing—three combat boxes, 54 bombers—had a hit probability of 98.5 percent for one bomb, or 84.5 percent with two bombs. If two combat wings were used—six boxes, a total of 108 bombers—then the probability of one hit was 99.9 percent, two hits 96.5 percent, and three hits 89 percent.[36]

This meant that, if the goal was to put a single power station out of commission for the duration of the war, a raid by two full air wings simultaneously dropping 356.4 tons of bombs, carried out in ideal daylight weather conditions, would provide an excellent probability of success—but would not guarantee it. Against the fifty-six proposed targets—that is, against the German power system as a whole—nearly 20,000 tons would have to be dropped in fifty-six raids by two wings. Statistically, at least six attacks would be failures, and several other partial failures would occur.

All of this is based on the assumption that bomb-damage estimates are correct—that three hits would put a station out for the duration. And it rests on assumptions concerning blast damage and repair capabilities. This last was a major point of disagreement between the Americans and the British. "Bomber" Harris, who commanded the British strategic-bombardment program, firmly believed that industrial targets could easily recuperate from an attack. Harris felt that economic intelligence was extremely poor in general and that American intelligence vastly underestimated German recuperative powers in particular. Skilled workers and copious spare parts could, he argued, bring the stations on line in a matter of days or weeks: the German power grid would never be completely disrupted. So the British air commander, Harris, bitterly opposed American attacks on specific German economic targets. He believed that only a direct attack on the social structure—terror bombing—could break the back of the Germans. Precision bombardment, he felt, even if it could be achieved, could not break the back of the German economy.[37]

The postwar data supported British misgivings about the bombing of industrial sites. As the U.S. Strategic Bombing Survey reported:

> Destroyed capital equipment can be replaced out of reserves, destroyed factory space can be made good by drawing on empty factory space, lost man-hours can be made good by making working hours temporarily longer and by taking on more labor. Each and all of these methods were, of course, resorted to by the Germans to offset the effects of air raids. The strain imposed on the economy by an air war may, under such conditions, merely slow down or temporarily halt the process of expansion, without causing an actual diminution of production in the vital industries. An expanding and resilient economy, moreover, is not merely better adapted to absorb the shock effects of a bombing offensive; it can also afford to take more far reaching measures to immunize itself against future attack.[38]

Hansell's assumptions on how long these sites would be out of commission proved to be vastly optimistic. To destroy the electrical grid, the sites would have to be visited over and over again, at least weekly. About seven hundred planes would have to be permanently assigned to this single task, not counting casualties and other attrition. And this assumes that the weather would always be optimal, something that never occurs in Europe, or elsewhere for that matter. In 1945, for example, 80 percent of all raids took place in bad weather.

The Allies sought to develop technical means for coping with poor flying conditions. The most important innovation was H2X, an airborne ground-tracking radar that would, it was hoped, permit precision bombardment in spite of clouds or darkness. Beyond distinguishing between land and water, H2X could not make out any major land features. When it was used, accuracy declined dramatically, with the circular error of probability rising to a full two miles, raising the number of sorties required to guarantee target destruction to astronomical, and totally improbable, levels.[39] With H2X, the area that would have to be saturated would be 130,612 times greater than what would be needed in good weather—and the number of aircraft required would have to be multiplied by similar amounts.

By 1945 these numbers had improved somewhat. In large part, this was the result of more experienced aircrews and bombardiers—of taking the equipment to its maximum possibility. But it really did not matter. The level of accuracy achieved was still not enough to create an economy of force. The number of bombers necessary to suppress industrial production, and maintain that suppression, far exceeded the number of bombers that could be produced. And even the production of the existing bombers and engines had required a substantial diversion of resources from other weapon types.

In the end, the dispute between the Americans and the British over the efficiency of precision bombing became purely theoretical. The Americans were nearly as imprecise as the British. The American bombing campaign relied as much on area saturation as the British did. In bombing economic and industrial sites, inaccuracy led the Americans to bomb surrounding residential areas. In bombing residential areas, the British destroyed industrial sites. Only the intent was different—and the time of day. The outcome remained very much the same.

Aerial bombardment did not break German morale, as the British had predicted. As with Londoners during the Blitz, this was due to the natural rage of the public against the bombers. In addition, the bombardment created a sense of shared fate. During World War I, soldiers were bitter at the disparity between their suffering and life on the home front. Strategic bombardment eliminated this morale problem because civilians suffered as much as soldiers. Equally as important, bombing was democratic—government officials and their families were as vulnerable to bombing as were ordinary citizens. Finally, the savagery of the attacks convinced the public that any talk of armistice or capitulation was pointless—the sort of people who could bomb helpless civilians night after night promised only a Carthaginian peace. Better to fight on.

From Incendiary to Atomic Bombs: The Japanese Conclusion

The Japanese case was a more intense test for airpower than the European. For one thing, the introduction of the B-29 bomber increased carrying capacity dramatically. Where the average load in Europe was about 3.3 tons, the B-29 could carry a 10-ton payload for 1,500 miles at 300 mph at 30,000 feet. It carried twelve .50-caliber machine guns as well as a 20-mm cannon. Above all, the B-29 was a stable platform for the Norden bombsight. These were its virtues. Its defect was that it was a fuel hog, requiring 6,000 gallons of fuel for each 1,200-mile flight. It also required a 1.5-mile runaway.

The Pacific war was the first one in which the use of airpower conditioned grand strategy. Accepting the theory that the key to victory was the ability to bomb Japan into surrender, Chester Nimitz's plan was to bring the B-29 within range of Japan. The range of the B-29 determined where air bases would have to be. The ideal location, in the Soviet Union, was impossible for political reasons—the Soviets were not at war with Japan. Chinese bases were tried, but this proved to be a logistical nightmare. If Japan was to be bombed, bases within 1,500 miles of the Japanese mainland would have to be found. By seizing Saipan, Tinian, and Guam, a cluster of islands about 1,500 miles from Tokyo, the United States would have bases from which to conduct daylight strategic bombardment of Japan. Later, Iwo Jima was taken simply to provide an emergency landing strip for damaged B-29s returning from Japan.

The first B-29 attacks on Japan were no more successful than were the first raids on Europe. The inaugural attack—staged from China—was against the Yawata steel mill. While losses were light, most of the bombs dropped at thirty thousand feet were scattered. Only 10 percent of the bombs landed anywhere near the target.[40] In eight raids against the Nakajima Aircraft plant in Mushashino near Tokyo, 835 sorties had destroyed less than 4 percent of the roof area.[41]

Daylight precision bombardment had failed again, foiled this time by unpredictable winds, poor intelligence, and general unreliability. Curtis LeMay, who commanded the B-29s in the Pacific and was a strong advocate of daylight precision bombardment, had to face the same two realities that determined the course of the European air war. First, precision bombardment was a myth. Second, the only tactic that made it even conceivable—low-level, daylight, clear-weather bombardment—raised losses to utterly unacceptable levels. LeMay was transferred from China to Saipan, where he replaced Haywood Hansell, the archprecisionist. LeMay switched to nighttime incendiary bombardment in March 1945, postponing, if not abandoning, an American dream.[42]

The primary virtue of incendiary bombardment is that its area of destruction is much greater than that of high-explosive bombs. Assuming that the target consists of combustible material—which, given lightweight construction in Japan, was certainly the case in its cities—a single incendiary device can set off fires that can destroy city blocks. A *saturation* attack by incendiary devices can destroy entire cities. Like Harris, the chief of the British bomber command, LeMay was driven to undertake a counterpopulation strategy—terror bombardment—by recognizing the limits of precision bombardment.

On March 9 and 10, 1945, B-29s staged nighttime incendiary raids on Tokyo. LeMay stripped the bombers of their machine guns, since the Japanese Air Force had been decimated and was virtually useless at night anyway. This allowed each bomber to carry eight tons of ordnance and still travel the three-thousand-mile route. This meant, in effect, that one bomb could be dropped every fifty feet, and each bomber was able to burn out sixteen acres.[43] More people died in Tokyo during those two nights than in any other bombing raid of any war before or since—including Hiroshima and Nagasaki.

The firebombing of Tokyo represented the high point of strategic bombardment—the most efficient use of the manned bomber in a conventional mode. Without fighter escort or onboard defensive weapons, enjoying command of the air without any parasitic load, the B-29s concentrated on being platforms for the delivery of ordnance—that is, airborne trucks. The type of ordnance selected was suitable to the target, and the quantity carried was sufficient to carry out the task. The result was the destruction of the center of Tokyo, the slaughter of nearly one hundred thousand civilians, and the disruption of economic life in the city. But Tokyo survived.

LeMay's strategy was one way to compensate for the inaccuracy of bombs—saturate the area so that the intended target would have a large statistical probability of being hit. It was the clearest case of the application of Douhet's and Mitchell's theories of strategic bombardment, and the closest we have come to a confirmation of that theory. It was also the clearest case of the pure air offensive. The Japanese skies belonged to Americans, attrition rates were low, and bombs could be dropped without significant fear of ground fire. The damage inflicted was savage. Above all else, it was an air war with minimal parasitic elements. It was a war of platforms and projectiles, of pure statistics. In this campaign the inexorable dialectics of defensive parasites and senility had not yet set in.

By the end of World War II, the culture of strategic bombardment had taken firm hold in the U.S. Air Force, culminating in the creation of the Strategic Air Command. The Soviets were less sanguine about strategic airpower, given their tradition of tactical airpower. The Soviets decided to place their faith in unmanned rockets instead of in the manned bomber. This choice flowed from their traditional strength in artillery—artillery being a separate service in the Soviet Union, and all intercontinental missiles being part of that service.[44] For both the Soviet Union and the United States the strategic fixation left them relatively indifferent to the problem of tactical airpower—a neglect that would haunt them in Vietnam and Afghanistan, much as the threat of nuclear war would haunt the world during the Cold War. The atomic bomb was, after all, the midwife of the Cold War as well as the culmination of the Second World War. And it was the logical conclusion of the quest for strategic bombardment.

In the end, World War II was won by the army and navy, and not by the air force. However, the atomic bomb reaffirmed the importance of airpower—it was the ultimate weapon, terrible and irresistible. Neither the army nor navy could deliver the atomic bomb to its target, only the airplane could do that. With the advent of the atomic bomb, Douhet's blustering claim that the airplane had made all other forms of war obsolete finally appeared to be on target—in a completely unanticipated way. This totemic fascination and dread of atomic and nuclear war would shape culture in general and devastate military culture in particular.

10

Rethinking Failure

VIETNAM AND THE FAILURE OF AIRPOWER

During the Cold War, the world was gripped with a sense of the imminence of nuclear war. The confrontation between the American alliance and the communist world ran from Norway's North Cape to the Aleutians. There was constant friction along this periphery, as well as in the rear where the European empires were crumbling. On occasion, as in Berlin or Cuba, the friction seemed about to explode into nuclear war. On other occasions, wars did break out—conventional wars.

Some, such as Korea and Vietnam, were part and parcel of the Cold War. Others, such as the Arab-Israeli wars, had significant connections to it. Yet others, like the Falklands War, had hardly any connection. What all of these politically diverse wars had in common was that all of them were nonnuclear wars **and none of them was expected.**

The radically new weapons systems of the Cold War—thermonuclear weapons, the solid-fuel intercontinental ballistic missile launched from silos and submarines, satellite surveillance and warning systems, the first surface-to-air missiles—were created as part of the nuclear obsession. Many of these weapons had profound consequences on the development of conventional systems, of course. Strategic and surface-to-air missile technologies were turned into conventional tactical missiles; reactors for nuclear submarines found their way into aircraft carriers and cruisers; satellite systems designed to locate enemy missile silos evolved into satellite systems capable of monitoring conventional capabilities.

The problem was that the evolution lacked guiding principles. Until the 1980s, the transfer of technologies from nuclear to conventional was done without any overarching, systematic plan. **This was because until the 1980s there was no planned, strategic understanding of nonnuclear warfare.** Nor could there be. According to the prevailing wisdom, conventional warfare threatened to escalate into nuclear war. And if the enemy had nuclear weapons, he could not be compelled to surrender. Thus, because conventional warfare could no longer achieve the desired results, it was undesirable as an instrument of national power.

Though the obsession with nuclear war did not abolish conventional war, it did abolish serious thought about it. Even so, conventional warfare persisted. Between 1945 and 1990, the United States spent ten years in conventional wars. The Soviet Union spent eleven years fighting the Afghans. The best minds thought about nuclear war and guerrilla war—called unconventional war, which were wars about societies, shattering the social order and rebuilding it. The old thinking about warfare, about how to destroy the enemy army without concern for social destruction or reconstruction, seemed passé, irrelevant.

Airpower found itself in a particularly difficult position. Before the advent of nuclear weapons, all of the passion and technology of the air forces had gone into strategic bombardment. The question of the role of airpower in a nonnuclear war was left barely examined, since conventional wars were no longer expected to take place.

Korea was a conventional war—not only because atomic weapons would not be used but because the goals of the war were inherently modest, deriving from the limited nature of American grand strategy. U.S. strategy was containment, to stop the communist attack. Following Douhet's vision of war as a total and merciless assault on enemy society, the Air Force had a doctrine for *annihilating* the enemy—even if it lacked the means to do so—but it did not have a doctrine for a more limited use of airpower. And it was very much at a loss as to what its role was in this intermediate realm. The Soviets had always kept their distance from Douhet's theories and began the Cold War by seeking to integrate the air forces with the operational level of war, that is, working with ground and naval forces in the general prosecution of war. The U.S. Air Force, on the other hand, imagined a much grander role for itself. As a result, in the area of the operational, it was very much at a loss, and air commanders were forced to improvise, with varying degrees of success. It is not that the commanders did badly, but rather that their success was in opposition to their doctrine and their inclinations.

After Korea, relatively little was done to prepare the U.S. Air Force for conventional war. Weapons progressed, but doctrine remained stable—and sketchy at best. The Air Force saw itself primarily as part of the strategic balance, with its real focus on deterring Soviet nuclear power. This fit in nicely with the view of the Eisenhower administration, which regarded the Korean War as a disaster in which the United States permitted the communists to define the war to their advantage.

The prevailing doctrine—massive retaliation—meant that the United States would not respond to a conventional challenge with conventional forces but was prepared to use nuclear force, hoping that this willingness would deter communist adventurism. Berlin, and even more so Cuba, seemed to be persuasive examples of how America's strategic airpower could compel the Soviets to capitulate politically without having either to use nuclear weapons or to engage in a conventional war. Obviously, this placed the Air Force, and particularly the Strategic Air Command, at the center of U.S. strategy. This also permitted the Air Force an Olympian contempt for lesser matters, such as the growing communist insurgency in Southeast Asia.

However, if the Air Force was pressed to devise a doctrine for dealing with limited war, the operational principle had remained unchanged since World War II: the Air Force would act as a nonnuclear strategic force, using conventional ordnance and strategic aircraft to attack the enemy's social and economic base, thereby compelling the enemy to submit. Nonstrategic aircraft would support ground operations, including close air support and interdiction strikes to the enemy's rear. That this had not worked in World War II did not change the doctrine. The deep conviction of the Air Force was that, given the proper tools, strategic bombardment would have to work. And pointing to the significant advances in aircraft since World War II, it did not believe that what had failed once would fail again.

Cheap Victory: Airpower's Promise in Vietnam

The early planners of the Vietnam War argued that Vietnam was geopolitically important to the United States, that it was an essential part of our containment strategy against the communists. In addition, even if Vietnam were not important in and of itself, the failure to defend Vietnam would be taken as a signal of American weakness, frightening allies in more strategic places and emboldening the communists globally. The psychological and political effect of Vietnam's fall on the defensive archipelago stretching from Japan to Indonesia and Malaya and through the rest of Asia from Thailand to the Levant could be devastating, even if the communization of Indochina would not destroy the American geopolitical position in Eurasia.

It is difficult to conduct a war in which avoiding defeat is the only interest, as it makes it difficult to calibrate the amount of forces that need to be allocated. A vital interest, such as Western Europe, would require the use of virtually all available force. What should be devoted to defending a nation whose functional value was unclear?

The American inclination was to total war, the commitment of all available forces. It would have been deranged to make such a commitment in Vietnam, as forces had to be retained for the defense of much more vital American interests, such as Europe or Iran, or for an invasion of Cuba. The forces available for use in Vietnam were determined by considerations having nothing to do with Vietnam, and the question of whether the total of those forces would have been enough or appropriate was never addressed. The strategically defensive nature of the war confused matters even further. Rather than proposing a rapid offensive *into* North Vietnam, the strategy was to hold South Vietnam *against* North Vietnamese attack. This meant that North Vietnam would determine the tempo and duration of the war.

One way to limit the American commitment was to have the South Vietnamese Army do much of the fighting, advised or commanded by American advisers. This approach was reinforced by the American conviction, particularly by civilians in the Kennedy administration and by Special Forces theorists in the Army, that Vietnam was less a war than an exercise in nation-building. Encouraging the South Vietnamese Army to defend Vietnam would help build a strong national institution, regardless of how well the soldiers fought. It was an interesting theory, and if it had worked, it would have been an outstanding solution to the American problem of holding down the American combat presence in Vietnam.

A purely defensive war is rarely a wise policy. On the other hand, taking the offensive and threatening the continued existence of the North Vietnamese might have proven intolerable to the Chinese. The United States was not prepared to cope with another war with China. Therefore, it wanted to find a way to stop the North Vietnamese government from supporting the Vietcong in the South without threatening the existence of the North Vietnamese regime. Airpower provided the United States with the means of inflicting pain on the North without threatening the North's communist government. It appeared to be offensive action without the risks and without the massive costs that a ground invasion would entail.

The key assumption was that airpower would be so effective and frightening that the North would have no choice but to abandon its political goals and make peace on terms favorable to the United States. The strategy was to interfere with North Vietnam's war production, interdict the flow of supplies to the South, destroy concentrations of communist troops, protect South Vietnamese Army installations from being overrun, and above all, inflict a high level of pain on the North Vietnamese regime and population. A secondary assumption was that this could be done with little cost to the United States.

An air campaign was extremely attractive to civilians who supported intervention in the war but did not wish to expend great resources on the effort. It also appealed to Lyndon Johnson and Robert McNamara, both of whom were seduced by airpower. The Air Force was also pleased by the possibility of showing its stuff, even though it was extremely concerned by the limits being placed on its ability to execute a genuine strategic campaign. The regular Army, however, was downright skeptical. Deeply influenced by MacArthur's dictum that the United States should never again fight a land war in Asia, it feared that, despite the claims of the Special Forces, the Air Force, and civilian analysts, the Army would be left holding the bag if the air campaign failed.

And fail it did. In this case, however, the inability of airpower to subdue the enemy was not hidden, as it was in World War II, by the success of the ground and naval war. At the same time, the strategy behind the war in Vietnam was so poorly thought out that it is difficult to see how airpower could be singled out for blame, or how this war could have been regarded as a fair test of airpower.[1]

Secretary of State Dean Rusk and the Joint Chiefs of Staff, for example, advocated a rapid, brutal campaign against key economic targets in North Vietnam. They assumed that this would rapidly undermine the North Vietnamese economy, leading to its collapse and forcing the North to make peace. It was thought that when the North Vietnamese saw what was happening, they would understand that their only choice would be to become vassals, economically dependent on either the Chinese or the Americans. They would quickly reach a deal with the United States, abandoning the National Liberation Front in the South in order to retain their economic autonomy in the North.[2]

Arguing against this classic airpower view and in favor of a redefined role for airpower were Robert McNamara and his analysts at the Defense Department. McNamara feared both success and failure. If the bombardment succeeded, it could force the Vietnamese to turn to the Chinese or Soviets; if it failed, it would lead to involvement beyond the means and interests of the United States: "The likelihood and political costs of *failure* of the approach, and the pressures for U.S. escalation if the early moves should fail, require serious examination."[3] The logic of McNamara's position should have led him to argue against intervention. Instead, he argued for a strategy that could not, by definition, bring the war to a rapid conclusion. Unwilling to risk breaking the North Vietnamese, McNamara chose a strategy in which the United States could never exert enough force to bring the war to a successful end. Without a meaningful offensive capability, the United States would be forced back onto the defensive—an endless defensive nightmare.

McNamara's strategy prevailed. Fear of Chinese intervention took precedence over all other considerations. Of course, it is not at all clear that a bombing campaign could have defeated the North—it may have been hopeless anyway. But McNamara's strategy was based on the assumption that the United States could not afford to try. Thus, rather than an all-out strategic bombing campaign designed to smash the North's economy, a gradual escalation was undertaken, in which the pressure on the North was slowly increased. The idea was to convince the North Vietnamese that the United States was prepared to go to any extreme to achieve its strategic goals, without frightening them into asking for direct Chinese or Soviet assistance. The net result was to convince the North Vietnamese that the United States was *not* prepared to exert the necessary force or to take the necessary risks.

Given the importance to North Vietnam of its goal—of national unification—it would have taken an enormous amount of pain to dissuade Hanoi from its course. Since McNamara's strategy was to inflict pain in gradually increasing increments without ever threatening the existence of the regime, the North was able to adjust physically and psychologically to its circumstances. The result was increased aggressiveness on the part of the North.

This is a point that has been made many times, particularly by advocates of the Joint Chiefs of Staff–Rusk strategies. For example, when asked if the Vietnam air campaigns represented strategic air thinking, Air Force general Curtis LeMay replied, "Definitely not. It wasn't until the last two weeks of the war that we even approached it. When we turned the B-52s loose up north—that started what would have been a strategic campaign, and would have been completely over in a few more days if we had just continued it."[4]

LeMay, not without reason, felt that the failure of the bombing of North Vietnam, far from being a refutation of strategic airpower, was actually its confirmation. Had the United States actually employed strategic bombardment, as it did during the Christmas bombing of Hanoi in 1972 (Operation Line Backer II), its uses would have been amply confirmed.

The defect in LeMay's thinking, and in the planning of Rusk and the Joint Chiefs, was to assume that the economic foundations of North Vietnam's war machine were to be found on the ground in North Vietnam. Nothing could have been further from the truth.

This is not, as some have argued, because North Vietnam was fighting a simple, peasant war. North Vietnam was mounting a modern, complex, and sophisticated war effort. It was because the bulk of the matériel was being produced *outside* North Vietnam—in Canton and Smolensk, in Pyongyang and Prague—and transported into the North, and from there into the South.[5] The North's own economy was a peasant, agricultural economy that could not be broken by airpower. So, in a sense, the North Vietnamese had the best of both worlds: a military-industrial base beyond the reach of the Americans and a peasant economy impervious to air attacks (excepting possibly an attack on the Red River dams).

No strategic air campaign was possible in Vietnam. If such a campaign had been undertaken, it would have entailed an attack on China and the Soviet Union. But this was not feasible. First, it would have violated American grand strategy, which was to contain communism, never to attack it directly. Second, it would have been too dangerous. On the part of the Soviets, it could have triggered a strategic bombing campaign against the United States proper. On the part of the Chinese, it might have triggered a ground intervention that would either have swamped American forces in Vietnam or required a World War II–style reinforcement, which would simply not have been worth the effort.

Strategic bombardment is viable in a condition of total war, such as World War II, or where the enemy is so weak and isolated that it has no suitable response, as in Iraq. But in a case where the target is neither weak nor isolated, and where total war is something to be avoided, strategic bombardment is not a rational choice. The weakness of strategic bombardment is that, unless it rapidly eliminates the ability of the enemy to respond, it frequently inflicts costs on the bomber greater than the benefits being reaped. It is, therefore, rare to find a case in which the strategic goal is worth the effort of a strategic bombing campaign, or in which a suitably weak and isolated victim can be found. World War II was a case of the former, the Persian Gulf War of the latter.

The problem was that the Air Force thought it knew how to destroy the enemy's economy and society. What it did not know how to do was destroy the enemy's military. The Air Force was incapable of isolating the battlefield either tactically or operationally. It was incapable of cutting either lines of communications or lines of supply. It could not stop the North Vietnamese from sending their Army south, feeding it, or commanding it. The Air Force could devastate enemy troop concentrations if intelligence about their location was available. It could also add firepower to tactical engagements. But the Air Force found itself incapable of systematically assaulting the operational capability of North Vietnam. It simply didn't know how.

Blindman's Bluff: War Without Doctrine

Supplies were entering North Vietnam via the port of Haiphong and on two rail lines from China. There was a great deal of concern about bombing Haiphong Harbor and hitting foreign vessels—or bombing the rail lines near the border and bringing China into the war. Moreover, World War II taught us that rail lines could too easily be repaired.

Fortunately, geography provided a solution. The two rail lines from China both converged on the same rail bridge over the Red River near Hanoi—the Paul Doumer Bridge. The rail line from Haiphong also crossed this bridge. From the Paul Doumer, rail traffic to the south proceeded on a single rail line, which crossed the Chu River over a bridge at Thanh Hoa. Supplies would move down this rail line, to be off-loaded somewhere around Vinh, and then brought down to the South via the Ho Chi Minh Trail. Alternatively, supplies would leave the Doumer via roads into Laos, connecting with the Ho Chi Minh Trail from there.

The entire logistical structure of the North Vietnamese Army and the Viet Cong could have been shattered by destroying two key bridges—the Paul Doumer and the Thanh Hoa—and keeping them from being repaired. With these destroyed, rail and road traffic would have been cut, and outside supplies would have been entirely prevented—or at least seriously delayed—from reaching the fighting fronts in the South. Attacking these bridges would not trigger any outside intervention, nor would it require the use of ground forces. It was a perfect mission for airpower. It would not be an exaggeration to say that the entire war turned on the success of this mission.

On April 3, 1965, seventy-nine U.S. aircraft, including forty-six F-105s, twenty-one F-100s, two RF-101s and ten KC-135 tankers, attacked the Thanh Hoa Bridge. One hundred twenty 750-pound bombs were dropped along with thirty-two Bullpup guided missiles. Hits were scored, but only minor damage was achieved. Two planes were shot down. The next day, another 384 750-pounders were dropped. Pilots claimed that 300 bombs struck the bridge—unlikely, since the bridge was easily repaired. Three more aircraft were shot down, two of them by the first MiG attacks of the war.[6]

The problem was twofold. First, the bridge was of superb French construction, able to resist explosives in excess of a thousand pounds each. Second:

> The placing of a bomb on a 16-foot-wide, 500-foot-long steel bridge from a fighter aircraft traveling over 500 mph while being fired on by myriad AAA weapons became a monumental task! Day and night strikes against the bridge, using visual as well as radar bombing techniques, had succeeded only in shaking the steel girders.[7]

The U.S. Air Force and Navy together lacked the means for destroying two bridges permanently—although the Paul Doumer was taken out for a time during the 1967–68 bombing campaign, in a mission that required 177 sorties and over 300 tons of bombs. One of the foundations of American strategy was unattainable because of technical limits. Airpower simply was not up to the job.

If large, fixed installations could not be permanently destroyed, it was a thousand times more difficult to interdict the flow of supplies down the complex and hidden jungle tracks of the Ho Chi Minh Trail, as the Americans called it. The following is a description by North Vietnamese senior general Van Tien Dung:

> The strategic route east of the Truong Son Range was the result of the labor of more than 30,000 troops and shock youths. The length of this route, added to that of the other old and new strategic routes and routes used during various campaigns built during the last war, is more than 20,000 km. The 8-meter-wide route of more than 1,000 km, which we could see now, is our pride. With 5,000 km of pipeline laid through deep rivers and streams and on the mountains more than 1,000 meters high, we were capable of providing enough

fuel for various battlefronts. More than 10,000 transportation vehicles were put on the road.[8]

The Ho Chi Minh Trail was designed to have neither choke points nor vulnerable clusters of transport vehicles. Much of the supplies were brought down on the backs of infiltrating troops or on bicycles. Trucks came down through complex, circuitous routes, singly, so as not to provide aircraft with targets.

The only way to cut the Ho Chi Minh Trail was to position troops across the Laotian panhandle, linking defensive positions across the demilitarized zone with a line running along the Quang Tri–Sano highway. This would have meant the violation of Laotian "neutrality," already a complete fiction. More important, it would have required the complete rethinking of American strategy. Any interdiction deployment would have had to be able to resist a direct assault by the North Vietnamese Army. The strategy would have had to assume that the North Vietnamese would have engaged in a battle of annihilation to save their southern forces, and that American troops, strung out along a defensive line, would have been vulnerable in such a battle. The distance from Quang Tri to Sano was about 150 miles. For a static defense in depth, this would require a force of six divisions, along with a strategic reserve of at least two more divisions. In other words, the bulk of American combat forces (the United States had deployed the equivalent of eleven divisions in 1968)[9] would have had to have been deployed defensively, leaving the guerrilla war in the rear to South Vietnamese troops. It would have turned the Vietnam War into a replay of the Korean—in retrospect, not a bad idea.

The failure of the air assault on the North, which was intended to force the communists to cease attacking the South, left the U.S. Army in the position it had feared most: it became responsible for suppressing communist forces in the South. The Army could not take the offensive outside of South Vietnam but had to wait until North Vietnamese men and supplies actually entered the South. Militarily, this dissipated U.S. power in countless minor engagements. Politically it turned the U.S. forces into policemen. The Army could not carry out its responsibilities without massive reinforcements of troops from elsewhere in the world. By 1970, it was no longer clear that the United States could carry out its missions in other parts of the world.

In 1964, had anyone described the burden that Vietnam would impose by 1967, either the war would have been avoided or a very different strategy undertaken. The Johnson administration did not want

South Vietnam to fall. At the same time, it was not prepared to devote the resources necessary to prevent its capitulation. U.S. officials bridged the gap with fantasy. One fantasy was that the South Vietnamese Army was sufficient to carry out the mission. The more serious fantasy was that U.S. airpower could compel the North Vietnamese into the kind of political settlement the United States wanted.

The Air Force could not break the center of gravity of the North Vietnamese war effort. As we have seen, the Air Force was unable to shut down the rail and bridge system in the North. Nor was it up to the task of cutting the Ho Chi Minh Trail. Brig. Gen. Soutchay Vongsavah of the pro–American Royal Laotian Army, said, "Despite the full weight of American bombing operations, Ho Chi Minh Trail truck traffic increased day by day and the NVA [North Vietnamese Army] continued to improve the road network, keeping passages open and making roads suitable for high-speed traffic."[10]

Airpower's supporters were the worst enemies of its reputation, accepting assignments that no one could reasonably expect airpower to carry out. Mission after mission was flown in a vain attempt to block the increasingly complex and sophisticated trails. It was as if a gushing wound were to be tied off capillary by capillary, rather than at an artery, with the ties dissolving at the end of each day, requiring the surgeon to begin anew the next. It was an impossible task.

Where airpower did make a difference was in providing support for ground forces engaged with the enemy. Here, airpower behaved less as the original theorists had envisioned and more as airborne artillery. Bombs were dropped from low altitudes, on targets designated by forward air controllers. Like artillerymen, these men were adding their firepower to engagements they could not themselves see. In doing so, they increased the rate of attrition of the enemy, in many cases turning the tide of battle.

But tactical support was not what strategic airpower had promised. According to Douhet's theory, that tactical combat need not take place if the war-making capability of the enemy was destroyed in the factories or on the roads. As we have seen, part of the problem was geographical— Vietnam's factories were located in other countries. But the central problem was technical. The United States failed to isolate the South from the North because its bombers were too inaccurate to permanently destroy two bridges. This inaccuracy was the root of the American defeat in Vietnam. And it was not an isolated problem: it was at the increasingly rotten core of modern warfare.

Sudden Dawn: Precision-Guided Munitions Appear

On April 27, 1972, something extraordinary happened. It would be too late to change the outcome of the Vietnam War, but it was extraordinary nevertheless. Earlier that month the United States had resumed its bombing of North Vietnam. The goal was primarily political—to force North Vietnam to reach a settlement with the United States prior to the November presidential elections. Nixon intended to show the North Vietnamese that the United States could, if it wished, destroy North Vietnam's command and logistical structure. But he also wanted to make it clear that the failures of the past seven years had been rooted in American restraint and not in any technical incapacity.

The massive B-52 bombing raids over Hanoi in 1972 devastated large portions of the city and strained the North's air defense capacity beyond its limits. But by themselves, they did not touch the North's ability to wage war. They did not shatter the North Vietnamese command structure or its ability to maintain control of its forces in the field. Nor did they have the ability to break apart the North's logistical system.

What was deadly, and utterly new, was the ability of the United States to attack and destroy the North's two key bridges. In 1972 the Paul Doumer and the Thanh Hoa were finally brought down together—and finally, if too late, the South was isolated from the North.

As we have seen, the Thanh Hoa Bridge had been a particularly important and resilient target. It had been bombed numerous times prior to the bombing halt in 1968 without being destroyed. Attackers were simply unable to place sufficient tonnage at precisely the right spot to do serious, permanent damage. On April 3 and 4, 1965, 504 750-pound bombs were dropped in over one hundred sorties, with five planes lost. On the assumption that the bridge would on average be closed for a week as a result of such a strike, it would cost the United States over 250 planes a year to keep it shut down, with a full air wing permanently tied to the mission—an impossibility. Indeed, attacks on the Doumer Bridge in Hanoi, an even more important target than the Thanh Hoa, were simply abandoned in 1967.

On April 27, 1972, twelve F-4 Phantoms of the Eighth Tactical Fighter Wing in Thailand were ordered to strike at the Thanh Hoa. Eight aircraft carried two-thousand-pound bombs; four carried chaff, designed to foil North Vietnamese radar and protect the attacking planes from radar-guided surface-to-air missiles and antiaircraft guns. The explosives in the bombs were no different from the explosives used in other raids. What was different was that the bombs, once released, were not merely subject to gravity. Rather, they were "smart" bombs that could correct their course while falling.

There were two kinds of guided bombs used on April 27. The first was the electro-optically guided bomb. This was a conventional high-explosive bomb with a small TV camera in its nose. During the attack run, the pilot would point the plane (and thus the bomb's TV camera) at the target. The bomb's camera would transmit a picture to a screen in front of the weapons system officer sitting in the backseat of the F-4. The weapons officer would select a point on the target with high contrast—a small area on the target clearly defined because of distinctions of light and dark—which a small, primitive computer would lock onto. After being released, the bomb, locked on the contrast point, would direct itself using vanes—small wings and tails that would provide guidance—and using inertia and gravity for energy. Once the bomb was released, the F-4 could leave the area—it could fire and forget. Not only would the plane probably not be shot down, the bomb would also probably hit the target at the precise point that had been designated.[11]

The second type of munition was the laser-guided bomb, which required the illumination of the target by a low-power laser beam.[12] This beam was emitted by a pod carried under the fuselage of the attack aircraft—or some other aircraft accompanying the bomber. The weapons officer would point the pod at the target (note that the plane itself did not necessarily have to be pointed at the target, only the pod, which could rotate), illuminating the point that was to be struck. The bomb, like the electro-optically guided bomb, guided itself to the target using inertial energy and controlling vanes.[13] A great advantage of the laser system (designated Paveway) was that more than one aircraft could use the point designated by the laser beam. Since the target would have to remain illuminated until the bombs struck, this meant that only one aircraft had to loiter around the target (a dangerous practice, even if the plane was heading away from the target), while the others could shoot and scoot.

Each system had advantages. The laser-guided bomb had an easier time locking onto the target, since there was constant, clear illumination. On the other hand, it was not simply a fire-and-forget weapon, and just as important, it required clear vision, without clouds or smoke, to work. The electro-optically guided bomb had a tendency to lose its lock if the contrast point was obscured or if perspective shifted, but it was a fire-and-forget weapon. The weapon of choice, therefore, depended on weather, lighting, and other atmospheric considerations. It also depended on how well the target was defended, and how dangerous it would be to loiter.

In the raid on the Thanh Hoa Bridge, weather conditions were cloudy, so laser bombs could not be used. Four F-4s dropped chaff to screen the attacking aircraft. The rest, eight F-4s, dropped five two-thousand-pound electro-optically guided bombs. The bridge was rendered unusable. No U.S. aircraft were lost.[14]

On May 13, 1972, the Thanh Hoa was revisited. Fourteen strike aircraft dropped nine three-thousand-pound LGBs, fifteen two-thousand-pound EOGBs, and forty-eight conventional bombs. The western span of the bridge was completely knocked off its struts, while the rest of the bridge was rendered unusable. Again, there were no losses to the attacking planes.[15]

As effective as these attacks were, it was the attack on the Doumer Bridge that signaled the true meaning of the revolution in warfare. The Doumer Bridge had not been attacked since 1967. U.S. losses had always been high, as the attack brought aircraft into range of Hanoi's strong air defenses. Even though it had been damaged once, it was a tough target. On May 10, 1972, this changed. Sixteen F-4s armed with twenty-two LGBs and seven EOGBs (all two-thousand-pounders) attacked the Paul Doumer Bridge. The sixteen attack aircraft were accompanied by eight F-4s dropping chaff, fifteen F-105G Wild Weasels designed to destroy the enemy radar that guided North Vietnamese surface-to-air missiles and antiaircraft guns, and four EB-66 electronic-countermeasure aircraft to jam the remaining radar and communications facilities. Of the twenty-nine bombs released, twelve hit, four were probable hits, and thirteen either missed or could not be observed. Several spans were destroyed or damaged. The next day, four more F-4s came in to mop up with two three-thousand-pound and six two-thousand-pound LGBs. There were no losses, and three more spans were destroyed. The Paul Doumer Bridge never again carried traffic until the United States withdrew fully from the war.

A number of extraordinary changes had occurred from 1965 to 1972. First and most obvious was the extraordinary change in accuracy. In the 1965 raid on the Thanh Hoa, seventy-seven aircraft were used on the first day. In the 1972 attack on the same bridge, eight attack aircraft were used. In 1965, 45 tons of bombs were dropped (with another 144 tons dropped the next day), whereas only 5 tons were dropped on April 27, 1972, with another 28.5 tons dropped on May 13. In 1965, 189 tons could not render the bridge unusable, while in 1972, 36.5 tons (less than 20 percent of the amount used in 1967) was sufficient to close it.

Another startling contrast can be seen. In 1965, of the seventy-nine aircraft involved in the attack, only seven were assigned to flak suppression, two to reconnaissance, and none for chaff or electronic countermeasures. During the attack on the Doumer Bridge, sixteen planes dropped bombs, while twenty-seven were devoted to flak suppression and antiradar measures. The earlier, unsuccessful raids had been overwhelmingly devoted to the offense. The percentage of attack aircraft had shifted from 91 percent to 37 percent; in fact, the bulk of the aircraft that flew missions in 1972 were intended to protect the attackers.

Airpower failed strategically in Vietnam but succeeded tactically—it greatly increased the power and effectiveness of ground forces.[16] But in the end, where airpower showed the greatest promise was on the operational level: the attacks on the North Vietnamese transportation system in 1972 indicated that airpower would be able to prevent enemy forces from carrying out their mission—by interfering with the movement of men and material, by affecting the ability to command and control troops, and by breaking the cohesion of large concentrations of forces.

In other words, the late stages of Vietnam opened a new possibility for airpower, a possibility substantially different from the one imagined by Douhet and Mitchell but, in a way, as significant. It was a vision closer to Clausewitz's orginal sense of the purpose of warfare—to render the other side incapable of further resistance. These new potentialities at the end of the Vietnam War would give rise to a thoroughly new way to think about airpower. But first, the long, miserable experience of Vietnam would have to be thought through.

Wha' Hit Me?—First Thoughts on Airpower's Failure

Defeat in Vietnam stunned the military more than the nation as a whole. Americans were exhausted by the war and simply pleased that it was over. For the armed forces, their entire identity, their sense of self, was challenged by the loss that had occurred. How could an army of light infantrymen, without airpower, usually without any artillery support larger than mortars, have overwhelmed the previously invincible U.S. Army, Marine Corps, and Air Force.

The first response was to claim that the United States had won the military war in Vietnam but had lost the political war at home. The military had enjoyed a string of successful engagements in Vietnam—

indeed, there was not a single major battle or campaign in which it had been defeated. From the first battle of the Ia Drang Valley onward, American forces, usually outnumbered and fighting with inferior intelligence, consistently engaged and defeated North Vietnamese forces. Even on the defensive, as at Khe Sanh and during the Tet Offensive, American forces held, repelled, and destroyed attacking forces.

Defeated military forces, particularly proud forces with strong traditions of victory, tend to respond to defeat by deflecting criticisms to the domestic front. The German Army, defeated in World War I, blamed Jewish revolutionaries for undermining political support at home. The French Army blamed their defeat on the politicians of the Third Republic. The Soviet Army blamed their early defeats on Stalin's purges and mismanagement. The German Army in World War II blamed Hitler's bungling.

Something very similar happened after Vietnam. According to the most prevalent theory, the antiwar movement was the culprit—draft-card burning, dope-smoking, communist-sympathizing perverts had undermined popular support for the war. There was a strong element of class anger here. The bulk of the enlisted men in Vietnam were drawn from the lower and lower-middle classes. The officer corps was drawn from the lower-middle and middle classes. At best, the enlisted men were high school graduates, while the officers were graduates of large state universities. The antiwar movement was perceived as upper class and centered around prestigious universities. Hence the perception grew that the elite had first taken the country into war and then, encountering difficulty, thrown the lower classes to the wolves.

As in all "stab-in-the-back" theories, there were elements of truth here. The bitterness of enlisted men returning home late in the war to contempt and indifference was surely justified. But this theory failed to explain two things. First, prior to 1968, the antiwar movement was small and marginal. Most Americans held protesters in contempt and supported the war. Indeed, from 1964 to 1967, the military received virtually any resource it required. Yet, during those years, it was still unable to bring the war to a successful conclusion, or even to erode the communists' will to resist. Second, in the post-1968 period, the antiwar movement—understood as the radical, campus-based youth movement—grew relatively little. The growing antipathy toward Vietnam derived not from increasing support for the student movement but from frustration at the military's inability to end the war and a sense that the time and resources being committed to it were out of proportion to American interests. We must remember that Richard Nixon was elected president in 1968 *after* the Tet Offensive.

A more sophisticated explanation was needed, and one was ready to go, having actually been bandied about since the beginning of the war. The argument was made that the United States lost the war because unreasonable political constraints had been placed on the military. The United States could not invade communist sanctuaries in Cambodia and Laos, could not invade North Vietnam nor unleash an unlimited bombing campaign on the North. Thus, this theory went, the United States could not win the war because it was fighting with one hand tied behind its back.

There is much to be said for this argument, particularly for the contention that the war was lost because American forces were kept on the strategic defensive, unable to move against the North's logistical system outside of Vietnam. But the military shares responsibility in this matter. At no time before 1968 did the Army's command at any responsible level inform the secretary of defense, the National Security Council, or the president that the war was unwinnable under the current operating strategic constraints. This was not because they were dishonest—although careerism did skew their professionalism somewhat. Rather, it was because they could not imagine defeat or even stalemate.

During Vietnam, the key measure of military force was firepower—the ability to bring explosive force to bear on the enemy. American artillery, helicopter gunships, AC-130 "Puff, the Magic Dragon"s, and ubiquitous aircraft gave the Americans superior firepower in all engagements. Superior firepower that cost a great deal to produce, supply, and maintain. The North Vietnamese lacked this firepower, and it was therefore assumed that, in the long run, they would be worn down by attrition and negotiate as the only way to save the army.

What the Americans forgot was that the North Vietnamese had several advantages over the United States that cost them very little. The first and most important was intelligence. The very act of bringing firepower to bear, particularly setting up artillery firebases, is an intelligence giveaway. The North Vietnamese Army could not help but know where the U.S. artillery was, what kind it was, how much, and what its range was. Intelligence also permitted them to know how to avoid American firepower. Large-scale infantry operations—search-and-destroy missions—were similarly obtrusive. North Vietnamese scouts could easily detect the buildup of U.S. forces in an area, and excellent intelligence through sympathizers in the South would frequently supply them with operational outlines. Avoidance and engagement on favorable terms was made possible by this intelligence.

The United States, by contrast, had relatively poor intelligence on the operation of enemy forces. The North Vietnamese were a light infantry force, unencumbered by heavy weapons, tied together by a superb command and control system that allowed their formations to break into smaller groups and regroup at opportune moments. The Americans simply did not know where they were. Thus, on one side were the North Vietnamese with inferior weapons but superior intelligence and operational doctrine. On the other side you had the Americans with superior weapons but an inferior war fighting *system.*

Weapon for weapon the United States should have won the war. But wars are not fought by weapons. They are fought by systems of war, which bring systems of weapons into contact with systems of weapons. The superiority of American weapons confronted the inferiority of American strategy, which in turn was bound up with the weakness of American intelligence. Defenders of the military were altogether correct when they pointed out that the United States tended to win firefights and almost always won battles and campaigns. What they missed was that the engagements that the Americans won were rarely the ones that they should have been fighting. The ability of the North Vietnamese Army to avoid combat except on its own terms meant that the measure of success was not to be found in the body count of any particular engagement but in the count of how many engagements desired by the Americans actually took place.

Because combat between the United States and the North Vietnamese normally occurred at the time and choosing of the North Vietnamese Army, it usually took place when U.S. troops were at a disadvantage. At that moment, the only way the United States could bring additional firepower to bear was to call in air strikes.[17]

Airpower became the final guarantor of the American position—a decisive tactical asset. But it never became a decisive *strategic* asset because none of the battles the United States fought was strategically essential.

American aircraft were generations ahead of anything the North Vietnamese had. That was their problem. They flew too high and too fast over the battlefield. In the South, they were too sophisticated to deal with small-unit engagements, while in the North they lacked sufficient sophistication to shatter logistical and communications facilities. This was the case at least until near the end of the war. Some airmen were prepared to acknowledge this. But even those who argued that American aircraft were inappropriate to their mission did not take the next step, which was to argue that it was not airpower but American air doctrine that had failed. Simply put, no one had any idea what airpower was supposed to do in a conventional war, just as no one had thought about what weapons ought to be used.

There's Gotta Be a Better Way: Thinking About Reform

After World War II the design philosophy behind American weaponry had emphasized high quality regardless of cost, even if it meant lower quantity. Several experiences during the war moved the United States in this direction. Shortly after Pearl Harbor, the United States discovered that the Japanese Zero was superior in dogfighting to any American fighter plane. This placed the United States at a tactical disadvantage as well as a strategic disadvantage if the Zero were able to gain air superiority over an American carrier fleet or an embattled island.[18] In the last days of the war, American bombers encountered German jet fighters that were far too fast and maneuverable to be shot down. The Air Force realized that had these been introduced earlier and in greater number, the U.S. bombing campaign would have been a failure. On the ground, the Army was shocked into a sense of mortality when it discovered that the Sherman tank's 75-mm gun could not penetrate the armor of the German Panther. Had the Panther appeared earlier and in quantity, American victory would have been in real doubt.

In the air and on the ground, the United States came away with a deep sense of technological vulnerability. The American strong suit was the ability to rapidly mobilize its industrial plant for war production. It had built rather ordinary weapons with extraordinary efficiency, concentrating its innovative skills on production techniques rather than on the design of cutting-edge weapons systems. In a way, the nature of the war forced them in this direction. Because the United States needed to arm the British and Soviets as well as its own army, mass and speed were imperative. In addition, American industrial philosophy had always been based on mass. Henry Ford had never made the world's finest car, but rather the world's most efficiently produced car. What the United States did best, therefore, was build lots of usable, but rarely outstanding, weapons.

In pursuing this course, however, the Americans opened themselves to defeat by the enemy's more sophisticated weapons. In each of the cases mentioned above, the United States felt that it had been saved by rapid innovation and development in response to these weapons, by greater troop strength, or by the sheer proximity of victory. The United States saw World War II as a cautionary tale.

The lessons of World War II were reinforced by the realities of the Cold War. With the Soviets, the Americans encountered an enemy who was numerically superior and who had also acquired during World War II a massive industrial base able to produce weapons. Moreover, the Soviets could initiate hostilities at the time and place of their choosing and establish the tempo of the war. Not only was it unclear that the United States could outproduce the Soviets, tank for tank, plane for plane, but it was unclear that, in the event of a war in Europe, the United States would be given time to mobilize its industrial base.

The World War II pattern of mobilizing men and industry *after* the outbreak of war had suddenly become inappropriate. If the possibility of sudden and unexpected war was enough to persuade the United States to field a standing army, it followed that that army had to be ready to go to war—an event unprecedented in its entire history. So the question became how—without dedicating the entire industrial plant of the country to mass-producing weapons—should a standing army be armed?

Postwar analysis showed that the Soviets had developed excellent homegrown weapons, including the T-34 tank, the Stalin Organ multiple-rocket launcher, and a wide variety of superb artillery pieces. They captured a wide array of German technology, and thus there was every reason to believe that they had, in addition to more weapons, better weapons. Analysis also proved that, because of demography and geography, the United States was incapable of numerically matching Soviet forces on the ground.

Military planners who thought beyond the knee-jerk response to attack, atomic weapons, recognized two practical counters to the Soviet threat. The first was to produce technically superior weapons across the board. The second was to train the standing army and reserves so well that they could compensate for Soviet numerical superiority on land and in the air. This was not the case at sea, where both American quality and quantity reigned.

Where World War II was built around mass production, the Cold War was built around technical excellence. The operating principle was that each weapon should incorporate the most advanced technology at the moment of its development and that newer versions should be constantly updated. Weapons were to remain on the technological cutting edge at all times, even if this meant that fewer weapons could be built. In general, at this time military funds were not in short supply, and the defense budget could be expanded if the need or opportunity to produce new weapons arose.

The Soviets adopted a very different strategy, partly out of necessity and partly out of choice: to make a large number of adequate—as opposed to cutting-edge—weapons. Their general experience of war was to absorb a massive enemy offensive and impose a war of attrition and, at the end, have sufficient weapons remaining to drive the enemy from their territory.

The argument in favor of massed weapons found justification outside of Russian history, in mathematics. In 1914, Frederick W. Lanchester, a British engineer, published a set of equations intended to describe modern combat.[19] In pre-gun warfare, he argued, war was a linear function. One combatant equaled one combatant. Thus, if side A had two thousand swordsmen and side B had one thousand, side A would defeat side B, with a thousand men surviving. Theoretically, excellence and failure, bravery and cowardice, would be equally present in both armies and thus canceled out.

In modern warfare, where sides are separated during combat and men can concentrate fire on targets, combat is not a linear function but a squared function. If two thousand riflemen were to face one thousand, the surplus one thousand would be free to concentrate fire on part of the enemy line. In the first moment of combat, the extra men would be able to impose disproportionate losses on the smaller enemy. In the next moment, the advantage would increase even more, until attrition wiped out the enemy, with much smaller losses for the larger side.

Because of this ability to concentrate fire, Lanchester argued that the number of riflemen was a squared function. This simply means that 2,000 riflemen have a value of 4 million, while 1,000 riflemen have a value of 1 million—a 4–1 advantage to the larger side, rather than the 2–1 advantage the numbers would seem to indicate. A force of 3,000 facing a force of 1,000 would have a 9–1 advantage. In the first example, Lanchester argued that the larger force would lose only 500 men while annihilating the other side. In the second example, the larger force would lose only 333 men while annihilating the other side. Thus, the larger the side, the faster it will annihilate the enemy and, therefore, the fewer will be its own casualties.

These equations, which are the foundation of modern war-gaming and military modeling, clearly show that in war, more is much, much better than less. But what about quality? What about training and technical skill and the other qualitative differences? Lanchester was obviously aware of these considerations, but he argued that while they ought to be included in the equation, they belonged there only as a single coefficient—they should not be squared. Lanchester had good reason to make this argument. Qualitative improvements still leave a rifleman with only one weapon. So, even if the rifle is so accurate that it hits the target twice as often as another model, it still fires only once each time frame, whereas five guns fire five times as often at the same target.

Assume that the smaller side consisting of a thousand men were twice as good at firing rifles as the larger, which had three thousand men. Instead of a ratio of N^2 to N^2 (N = number of riflemen) for each period of time, the ratio would be N^2 to $2N^2$, or $3,000^2 = 2 \times 1,000^2$. This would translate into 9 million versus 2 million. Doubling effectiveness would only increase the ratio from 9–1 to 9–2. But, and this is the key, by doubling the *number* of riflemen without improving quality at all, the ratio would improve to 9 million to 4 million, or 9–4. In other words, doubling quantity is equal to quadrupling quality.

The obvious conclusion to be drawn from the Lanchester equations is that money should be spent on buying as many weapons as possible while spending the least possible amount on quality—you get more bang for your buck from many weapons than from a few good ones!

This was the conclusion drawn by the Soviets, who turned out much less advanced aircraft than they were capable of so that they could keep production high. Their assumption was that numbers counted more than quality, and that the ability to evade threats, absorb losses, and concentrate firepower was worth more than a few outstanding aircraft. Whereas American planes were loaded with expensive radars, communications equipment, electronic-warfare devices, computers, and the like, Soviet planes were simple and rugged and relied much more on ground control.

American planes were designed so that expensively trained pilots could fly long-range missions, taking their early-warning, mission-planning, and other systems with them. Soviet pilots were much less well-trained, and their aircraft much less equipped for long-range missions. It was expected that Soviet planes would do most of their flying on or close to the forward edge of the battle area (FEBA), where Soviet ground controllers, using their radars, could control the aircraft.[20] The ground controller could detect enemy aircraft and vector the pilot to intercept them. Obviously, this was less efficient than the American system—but it was a lot cheaper per plane.

So, in the postwar period, the United States and the Soviet Union went in two completely different directions. The Soviets accepted fully the Lanchester view of the world; the Americans were ambivalent. While their war-gaming was drawn from Lanchester models, their modeling of any given weapon system focused on the quality of an individual specimen rather than on the behavior of large numbers of fielded models. One-on-one engineering models would always favor the technologically superior model, while general combat models might or might not, depending on the underlying assumptions.[21] From 1945 onward, the United States remained practically committed to quality and intellectually intrigued with quantity.

In the wake of Vietnam, however, when the United States was defeated by simple masses of peasants, the foundations of our defense policy suddenly seemed unconvincing. Some of the best and most thoughtful defense analysts began to make the argument that simplicity, reliability, and quantity were more important than technical sophistication. If the Lanchester equations were correct, then the money expended on increased technology would be better spent on more weapons.

Norman Augustine, chairman of the board of Martin Marietta, framed the problem of dramatically increasing costs per aircraft nicely when he wrote in one of his famous laws that "in the year 2054, the entire defense budget will purchase just one tactical aircraft. This aircraft will have to be shared by the Air Force and Navy, three and one-half days per week, except for the leap year, when it will be made available to the Marines for the extra day."[22]

This grim prediction was backed up by figures produced by other reformers, who argued that, in constant dollars, the cost of fighters had increased a hundredfold since World War II. The World War II–era P-51 fighter cost $300,000 each, while the leading-edge F-15 cost $30 million. Whereas the United States was able to produce up to six thousand fighter planes a year during the 1950s, contemporary fighter production has fallen to three hundred to four hundred.[23]

In other words, the United States was building one plane where it used to build fifteen. According to the Lanchester square law, in order to make up for the decline, the quality of new aircraft would have to be at least 225 times greater than older aircraft. The reformers argued that increasing the number of weapons was inherently easier and cheaper than increasing their quality, and that struggling to increase quality would inevitably decrease quantity.

In a way, the traditionalists (the old radical airpower advocates, grown powerful and comfortable) did not disagree with the radicals that it was more and more difficult for aircraft to carry out their mission. Increasingly sophisticated warning systems, both on the ground and airborne, increasingly intelligent surface-to-air missiles, and increasingly effective ground air defense systems made aircraft survival more and more difficult. Air planners on both sides were reasonably focused on the classic American mission—mounting an air strike against a well-defended target far away from the continental United States. And both sides were asking the same question: how, or whether, this mission could still be carried out.

The disagreement between the traditionalists and the reformers was over whether the solution lay with the postwar tradition of over-matching the defense with advanced offensive technology, or with the U.S. World War II strategy and Soviet Cold War strategy of having redundancy. The latter meant lots of good weapons, and the ability to absorb casualties and keep on coming.

In a sense, the Soviets showed the direction even as the Americans debated the issue. During the 1950s and early 1960s, the Soviets had produced such planes as the MiG-15, 17, 19, and 21, as well as the Su-7 and Su-22. All of these conformed to the Soviet model of being cheap, easily maintained, easily flown fighters with minimal onboard avionics, controlled from the ground. Huge numbers of all of these aircraft were produced. But the Soviets discovered in Vietnam, and especially during the Arab-Israeli wars, that Western aircraft were able to establish air superiority even in the face of far more numerous Soviet-produced aircraft. Indeed, the real threat to U.S. aircraft in the 1960s and 1970s came from Soviet-manufactured SAMs rather than from MiG interceptors.

By the 1960s, the Soviets were beginning to deviate from their original design philosophy. The MiG-23 differed greatly from older Soviet aircraft. It had a variable-geometry wing, like that of the F-111 or the F-14, which meant that it could take on different shapes to maximize performance. It also had a pulse Doppler radar, similar to that carried on the F-4 Phantom, and a laser range finder.[24] The MiG-23 was clearly a challenge to the F-4, which had dominated the skies over Vietnam as well as over Israel's battlefields after 1967. During the 1970s, the Soviets developed the MiG-29, comparable to the F-15/F-18 generation of American fighters. The MiG-29 is capable of pulling up to ten g's and is controlled by a sophisticated fly-by-wire system—a computerized control system rather than a mechanical one. It carries a look-down/shoot-down radar, able to detect low-flying craft in the ground clutter, as well as an infrared search-and-track system and a wide range of infrared and radar homing antiair missiles. Later versions of the plane carry air-to-ground missiles.[25]

The MiG-29 and the even more sophisticated Su-27 all-weather fighter represent the Soviet recognition that larger, low-technology air forces cannot hold their own with smaller, high-technology air forces.

The Soviets were now spending around $30 million per plane and slashing aircraft production into the hundreds annually.

Whatever doubts the Soviets had about quality versus quantity were dispelled in 1982, when Israeli F-15s and F-16s shot down eighty-five Syrian MiG-21s and MiG-23s without losing a single plane themselves. The destruction of yet another Arab air force at the hands of Western aircraft prompted the Soviets to emulate Western design philosophy at the same moment that the reformers were challenging that philosophy in the United States.

What drove the American reformers was the very real sense that aircraft were becoming senile, in the sense that we have been using the term. As we have seen, senility is a condition in which a weapon is not obsolete—it can still do the job. But the threats it faces are so great that expensive countermeasures have to be taken. Those countermeasures mean, in the case of aircraft, not only that not enough aircraft can be purchased but also not enough submarines, artillery pieces, or aviation fuel. An obsolete weapon is thrown away. A senile weapon continues to function by imposing an overwhelming burden on everything around it.

The advanced fighter planes were designed in the late 1970s to survive in threatening environments. To drop eight tons of ordnance on the enemy and return, strike aircraft had to be accompanied by fighter planes—fitted out with expensive radar, long-range television, and radiation sensors as well as antiaircraft missiles able to engage the enemy at various ranges. In addition, electronic-warfare aircraft able to suppress the enemy's ability to sense the presence of danger were needed, as were air-control planes, such as the Hawkeye and the AWACS, which could sense for the entire strike force, using a wide array of spectra, intercepting enemy communications, and able to integrate this data and command the air battle. **All this so a fighter could drop eight tons of munitions on a target.**

Each American fighter plane was, in effect, a multisensor platform, an electronic-countermeasure system, a communications center, as well as a weapons platform. Given the complexity, a fighter would inevitably need a great deal of maintenance, and the reformers argued that this meant that many aircraft were not ready to fly a mission at any given moment. One of the reformers, Pierre Sprey, claimed that the F-15 was mission capable only 35 percent of the time.[26]

The reformers understood that the emerging trend lines were disastrous and longed for a simpler day in which the eight tons of explosives were carried by sturdy, easily maintained aircraft, readily available for missions. Above all, they longed for a day when an air force could afford redundancy—enough planes so that they could absorb losses and keep fighting—something that the low production rates for high-cost aircraft seemed to make impossible. Whatever their nostalgia, they were not fools. The reformers did not fantasize that their simpler aircraft could carry out the missions planned for more expensive aircraft. They redefined the tasks that airpower was intended to carry out. Their solution was elegantly simple: they argued against missions that required high technology. For example, they argued that dogfights ought to take place within visual range—thus eliminating the need for sensor systems that were both expensive and subject to failure. They also redefined the air-ground mission. The reformers understood the threat posed by advanced air defense systems and the fact that extremely costly aircraft and weapons would be necessary to suppress them. Therefore, they argued that the deep air strikes against defended targets would have to be abandoned and that, instead, the Air Force should concentrate on supporting ground forces on the FEBA.[27]

The reformers understood with tremendous clarity the process of senility that was under way with airpower. They understood that to carry out the mission that airpower had claimed for itself since Douhet, the United States would, over time, have to devote resources to airpower disproportionate to the benefits received. Thus, at almost precisely the same moment that the Soviets succumbed to the Western model of airpower in the 1970s, the reformers tried to stop the process in the United States. What they failed to understand was that the process was unstoppable—that it was driven by both the logic of American geopolitics and the logic of weapons development.

The geopolitical problem was that the United States was committed to a policy of force projection. The United States fought overseas, usually in places in Eurasia, where it would fight outnumbered and where ground-based sensor and data support would not always be available. It needed aircraft with substantial range—which meant that we needed self-sufficient aircraft, equipped with sensors, data-processing systems, and weapons-guidance systems.

The tactical situation would invariably require more than merely increased firepower on the forward edge of battle. It would demand that the United States reduce the ability of the enemy to maneuver his forces. American forces *had* to avoid wars of attrition, and therefore, they needed to use their airpower to limit the operational capabilities of their enemies.

There was no way that the United States could free airpower from this role. It was an essential part of American force-projection strategy, as well as America's fundamental force multiplier. The reformers were undoubtedly correct in arguing that the kind of Air Force that could deliver air superiority throughout a theater of operations would, before long, become an unbearable burden on national resources. But until that point was reached, the logic of weaponry dictated that aircraft increase in complexity and cost, even as less expensive threats to aircraft were invented and proliferated. The Air Force would be considered increasingly sophisticated as it struggled to stay alive and function amidst all the threats facing it. In reality, the vaunted sophistication would be nothing more than temporary and desperate contrivances designed to keep the aircraft capable of carrying out its basic mission. The question will never be, can airpower carry out the mission? The question will be, can airpower carry out its mission without usurping such enormous resources that the war-making system as a whole can remain balanced and effective? Reform will only become possible when that question is answered with a resounding no. But real reform also requires a second criterion—an alternative means of carrying out airpower's mission. Both of these things must be in place before reform can be undertaken.

In the short run, Desert Storm proved the reformers wrong. In the Persian Gulf, whatever the underlying defects of airpower, the allied air forces operated with near perfection. By absorbing the costs in the design of its aircraft and force structure, the United States was able to suppress Iraq's air defense system and strike at the enemy's command, control, communications, intelligence, and logistical systems, as well as pound Iraq's ground forces. The aircraft the reformers hated the most— the F-15, the F-111, the Wild Weasels—were there and performed as they were designed to do. Given what they cost, it was the least they could perform. They broke the back of the Iraqis before the ground war began.

For the first time, airpower could make the claim that it was decisive. But we have already seen the underlying corrosion in airpower. Certainly, weapons will continue to fly through the air, but whether they will be carried by manned aircraft is the most pressing question faced by warfare. We suspect that, in retrospect, Desert Storm will be seen as the high point of the use of manned aircraft—its most perfect application. From there on out, conventional airpower will come under greater and greater pressure, to give way to new forms of weaponry and a new logic of war. We can already see this new logic as an embryo, hidden in the success of Desert Storm.

11

Dawn Breaks

DESERT STORM AND THE FUTURE OF AERIAL WARFARE

The outcome of the Persian Gulf War surprised many observers, and particularly the reformers. Two things were especially startling: the first was the extremely low number of allied casualties on the ground; the second was the effectiveness of the air force in suppressing Iraqi air defenses and destroying the ability of the Iraqi high command to communicate with and control its forces. The latter surprised many analysts, who had become used to overblown claims for airpower and expected a replay of Vietnam. The former surprised everyone. Historically, land warfare involving forces as large as those facing each other in the Persian Gulf has resulted in substantial losses even for the victors, if for no other reason than the natural friction of war. Without exception, experts predicted higher casualties for ground forces.

Trevor Dupuy, a noted independent analyst, predicted that ten thousand allied casualties in ten days of fighting would be reasonable.[1] Joshua Epstein, from the Brookings Institute, ran simulations indicating that the number would reach about twelve thousand in a three-week war.[2] Official calculations, which were made public, ran even higher. *U.S. News,* for example, reported without contradiction that the Joint Chiefs and the National Security Council thought that casualties would range between twenty thousand and thirty thousand.[3] Added to this list ought to be George Friedman's own prediction that the war would go on for two years with massive casualties.[4]

What led virtually everyone to overestimate losses? First, from World War I onward, war had been thought of as an exercise in mass, where the advantage went to the side with the bigger battalions. Obviously, many observers, particularly in the United States, bitterly disagreed with this notion. At the same time, the idea had penetrated so deeply into the defense culture that it was nearly unshakable. All of the war games on which planning strategies were based were driven by statistical assumptions concerning the importance of mass—perhaps because it was easier to quantify mass than other aspects of war. Just as important, the hypothetical results of these war games were consistent with historical experience. Hundreds of thousands of men were confronting hundreds of thousands of men in multidivisional ground combat. It was not rational to assume that a clash of this sort would yield tens of thousands of casualties on one side, and only a few hundred on the other.

The fundamental miscalculation by analysts, Iraqis, *and* the U.S. command was to underestimate the transformation that had occurred in U.S. airpower. Had airpower's effectiveness remained at about Vietnam-era levels, U.S. casualties would have indeed been high—and the outcome of the war perhaps in doubt. Simply put, everyone, including Norman Schwarzkopf, expected this air campaign to go the way of previous ones. Schwarzkopf dismissed the idea that airpower could do most of the fighting with the memorable phrase "That's a bunch of hoo-ha."[5] No one expected what happened, and one of the more amusing aspects of the war was watching specialists trying hard to pretend that they were not astounded.

To argue that airpower was not decisive, it would be necessary to argue that, even without the successful six-week air campaign, the ground assault could have been concluded in a matter of days with almost no casualties. Fantasies die hard, and the ideas of the reformers remained intact. Former senator Gary Hart, a leading member of the reformist school, testified before the House Armed Services Committee following the war:

> Some reports of fantastic weapons performance already seem highly inflated, if not grossly exaggerated. Nevertheless, Mr.

Chairman, some conclusions are obvious. We won. We won with very few casualties. And we won largely through maneuver warfare, a central theme as Colonel Boyd [another important reformer] has said within the military reform movement.[6]

But what made this maneuver possible? Would it have been possible to engage in multidivisional maneuver with highly exposed flanks and lines of supply had the Iraqi Army not been paralyzed? And would that paralysis have been possible without the very weapons and doctrines opposed for over a decade by the reformers? It might well be the case that the allied forces would have defeated the Iraqi Army under any circumstances. But with Iraqi command, control, communications, and intelligence intact, the victory would have been time-consuming and costly, to say the very least.

Air Clausewitz: Colonel Warden Thinks about Precision

Between the end of the Vietnam War and Desert Storm, a revolution in the theory of aerial warfare took place. The colonels and generals who defined the Persian Gulf air war had fought in Vietnam as junior officers. They had all participated in Vietnam's great tactical successes—and in the greater strategic defeat. They had all been schooled in the traditions of strategic aerial bombardment handed down from Douhet and Mitchell—and they had all experienced its failure in Vietnam. They had also sensed that the bombing of North Vietnam at the end of the Vietnam War, Linebacker II, was very different from the rest of the war, that it was cut from a different fabric. The conventional explanation—that we had finally bombed the hell out of them—was understood to be insufficient. What made the end of the war different was the sudden birth of the thing that had long been promised and never delivered: *precision!* Two poles, defeat and precision, compelled a doctrinal revolution that led directly to Desert Storm's success.

A generation later, pilots who had flown combat missions in Vietnam planned a new war in the Persian Gulf. Some drew the conclusion that our failure in Vietnam had been the result of the misapplication of a valid theory, that the Air Force had been prevented by politicians from applying sound theories. Other, more thoughtful officers came to the conclusion that the theory itself had been at fault—that failure in Vietnam was the responsibility of the military as much as of the political leadership. Their planning for Desert Storm began with a commitment to understand why they had failed in Vietnam and with a determination not to repeat that defeat.

In the Persian Gulf, the Air Force planners reacted against the radical presumptions of the earlier theorists—that airpower was supposed to assault the industrial base or even the social fabric—and returned to a more traditional, Clausewitzian understanding of war. For Clausewitz, the purpose of military force was to destroy the capability of the enemy's military to fight. In Desert Storm, strategists struck against Iraq's immediate ability to make war: they struck the nerve system of the Iraqi military in the hope of paralyzing it. No limits were placed on attacking Iraq's military, but also no weapons or resources were wasted on attempting to break Iraqi society.

Air Force planners also rethought the connection between violence and the will to resist. In Vietnam, destruction was inflicted gradually, in the hopes that the enemy would find himself compelled, by increasing pain, to negotiate. It was assumed that the Vietnamese would understand that beneath American restraint there was tremendous might and resolve. In fact, the Vietnamese took American restraint as a sign of weakness, as the inability of the United States to bring its power to bear, and adopted a strategy of outwaiting the Americans. In the Persian Gulf, once hostilities commenced, power was applied massively, constrained only by the logic of war. The Iraqis were not invited to negotiate; they faced a sudden, overwhelming, and coherent *spasm* of violence. The utter destruction of the Iraqi armed forces in the shortest period of time was clearly the intent—a result that was psychologically devastating.

In Vietnam, a massive air force fought without a coherent doctrine. In Desert Storm, a mature doctrine showed itself. At the heart of this development was the publication of a book, obscure at the time, and still insufficiently appreciated by the general public: John A. Warden's *The Air Campaign*.[7] Warden belonged to that generation of military officers who had experienced defeat in Vietnam. For an Air Force officer, Vietnam presented a more radical challenge than it did for an Army officer. It was possible to accept the idea that the U.S. Army had failed to win in Vietnam because it had failed to devise successful strategies for victory, without raising the possibility that ground warfare was, in and of itself, ineffective. In the case of airpower, however, its failure to live up to the tremendous promises its proponents had made at the beginning of the war did in fact raise the possibility that airpower was itself a failure or, at the very least, that it was not nearly as useful as its supporters had claimed.

Warden, like his Army counterpart Harry Summers, recognized that the military's failure in Vietnam could not be ascribed simply to civilian incompetence—something was fundamentally wrong with the military's thinking about war. Summers's argument, that the Army had abandoned traditional military doctrine in Vietnam and that this had led the Army to disaster, was oddly echoed by Warden. It was odd because the Air Force was radically innovative in all ways—including doctrine. What Warden recognized was that traditional military thinking applied to airpower as well.

Warden's approach was profoundly conservative, and very different from Douhet's or Mitchell's. In the first place, Warden argued that airpower could be decisive in a battle or campaign under certain circumstances. But this was far from arguing that airpower could simply and regularly supplant other types of military power, as both Douhet and the American advocates of strategic airpower had done. Warden understood that the measure of the effectiveness of airpower was its effect on the ground battle, on the campaign as a whole. Perhaps most important, he argued that the purpose of airpower was to strike against the operational center of gravity of the enemy's forces: "The term *center of gravity* is quite useful in planning war operations, for it describes that point where the enemy is most vulnerable and the point where an attack will have the best chance of being decisive. . . . Clausewitz called it the 'hub of all power and movement.' "[8]

Just as Harry Summers reintroduced Clausewitzian thinking to ground warfare in his *On Strategy*, Warden brought Clausewitzian theory to air warfare. Where previous air doctrine sought to bypass the limits of ground warfare, Warden sought to understand the uses of airpower as part of a systematic approach to warfare.

Douhet thought that the threat of annihilation, the threat that airpower would destroy a country's very social fabric, would force an enemy to capitulate. In this sense, nuclear weapons were the logical culmination of airpower theory. The U.S. Air Force was organized as an instrument of general annihilation, both to deter the Soviets and to overawe lesser enemies. The war for which the Air Force prepared never took place. Certainly the North Vietnamese had not been overawed. Following Vietnam, the Air Force had to invent, at long last, a theory of airpower that did not regard conventional war as irrelevant or strategic bombardment as the absolute weapon. Airpower had to be seen as part of a system of warfare that contributed to victory much like artillery, rather than as a self-contained or self-sufficient system.

Warden was far less concerned with destroying enemy societies than with breaking the command, control, communications, intelligence, and logistical capabilities that held enemy armies together. In this sense, he was concerned more with the operational level of war than with the strategic level.[9] Warden wanted to demonstrate the effect airpower could have on the enemy's ability to wage war or, more precisely, to operate its war-making system. For Warden, airpower was particularly suited to "decapitating" the enemy and isolating commanders from their forces by destroying the electronic sinews that bound them together.

Desert Storm was the coming together of an extremely conservative war-fighting doctrine and an extremely advanced war-fighting technology. The allied air forces made it impossible for Suddam Hussein to know what his own troops—let alone allied troops—were doing. They destroyed his ability to communicate with his forces and, therefore, made it impossible for him to develop a strategy for dealing with the threat facing him. They destroyed the ability of the Iraqis even to supply their armies with food and, through constant attacks, broke the morale of many of the fighting units.

Desert Storm was a much more modest campaign than its predecessors. It did not intend to break down Iraqi society; it intended only to break the Iraqi military and prevent national leadership from waging war. This it did by isolating that leadership and shattering its technical means for intelligence and communications. It is in this sense that Desert Storm's air campaign should be considered surgical. It struck against the exact center of gravity of the enemy with precise but overwhelming force.

Aircraft Performance and the Defeat of Iraq

Airpower was the key to victory in the Persian Gulf War, but what element of airpower held the key? The traditional model of air warfare is divided between the platform—the aircraft—and the projectile—the bomb. Of the two, the platform has always been seen as the more important. The platform has to lift the projectile, deliver it to the target area, and release it in precisely the right fashion, since, once released, the projectile is beyond control. Over time, far more research and development has gone into the platform than into the bomb.

Obviously, airpower did not behave in Desert Storm as it had during Vietnam: it was greatly more efficient. One might assume, in fact, that it had changed dramatically. **Although there have, indeed, been real improvements in aircraft since Vietnam, when considered carefully,** these improvements—save for stealth technology—have been incremental, not revolutionary. Changes in aircraft could not, by themselves, explain the revolutionary nature of the victory over Iraq.

Gross Performance Characteristics of Key Aircraft in Vietnam & Persian Gulf

Aircraft	Year/War	Role	Thrust/ Weight	Speed (Mach or mph)	Ceiling (feet)	Combat radius (miles)	Load (pounds)
F-100	1956/Vietnam	F/B	0.486	1.3	36,000	534	7,500
F-104	1954/Vietnam	F/B	0.567	2.2	58,100	900	7,500
F-4c	1961/Vietnam	F/B	0.381	2.0+	34,850	712	16,000
F-111	1964/Viet/PG	Bomber	0.182	2.2	56,650	3,000	30,000
B-52	1954/Viet/PG	Bomber	0.026	0.8	55,000	6,200	70,000
F-15c	1979/PG	Fighter	0.984	2.5+	60,000	720	23,000
F-16	1984/PG	F/B	1.111	2.0+	50,000	575	12,000
A-4	1954/Vietnam	Attack	0.379	0.8	49,000	920	10,000
A-6	1960/Viet/PG	Attack	0.153	0.8	41,600	1000	18,000
A-7	1968/Viet/PG	Bomber	0.339	0.9	na	951	15,000
F-14	1972/PG	Fighter	0.787	1.9	50,000	704	14,500
F/A-18	1983/PG	F/B	1.050	1.8	50,000	460	17,000

It is important to remember that many of the aircraft used in Desert Storm were already in use in Vietnam. Indeed, some of the most important aircraft in the Persian Gulf—the Navy's A-6 and A-7 and the Air Force's F-111 and the B-52—saw extensive service in Vietnam. Other aircraft were developed during the 1960s but became operational after Vietnam—the F-14 and F-15, for example. Very few aircraft—the F-16, F/A-18, and F-117A—were conceived, designed, developed, and produced after Vietnam.

The first things to consider about an aircraft are its gross performance characteristics: speed, range, and weapon load. In terms of these, there was, at best, marginal difference between aircraft of the Vietnam era and the Gulf War era.

- Aircraft remained basically manned, with cruising speeds well below the speed of sound. Engines had not improved to the point where they could carry sufficient fuel to maintain supersonic flight for extended periods. Supersonic flight was possible, but only for short bursts.

- The range most aircraft could fly without refueling remained limited to five hundred to a thousand miles. As in Vietnam, this optimal range decreased rapidly as the weapons load was increased and as energy-intensive combat maneuvers took place. Fuel consumption actually increased in later aircraft, making aircraft increasingly dependent on midair refueling for normal operations.

- The total maximum weapons load of the new aircraft remained fixed at about fifteen thousand pounds.

What did dramatically improve was the tactical performance of the new aircraft, represented by substantially improved thrust-to-weight ratios and other characteristics such as wing loadings. Once they arrived in the actual combat zone, the new planes could maneuver more efficiently and with greater agility than the older generation— but only for the short time allowed them by their fuel-consumption rates. This, in turn, improved survivability or, more precisely, permitted the traditional manned jet aircraft to continue to function on the battlefield.

Traditionalists assumed that this tactical improvement was the reason why American aircraft suffered such small losses. Indeed, losses were low considering the strength and sophistication of the Iraqi air defenses. The density of antiaircraft systems in Baghdad alone was seven times greater than the density of Hanoi's defense during the 1972 raids.[10] The Iraqis deployed Soviet SA-7, 8, 9, 13, 14, and 16 missiles, as well as ZSU-23/4 and ZSU 57/2 radar-guided antiair guns. The Iraqi Air Force consisted of some seven hundred combat aircraft, of which fewer than half were third-generation (comparable to Vietnam-era F-4s) or fourth-generation (comparable to F-15s) French F-1s.[11] The entire system was linked together with a French-designed, computerized command, control, and communications system—KARI. Thus, at least on the surface, the Iraqi air defense system was at least the equal of the North Vietnamese system.

The expectation was, as with casualties in the ground war, that the allies would suffer substantial losses in the air war. The Air Force Studies and Analysis Agency estimated that losses would run about 4 percent per combat sortie. Other forecasts ran as high as 10 percent.[12] So when losses turned out *much* lower than expected, there was a sense of a breakthrough in airpower's capacity to cope with air defenses. Yet the fact is that while losses were low, they continued a trend toward declining losses per sortie that has been under way since World War II.

SORTIE, TONNAGE, AND LOSS RATES FOR U.S. AIR FORCE

War	Tonnage	Sorties	Losses	Losses/ Sorties	Losses/ Tonnage	Tonnage/ Sorties
WWI	138	28,000	289	1.032%	2.094	0.005
WWII	2,150,000	1,746,568	18,369	1.052	0.009	1.230
Korea	454,000	341,269	605	0.177	0.0013	1.330
Vietnam	6,162,000	1,992,000	1,606	0.081	0.00026	3.093
Persian Gulf	60,624	29,393	14	0.048	0.00023	2.062

Source: Hallion, *Storm Over Iraq*. (Note that above figures exclude Navy and Allied sorties for the sake of historical consistency.)

The USAF loss rate per sortie in Korea was only 17 percent of that in World War II. The loss rate in Vietnam was 46 percent of the loss rate in Korea. The loss rate in the Persian Gulf was 59 percent of the loss rate in Vietnam. The decline is continuing, but the rate is slowing. In this sense, the Persian Gulf War, rather than being an unprecedented breakthrough, was the tail end of a steady decline in losses experienced by American pilots since the First World War.

Examining the data more carefully, we see that the improvement was not nearly as marked as might be suspected. When we look at losses per ton of bombs dropped, we find that the loss rate in the Persian Gulf was 0.00023 planes per ton of bombs delivered as opposed to 0.00026 in Vietnam. This is a critical yardstick, since it calculates losses against the quantity of explosives actually delivered to the target. **By this measure, which focuses on the primary mission and excludes peripheral and support missions, losses during Desert Storm were almost exactly the same as losses during Vietnam.**

Losses per sortie declined while losses per tonnage stayed the same. The reason for this apparent paradox is that tonnage per sortie declined dramatically in Desert Storm. In Vietnam the average bomb tonnage per sortie was a bit over three tons. In the Persian Gulf, it declined to a bit over two tons, one-third less. As we have seen, the tonnage capacity for attack aircraft and bombers remained pretty much the same since Vietnam. What could account for the somewhat mysterious decline in tonnage per sortie?

One explanation is that fewer bombs were needed—an argument

that has much to recommend it. Due to an improvement in avionics, the accuracy of aircraft in dropping bombs improved tremendously. Sensor capabilities on aircraft included phased-array radar for managing air-to-air combat, look-down/shoot-down radar, ground-scan radar, infrared sensors, and light-enhancement night-vision systems. Most important of all, onboard computers and display systems permitted data management to be either automated or displayed in such a way as to maximize pilot efficiency.

Consider this example from John Hallion's *Storm Over Iraq*, concerning the number of two-thousand-pound bombs that had to be dropped to hit a sixty-by-one-hundred-foot target from medium height:

War #	# Bombs	#Aircraft	Circular Error of Probability
WWII	9,070	3,024	3,300 ft.
Korea	1,100	550	1,000
Vietnam	176	44	400
P. Gulf	30	8	200

It should be remembered that two things remain unchanged. First, in the final analysis, the pilot is still depending on his ability to coordinate hand, eye, and aircraft in selecting the precise moment for releasing the bomb. Second, once released, the bomb's course is beyond recall, and any errors that may have crept into the calculation become uncorrectable. So, while improved, accuracy in dropping bombs from aircraft was still not sufficient to guarantee the destruction of key targets.

A more important cause of the decline in tonnage was the widespread use of precision-guided munitions. During Desert Storm, these weapons accounted for 6,600 tons, or 10.9 percent of all tonnage.[13] Measured another way, they accounted for 7.6 percent of the 227,166 projectiles dropped during Desert Storm.[14] If we accept the assumption that each precision-guided munition substitutes for thirty gravity bombs, the 17,162 PCM used would have had to have been replaced by 514,000 iron bombs—increasing total projectiles to 724,864, almost tripling the number of bombs used and at least doubling the tonnage. This relatively small number of weapons had an enormous impact on tonnage, sorties, and losses.

There is another reason for the decline in tons per sortie, a more ominous one—the increased need for support aircraft to carry out a mission. One of the arguments made by the Air Force in justifying expensive stealth aircraft was that stealth would dramatically reduce the number of aircraft needed to carry out the mission. The Air Force itself painted a grim picture in which more and more planes were needed to deliver the same amount of munitions. It claimed that to get a strike consisting of thirty-two attack aircraft through to the target, the mission had to include fifteen tankers, sixteen air-superiority fighters, eight air-defense-suppression aircraft, and four search-and-rescue helicopters—a total of forty-three support aircraft to deliver thirty-two bombers.[15] So many combat sorties have been devoted to fueling, suppression of surface-to-air missiles, air cover, and other nonbombing functions that the bomb load has become divided among many aircraft. Thus, the average tonnage per sortie has dropped considerably—the statistical sign of growing inefficiency.

This is not a new problem. It emerged late in the Vietnam War in raids over the North, when the ratio of support to attack aircraft tilted dramatically. During a naval air attack on Haiphong in May 1972, sixteen A-6s and A-7s were accompanied by nine F-4 escorts, four F-4 and two A-7 flak suppressors, and a reconnaissance aircraft and tanker, for a 16–17 bomber/support ratio. The only thing that kept the Vietnam War bombing efficiency numbers high was that a large number of raids were in the South, where neither surface-to-air missiles, radar-guided antiaircraft, nor MiGs were to be found, and where more efficient strike packages could be assembled. In addition, many raids were carried out in the North prior to 1968, when the air defense system began to mature. After 1968, the price for delivering bombs was high in terms of airframes and engines—sometimes every bit as high as they were in Desert Storm. Indeed, had the strategic destruction of Iraqi air defenses not succeeded, the cost in terms of excess sorties and casualties would have soared, one suspects, to new heights.

In evaluating the performance of conventional manned aircraft at the end of the Persian Gulf War, it appears that a developmental limit, relative to threats and missions, has been reached. A Persian Gulf–era aircraft did not perform demonstrably better than aircraft had done twenty years earlier in Vietnam. It could go no farther, no faster, and could carry no more munitions than before. In combat, it experienced losses that were, in absolute terms, better, but not remarkably better, than in Vietnam; in relative terms they were virtually identical. Tonnage efficiency was lower—it delivered less munitions per sortie than before.

The air victory in the Persian Gulf was not due primarily to better manned aircraft. Improvement in individual airborne weapons platforms did not defeat the Iraqis. Yet the U.S. Air Force did lay the groundwork for the defeat of the Iraqi Army. Why?

The key lay in the ability of the allies to break the Iraqi command, control, communications, and intelligence network. The destruction of this network caused the military to suffer a concussion. Like a concussion, it did not necessarily render the Iraqis permanently unconscious. It merely disrupted sensory perception, the ability to conduct coherent analysis and, ultimately, the ability to command and coordinate the limbs. The ability to strike at enemy troop concentrations from the air, the ability to deploy American airborne forces far to the west and to use American mechanized forces to encircle the Iraqis in a vast maneuver, the ability to choose the time and place for the opening of hostilities and to deny the Iraqis the initiative—all of this derived from the blow to the head.

The destruction of Iraq's air defense radar network not only meant that its air force could not defend its airspace or fire surface-to-air missiles effectively, it also meant that the air force could not carry out aerial reconnaissance, even from its own territory. Once the air war began, the simplest aerial surveillance became impossible. The Iraqi command had no idea what allied forces were doing and, therefore, could literally no longer engage in combat, except at the time and place chosen by the Americans. Some argue that the Gulf War was won by A-10s destroying Iraqi armor, or by maneuver warfare, or by the isolation of the battlefield from the air. What all of these analyses forget is that, without the semi-paralysis of the communications and intelligence functions, none of the others would have been possible, save at enormous cost.

Keeping Aircraft Viable: Helicopters and Stealth

The first mission of the Gulf War was to blast open Iraqi air defenses so that F-15s could attack their targets without fear of Iraqi surface-to-air missiles, antiair artillery, or fighters. Key radar stations in southern Iraq had to be taken out, so that the Iraqis would be blinded and not see the mass of aircraft pouring through to attack other facilities, including air bases and the rest of the air defense system. This first mission was not carried out by manned, fixed-winged bombers. They were too visible and vulnerable. Rather, the honor went to the stepchild of aviation—the lowly helicopter.

Forty minutes before H hour, Task Force Normandie took off from Al Jouf near the Iraqi border and crossed into Iraq. Three MH-53J helicopters with Pave Low laser-target designators led the way, followed by nine Apache gunships. The MH-53Js, operated by U.S. Special Forces teams, navigated using satellite-based ground-positioning systems (GPS), forward-looking infrared (FLIR), and night-vision systems. Using infrared, the Apaches spotted the target minutes before the main attack was to take place. Opening fire at 4.3 miles and closing, they fired twenty-seven laser-guided Hellfire missiles at one-to-two-second intervals. Fifteen hits were scored within two minutes. The radar positions were destroyed, and F-15s were able to carry out missions deep into Iraq.[16]

In the same way that infantry had to be used to clear the way for armor during the Yom Kippur war, helicopters opened a door so that the most advanced fighters available could carry out their missions. No one doubted that precision munitions such as the Hellfire could take out the radars. The problem was bringing the munitions within range of the target. Conventional aircraft would have alerted the air defenses there and throughout Iraq. The helicopter solved the problem. But their noise, low speed, low altitude, and limited range meant that, while they might be useful for specialized missions and for general close air support against tanks and infantry, their effectiveness was limited.

Throughout this century, and particularly in Vietnam, most aircraft were lost to groundfire, very few to air-to-air combat.[17] The same range of precision-guided munitions that was to destroy the Iraqis was available to the Iraqi antiair systems. Surface-to-air and air-to-air missiles were guided by infrared and radar, which were assisted by ground and air search radars of extraordinary range and sensitivity, linked together by complex communications networks. The traditional solution to ground radar is to come in at extremely low levels—nap of the earth, as it is called—as the Persian Gulf helicopters had. But the denser the air defense system, the more dangerous nap of the earth becomes. Being low when a thousand radar-guided guns open up on you is not healthy. Nor does it protect you against modern fighters equipped with look-down/shoot-down radar, which could see into the ground clutter below.

The conventional counter to sensors are countermeasures. From World War II on, pilots facing radar used chaff—strips of metal foil—to reflect radar pulses and thus disguise the location of the aircraft. In the postwar period, ever more complex and sophisticated electronic-countermeasure systems were developed. But they are expensive and require independent platforms—aircraft—to operate. In fact, specialized aircraft, such as the EF-111 Raven, were designed with the sole purpose of disrupting electronic sensors, both strategic and tactical. A far better solution to the problem of radar detection would be the construction of an aircraft that radar could not detect.

A large part of what made U.S. success in the Persian Gulf possible was the design of an aircraft that was relatively, if not absolutely, invisible to radar. The F-117A stealth fighter was designed to penetrate enemy air defenses unseen by either radar or infrared sensors—or even acoustical sensors. Indeed, painted jet-black and flying at night, it was to be invisible to the naked eye as well. The F-117A was the first operational example of stealth technology, and the second time stealth was used in combat operations. The first time was during the invasion of Panama, when it was less than wholly successful, as one might imagine with any new technology. Against Iraq, however, it was triumphant.

Stealth technology is an attempt to defeat sensor technology. The most direct threat to an aircraft is radar, which can see an aircraft at great distances and can also be used in guiding weapons to it. Therefore, a top priority is to eliminate, or at least to limit, the visibility of the aircraft in the radar spectrum and then in other spectra, from visible to infrared to acoustical.

The impetus for stealth could be found in Vietnam, and in the growing cost of penetrating Vietnamese airspace against the Soviet-built defense system. The cost became particularly painful during Linebacker II—the Christmas bombing of Hanoi—when the United States suddenly discovered how vulnerable the B-52 had become. On the first night of the attacks, three B-52s were shot down, while on the third, six went down. In just three days, 5 percent of all B-52s assigned to Southeast Asia were lost. Even after adjusting tactics, five more were eliminated.[18]

Arguments raged over what went wrong. The Air Force focused on tactics. The problem, however, was deeper than that. The subsonic, manned intercontinental bomber could no longer be certain of penetrating enemy airspace against radar-guided surface-to-air missiles. This meant that the manned-bomber leg of the nuclear triad (intercontinental ballistic missiles, nuclear submarines, manned bombers), which rested on the B-52's ability to penetrate Soviet airspace through thousands of miles and attack monitored and defended targets, was no longer capable of its mission. Indeed, concern over the survivability of the B-52 went back to 1960 and the downing of Francis Gary Powers's U-2, an event that drove home the growing power of radar-guided missiles.[19]

An urgent response was necessary, but in the wake of Vietnam, little money was available for innovation. One direction was antiradiation missiles, which would home in on radar emissions. Another solution was long-range missiles that bombers could launch far from the target and far from antiaircraft concentrations. A final solution was a more efficient bomber, one that could use speed and agility to penetrate enemy airspace.

The B-1 bomber, first proposed in 1970, was an attempt to consolidate all three things. It was intended to be equipped with advanced electronic countermeasures, to be armed with cruise missiles, and to operate at supersonic speeds—at both high and low altitude. Although it was an excellent idea, the B-1 did not work, particularly after the budget slashing of the Carter years. The cruise missiles and electronic-countermeasures equipment raised the weight to such an extent that the B-1's engines could not produce enough thrust for supersonic speeds. In addition, fully loaded at low altitude, the B-1 was not able to reach maximum altitude nor to be completely agile.

There are two basic approaches to making an aircraft stealthy. The first is to reduce the radar cross section—the area that reflects the radar pulse back to the receiver.[20] Every receiver has an inherent level of sensitivity, and a point where the return gets lost in standard background radiation, not unlike the sensitivity of an FM radio receiver that eventually loses the signal to background noise. The problem is how to lower the return so much that it is drowned out by background radiation—especially with a plane that is 95 percent aluminum and only 5 percent nonmetallic resins.[21]

The solution entailed rethinking the very shape of the aircraft. A flat surface has an extremely large cross section if it is at ninety degrees to the radar beam. If the surface can be tilted away from the beam, so that the beam hits it on an angle, the return is deflected. By creating a set of curved and swept surfaces, the return, as well as the probability of picking up the return, is reduced.[22] Thus, the stealth aircraft consists of triangular and trapezoidal facets, all of which are created less with performance in mind than with the manipulation of radar beams.

The second approach to achieving stealth is to use material that absorbs radar energy. Radar absorbing material (RAM), consisting of ceramic matrix compound, is applied to key surfaces, particularly those where angling cannot reduce return. This absorbs some radiation frequencies while letting others pass through. Aircraft windows are coated with indium-tin oxide, which reflects and diffuses virtually all radar with minimal reduction of visibility.[23] Finally, the engine outlet is designed to be invisible from the ground, diffusing the output and making it less visible in the infrared spectrum.

The goal is invisibility. Obviously, that cannot be achieved. In reducing the radar cross section, for example, the return is both reduced and diffused. At a certain distance, however, the return will rise above the background radiation, and even the narrowest beam will sweep over a receiver. The assumption with stealth is that the detection range will be so low that by the time the plane is detected, it will be out of harm's way.

A huge financial investment rides on this assumption. Stealth aircraft are so expensive that they require missions of only the highest national importance, or else missions in which their survival is virtually assured. As one might imagine, research on defeating stealth technology is well under way. The Department of Defense created a Counter–Low Observable Office in the late 1980s. Among other things, this office has developed a program code-named HAVE GAZE, which will use space-based reconnaissance satellites to deal with the stealth problem. The Navy Weapons Center at China Lake has also undertaken an interesting counterstealth program, which seeks to detect the ions produced by stealth engines as a by-product of its attempt to cool its exhaust.[24] The attempt to cope with stealth is, of course, now an international game. The Swedes, for example, claim to have developed a multistatic radar, using multiple radar installations, that can detect stealth aircraft.[25]

During the Persian Gulf War, the F-117A was assigned to the most difficult targets—critical air defense command posts and communication centers. It was able to penetrate densely defended and heavily monitored airspace so long as night lent it invisibility, and it was the only aircraft assigned to attack targets in downtown Baghdad.[26] Using laser-guided Paveway II and III weapons systems and the two-thousand-pound BLU-109/B warhead, which has a hardened case and a delayed fuse useful for attacking hardened targets, the F-117A helped shatter both Iraq's air defense and its command, control, communications, and intelligence systems.

Nineteen or twenty F-117As were sent to the Persian Gulf in August 1970 with another twenty to twenty-two arriving in late November.[27] Though they constituted less that 2.5 percent of all allied fighter and attack aircraft, they attacked 31 percent of the targets hit on the first day of the war. During the war as a whole, although the F-117A flew only 2 percent of all missions, it attacked 40 percent of all strategic targets.[28] The Air Force initially claimed that it enjoyed a 90 percent success rate, but this was later scaled back to about 60 percent. The biggest problem of the F-117, which is common to all laser-targeted systems, was bad weather or smoke from previous strikes.[29]

Col. Thomas J. Lennon, an F-111 pilot and wing commander, claimed that the F-111s were able to hit 85 percent of their targets using precision-guided munitions, equal to or better than the success rate of the F-117.[30] He may well have been right—but he missed the essential point. Using precision-guided munitions, virtually any aircraft that could approach within the munition's range of the target could achieve extraordinarily high rates of effectiveness. The problem was that conventional aircraft could not approach heavily defended targets and survive. Hence, only the F-117 was permitted to attack these targets.

The attack on Iraq's center of gravity required the use of an aircraft that cost over $100 million. The roughly forty F-117s ultimately involved in the Persian Gulf cost over $4 billion. As air defenses improve and spread, the ability of conventional aircraft to penetrate—even those equipped with electronic countermeasures—will decline, until only undetectable aircraft will be usable. And as the ability to detect improves, the cost of stealth will rise exponentially. **Thus, the success of the F-117, rather than showing the viability of the manned fighter and attack craft, represents the beginning of the end, an end posed by the very thing that made the F-117 so successful in the first place—precision-guided munitions.**

Precision-Guided Munitions and Victory

The success of the bombing campaign in the Persian Gulf was rooted in the radical revision of bombing probabilities. During World War II, nine thousand bombs were necessary to hit a given point; during Vietnam, three hundred bombs were necessary. During Desert Storm, one or two bombs were sufficient.[31] This dramatic, qualitative, fundamental shift in the probability of a projectile's striking a target had far less to do with evolutions in aircraft than it had to do with evolutions in projectiles. Indeed, the accuracy rates of a B-17 and an F-117 dropping a precision-guided munition would have been about equal—if the B-17 could survive. Aircraft evolution by itself did not lead to increased precision. Precision was built into munitions—which were delivered by aircraft. This raises the fundamental question of whether, and for how long, munitions will require manned aircraft for their delivery.

Precision-guided munitions refers not to a single weapon but to an entire class of weapons distinguished by their ability to strike targets with precision on a regular basis. They are not the same as intelligent munitions. Precision can be provided by the intelligence of the onboard sensors and computers, or—as was the case in the Persian Gulf—it can be provided by human intelligence using advanced targeting methods.

This is a critical distinction. The latter case requires that humans remain at the center of the targeting loop, guiding the weapon. To do this, they need to be in some sort of weapons platform—most likely an aircraft—which must go in harm's way. Indeed, regardless of how precise the munition, it might not be deliverable because the weapons platform cannot survive in the hostile environment protecting the target.

The *intelligent* precision munition can dispense with the human being in the loop or allow the human to operate far away from the action. Humans select the target, do some initial management of the trajectory, but it is the projectile that assumes the burden of directing itself to the target, recognizing the target when it arrives, and guiding itself on its final assault. Humans do not have to be in a position to observe the attack and, therefore, would not require a weapons platform that would put them in danger.

The distinction between precision and intelligent precision was obvious in the Persian Gulf. Precision munitions were very much in evidence, whereas intelligent precision munitions were only beginning to emerge. Indeed, as we have seen, the most important precision weapons had already been used to great effect in the Vietnam War and had changed relatively little in the intervening years.

In principle, precision weapons had existed since World War II. The Germans developed a weapon called the Fritz X, which could be released from high altitude and guided to its target with a joystick. Using his unaided vision, the bombardier would manipulate the joystick, which, in turn, would generate a radio signal picked up by a receiver in the tail of the bomb, which passed commands to the control system while a gyroscope stabilized the bomb. The Fritz X was first used in 1943 against the Italian battleship *Roma,* to prevent its defection to the Allies after the ouster of Mussolini. Dropped from eighteen thousand feet, well out of range of Italian antiaircraft guns, two Fritz Xs—three-thousand-pound, delayed-action, armor-piercing bombs—took nearly a minute to strike the ship. The ship was completely destroyed, with more than a thousand crewmen lost.[32]

Obsessed with gravity-bomb precision, the Americans did relatively little in developing precision-guided munitions during World War II. Prior to the war, U.S. designers had speculated about the possibility of television-, radar-, and infrared-guided weapons—all of which remained speculation until decades later. The United States did develop a weapon called AZON, which was actually used in the war, achieving some success in Burma.[33] Like the Fritz X, it was controlled by radio attached to a joystick. While only able to maneuver in the azimuth, AZON had the advantage of a flare in its tail, which made it easier for the bombardier to see it on its descent and hence to guide it.

AZON was modified into RAZON late in the war. RAZON, able to maneuver in all directions, was attached to superior sights for guidance and, ultimately, was linked to the Norden bombsight. Indeed, the Royal Air Force used RAZON more frequently than the U.S. Army Air Force did. RAZON grew to enormous size. The British had a 22,000-pound version, while the United States was using a 12,000-pound version called TARZON. TARZON was so big that it could not fit inside a B-29, but had to be carried protruding from its bomb bay. In Korea, TARZON worked well enough to take out six bridges out of thirty, but it did not seem reliable enough to constitute a solution to the problem of accurate bombardment.[34]

Controlled-descent weapons did not seem a priority in the early Cold War years, as nuclear weapons appeared to have made the issue of precision moot. Military research and development focused on the problem of delivering large munitions intercontinental distances, leaving explosive power to obliterate errors in placement. Thus, the development of large, conventional precision munitions was not pressed.

The Navy, however, remained interested in using precision weapons against naval and land targets. During the 1950s, it began to develop the Bulldog and Bullpup series of 1,000- and 250-pound guided bombs. Both used the same guidance principles as the AZON-RAZON-TARZON series—joysticks, radio guidance, and flares. But the bomb was simply not large enough to do the job unless it hit the target precisely. So in Vietnam, where it was employed early on, direct hits on a bridge did little damage. The search for greater precision resulted in the Paveway series of weapons—the most widely used unpowered precision-guided munitions during the Gulf War.[35]

Developed during the 1960s and designed to solve the problems of the Bullpup series, Paveway had showed its effectiveness in Vietnam. The heart of the Paveway system is the laser beam, drawn from a narrow band of coherent light. A bomb is fitted with a special Paveway kit, which consists of a sensor that can locate the laser beam, and a computer chip that locks onto the laser point and sends commands to wings and vanes that have also been attached to the otherwise ordinary iron bomb. When the bomb is released, it searches for and senses the laser beam illuminating a small portion of the target.[36] The pilot must drop the bomb somewhere within a "basket," a broad area close enough and properly angled so that the bomb can maneuver itself, using gravity and inertia, to the designated point. During Vietnam, 50 percent of the bombs hit their target precisely, with a circular error of probability of about twenty feet.[37]

The key to the Paveway is the targeting system—the system that produces an intense point of light on the target. The original method, which was used in Vietnam, was to have one aircraft use a pod attached to its fuselage to illuminate the target, with one or more other aircraft releasing munitions aimed at that point. The basic defect of this approach is that it requires multiple aircraft for a mission and leaves one of them in the area until impact. The second method, rarely employed, is to have ground designation. This is used in cases where special operations teams have gotten deep into enemy territory looking for hidden targets and then illuminate the target for strike aircraft. The alternative, widely used in the Persian Gulf, is to have the attack aircraft itself carry the laser designator. However, since the aircraft needs to leave the area quickly, the pod must be able to swivel, with the pilot or weapons officer still able to see the target, making certain that the laser remains fixed on it. This means that the laser needs to be linked to an infrared camera also able to swivel, which transmits images back to the cockpit.[38]

Originally designated PAVE KNIFE, this designation evolved into LANTRIN, for low-altitude navigation and targeting infrared system for night. Combining infrared imaging with computer-managed displays, LANTRIN permitted American pilots to attack Iraqi targets at night and in bad weather. The forward-looking infrared (FLIR) permitted pilots to see terrain and fly as low as one hundred feet above the ground. By combining navigation with laser designation, F-15Es were able to conduct near perfect strikes against key facilities outside the surface-to-air-missile belts where F-117s were sent. More than nine thousand Paveways were used, for a total of 5,700 tons.[39]

All of this, of course, reveals the weakness of the Paveway system. First, it needs a clear night to operate. Laser designation can be obscured by sunlight, clouds, smoke, or intentional smoke screens. Second, the aircraft must remain in sight of the target, to keep the designator on it, and if it is in sight of the target, it is in sight of air defenses. The reason Paveway worked well in the Gulf War was because the Iraqis' air defenses and command, control, communications, and intelligence systems were shattered. How it would have worked had it faced an intact air-defense force is another question. What should not be forgotten, of course, is that a Paveway II costs about $10,000 to $15,000 and can take out targets worth hundreds of millions. That is the key virtue of the Paveway and its heirs.

Paveway is a semi-intelligent weapon. It requires a human to designate the target but guides itself to that designated point. In this sense, it suffers from the basic defect of the unintelligent precision-guided munition—leaving an aircraft in danger. The next step, obviously, is a more intelligent weapon that can home in on an identified point without needing continual designation from a weapons platform.

Another step on this path was the electro-optically guided bomb (EOGB), actually a contemporary of the laser-guided bomb, and in some ways its stepchild. As discussed in the previous chapter, the EOGB was also used in Vietnam, particularly in daylight raids. Designated GBU-15, it had a television camera in its nose, which transmitted a picture back to the attack plane—always a two-seater—as managing the bomb was a short-term, full-time job. Like the laser-guided bomb, the EOGB is dropped into a basket, an area from which the bomb can be maneuvered to the target. When the EOGB is close enough so that its TV picture builds up sufficient detail, the weapons officer selects a point where there is good contrast, so that the bomb will lock onto it, as Paveway locks onto the laser point. After lock-on, the bomb guides itself to that target.

The EOGB's great defect was that it could only be dropped on clear days, which was even worse than being dropped on clear nights. It also had to be dropped from sufficient altitude so that, like the laser-guided bomb, it had room to maneuver on the way down. The effect of this is that aircraft were too exposed. Just as bad, using a television screen's contrast point meant that the aircraft was prisoner to television. It could only hit where there was sufficient contrast, which may or may not have been the most vulnerable point, or it had to hang around and guide the bomb all the way in, which was dangerous. Contrast changes with movement, and there was an excellent chance the bomb would lose its lock on the way in. Finally, since two-seaters were necessary, only the F-111 and F-15E could carry it.[40]

The electro-optically guided bomb was probably not ready for prime time. But for that brief period between the point when the weapons officer handed control over to the GBU-15 and impact, an autonomous intelligence showed itself. This slow progression to intelligence can be seen in another weapon first used in Vietnam that came into its own during Desert Storm—the AGM-65 Maverick. The first, and perhaps most striking, difference between Maverick and the laser and electro-optically guided bombs is that the former comes in several flavors. Some have shaped antitank charges, while others have a high-explosive warhead designed for antiship and land attack. The Maverick also comes with different guidance systems. During Desert Storm, Mavericks could be guided in three ways: by TV, electronically locked on the target point or self-locking on the center of the picture; by imaging infrared, acting as a television but using the imaging infrared spectrum to lock onto the target; or by lasers, as in the Paveway series.

The imaging models represent a major step forward. Once the pilot or weapons officer has locked Maverick's sensor on a target and fired, the aircraft is free either to flee or engage another target. Moreover, with natural light, TV, and imaging infrared to choose from, the weapon is usable in all weather and under all light conditions. The obvious next step will be to place multiple sensors on each warhead, so that a single missile will be suited for all missions.

Already used by the Israelis in 1973 and 1982 in small numbers but to great effect, the Maverick conducted a slaughter in Desert Storm. More than 5,100 Mavericks were fired, mostly the imaging-infrared version but also many TV versions, and it was estimated that about 80 percent of the missiles destroyed tanks. What was most deadly about the Mavericks is that they attacked the top of the tank, the least armored part, allowing them to penetrate Iraqi tanks simply because of their kinetic energy—the warheads were superfluous.[41]

The Maverick had many virtues. It could be, and was, fired from virtually every aircraft and helicopter; it cost only about $50,000; and it was extraordinarily accurate and deadly. Along with its helicopter-borne cousin, the Hellfire, it could decimate an armored force. It also had weaknesses. A tactical weapon, not particularly large, it was not useful against strategic targets. Its imaging-infrared and TV systems could be fooled by sophisticated countermeasures. Most important, the average range for firing a Maverick was about 3.5 miles, which meant that the F-16s and A-10s that used it had to place themselves in grave danger, at least against an enemy still capable of fighting.

A munition's intelligence must be combined with range and size. In other words, it must be able to guide itself over long distances, locate the target by itself, and have sufficient explosive power to destroy it. A munition must also be precise enough to be relied on operationally. In the Persian Gulf, that capability was present—albeit as a first approximation of what it will eventually become—in the Tomahawk land-attack missile (TELAM).

The Tomahawk was the only genuinely intelligent munition in the war, and it will, we suspect, serve as the prototype of all future intelligent munitions. What made it intelligent was that it guided itself precisely to its target over hundreds of miles, without requiring human intervention beyond programming and launch. Unlike self-guided bombs, such as the Maverick and Hellfire, the Tomahawk did not need an aircraft to fly in close to the target. Unlike laser-guided bombs, it did not need an aircraft to loiter in the area of the target, shining a laser beam.

Tomahawk: Cruising to Intelligence

A cruise missile differs from a rocket in that it has an air-breathing engine, whereas a rocket does not need air to operate. This means that rockets can go into space where there is no atmosphere, while cruise missiles cannot. It also means that cruise missiles behave more like aircraft than like rockets. The need to breathe air means that it needs to use hydrocarbon fuels and mix those fuels with air. It also means that, as with normal aircraft engines, there are limits on speed and maneuverability—on anything that would interfere with the flow of air. The cruise missile's great advantage over the rocket is that, like an airplane, it is capable of sustained powered flight. Because a rocket can fire its engine for only brief periods and derives the entire energy for the flight from that initial burst, it either has a very short range or it must leave earth's gravity to preserve the inertia of the initial surge. A cruise missile, on the other hand, can travel within the atmosphere at much lower altitudes because it receives steady levels of energy from its engine. A rocket that tried to do what a cruise missile does—fly hundreds of miles at sea level—would start bleeding energy once it ran out of rocket fuel and would be brought down by gravity in short order. A cruise missile, in other words, is nothing but an unmanned plane on a one-way mission.

The United States has run a cruise missile program since World War II, although most of its energy and resources have gone into rocket technology. The most important cruise missile developed by the United States prior to the 1970s was the Matador/Mace series, which were the first tactical nuclear missiles, and the Navy's Regulus, which was a high-altitude, supersonic, intermediate-range nuclear-delivery system.[42] The Matador and Regulus were built as a result of the American nuclear obsession, out of the fear that U.S. strike aircraft would not be able to penetrate Soviet air defenses. The Soviets also developed a family of cruise missiles, albeit for very different reasons. For the Soviets, the strategic imperative was to cut maritime supply lines between the United States and Europe. Therefore, their cruise missiles were primarily naval, designed to be carried by Soviet aircraft trying to attack American convoys.[43] Most operational cruise missiles had been abandoned by 1970, except for the Hound Dog, a short-range cruise missile carried by the B-52, which allowed the B-52 to lob a bomb toward its target without actually having to penetrate heavily defended areas around the target.

As soon as cruise missiles were abandoned, they had to be reinvented. Nuclear weapons were wide-area-blast weapons that traditionally did not require a great deal of targeting precision. During the 1960s, intercontinental ballistic missiles began to go underground, and during the 1970s, the underground missile silos were hardened. This meant that enemy missiles had to become more and more accurate to destroy them. The circular error probable of the Minuteman missiles was not sufficient to guarantee silo destruction, while submarine-launched missiles were even worse. One alternative, an extremely expensive one, was a third-generation missile like the MX. The other was the air-launched cruise missile (ALCM). This was designed to be a "self-guided, low-flying, terrain-following weapon capable of attacking ground targets with great accuracy at speeds of 500 miles per hour."[44] Launched from a B-52, it had a range of about 1,500 miles, which permitted the aircraft to launch from outside the densest antiaircraft systems.

The Navy also undertook a cruise missile program in 1972, using both submarines and surface vessels as the launch platform. During Vietnam the Navy realized that it needed a weapon that could penetrate increasingly impenetrable air defenses and hit a target with precision. The Navy's informal instructions to contractors were that they should "design a device able to be fired from a submerged submarine under way off San Diego, fly cross-country so low it has to pull up over mountains, and upon reaching Chicago, navigate to a point that puts it inside the base paths of Wrigley Field."[45]

The result was the Tomahawk.

The Tomahawk land-attack missile, which was designed to fit inside the torpedo tubes of a submarine, is only eighteen feet long and twenty-one inches wide, with eight-foot wings that unfold after launch. It is launched by a solid booster rocket—which gets it out of the tube with a twelve-second burn—after which a turbofan engine cuts in, permitting speeds between 381 and 571 miles per hour, with the lower speeds being more fuel efficient. Size constraints place severe limits on the amount of fuel that can be carried. The C series, which was used in Desert Storm and is the conventional-warhead model, has a range of about eight hundred miles.

What makes the Tomahawk so remarkable and so potentially intelligent is its three-part guidance system. Inertial guidance, similar to that used in intercontinental ballistic missiles, is the heart of the system. This determines the Tomahawk's trajectory by providing data, calculated by an onboard computer, concerning direction and speed.

Because there is built-in inaccuracy in any inertial system, the information it provides is supplemented by data from a terrain contour matching (TERCOM) system. Derived from an earlier system, ATAR, which had been built into the Matador, TERCOM measures with a radar altimeter the contour of the land the missile is passing over and constructs a topological map. The computer then compares this map, which includes both geological and man-made formations, to the one stored in its memory.

Using inertial guidance and TERCOM alone, TLAM's circular error probable would be about one hundred feet—which is unacceptable for targets requiring a high degree of precision. To boost the Tomahawk's precision to near perfection, another guidance system is used as it approaches the target area. Digital scene-mapping area correlator (DSMAC) cuts in in the terminal phase, using an optical scanning system to look for the target in the visible band. It compares the picture it is receiving to the one stored in its memory. On locating the target, it locks onto it and is guided to it with flawless precision, slamming into it with a thousand pounds of explosive, substantial kinetic energy, and burning fuel. This is obviously a potent weapon, but at $1.1 million a shot, it had better be.[46]

The Tomahawk missile represented a tiny percentage of all ordnance fired in the Gulf War and a small percentage of precision-guided munitions. But both it and the F-117 were indispensable. It carried out most of the attacks on the Iraqi electrical system and was instrumental in knocking out key air defense control facilities, as well as command and communications facilities in Baghdad.[47]

The Tomahawk had many weaknesses. It was slow, taking several hours to reach its target; its eight-hundred-mile range was a severe limitation. (It is about nine hundred miles from the Strait of Hormuz to the Kuwait-Iraq border. Had U.S. vessels been prevented from entering the Persian Gulf by mines or missiles, Baghdad would have been at extreme range from the Red Sea.) Its explosive power—a thousand pounds of ordnance—was sometimes inadequate (for example, it could not be used against hardened targets). It took a long time to program, requiring input of terrain data, terminal imaging data, inertial data, and so on. Moreover, it could only hit targets that the Navy had good pictures of and could only travel over terrain that had been carefully mapped. This meant that a savvy enemy could undermine its usefulness by constantly changing features on key installations so as to create significant differences between the picture in the computer's database and the image in its optical system. Similarly, terrain could be changed along expected axes of attack. (Some Tomahawks reportedly went astray because terrain had been altered by previous attacks by allied aircraft.)

Despite all of these weaknesses, the Tomahawk missile represents the future. Give it more speed, more range, better explosives, and a more advanced image-analysis program, and you have a weapon system that can carry out strategic bombardment without needing to put men into danger. Indeed, take a Tomahawk, load it with submunitions and have it fly to the target area, release the submunitions, which use their own sensors to home in on targets, and then fly home for another load. These things may be in the future, but not too far in the future.

Conclusion: The Persian Gulf as the Transition to the Future

Precision-guided munitions made possible an economy of force never before seen in an air war. Where hundreds or thousands of bombs had been necessary to destroy a given target in the past, a small handful would now suffice. Where dozens or hundreds of aircraft would have been necessary to carry out attacks in the past, only two or three were now necessary. This quantitative difference carried with it profound qualitative significance. Precision munitions, from laser and optically guided bombs to cruise missiles, transformed the expectations of military planners. They redefined the possible.

Every war has as its goal the disruption of lines of supply and communication. The traditional means were ground operations, which enveloped enemy formations. The airplane seemed to offer greater possibilities, but as was easily seen in Vietnam, both strategic and tactical isolation were impossible to achieve from the air. In Desert Storm, planning began with the assumption that it was possible to strike at communications and supply from the air and that by disrupting communications in particular the ability to execute the allied ground war would be dramatically enhanced. After all, the Iraqis held strong defensive positions with substantial forces. Their weapons may have been inferior to American weapons, but not, by all appearances, markedly so. Had Iraqi commanders been able to communicate with their forces in the field, and control their movements during ground combat, the ability of those forces to resist could not be underestimated.

Since the end of the war, a controversy has emerged as to how effective airpower actually was, particularly against command, control, communications, and intelligence facilities. This controversy reached its apex with the publication of the Air Force–commissioned *Gulf War Air Power Survey*. The survey concludes pessimistically:

> While some disruption and dislocation had undoubtedly been imposed on the Iraqi leadership by the attacks on L (leadership) and CCC (communications) targets, the regime's ability to function was neither paralyzed nor broken by the time the Coalition's ground offensive began on 23 February 1991. . . . In the absence of detailed information from the Iraqi side, a precise assessment of the extent of disruption and dislocation imposed by Coalition air strikes cannot be given.[48]

The *Survey* misses two important points. First, the question was not whether the regime continued to function. According to that standard of success, the air campaign must inevitably be considered a failure. The question, rather, is whether or not the regime could control its forces in Kuwait efficiently enough to wage war. Second, the claim that the only way to measure effectiveness would be to survey the Iraqis is nonsensical. Effectiveness has its best measure in the behavior of the Iraqi forces in Kuwait, who behaved as if they were *not* under central command. Certainly, Baghdad could issue a simple order for withdrawal and even arrange for units to form a screen. But there is a vast difference between ordering a withdrawal and *managing* a withdrawal, and war is about controlling as well as commanding.

The *Survey* actually understands this point quite well:

> While the leadership in Baghdad probably lacked the volume of communications and overall "connectivity" with the Kuwait theater necessary for real-time direction of extended offensive operations against mobile Coalition forces, links to the Kuwait theater had, apparently, not been completely severed.[49]

The statement that Iraq did not have the communications capacity to wage extended offensive warfare is the same as saying that Iraq could not wage war. Iraqi forces in Kuwait occupied a salient with an exposed western flank. If the allies could strike along that flank at will, Iraq would be defeated. The orthodox and reasonable counter to the allied offensive was a counteroffensive against its exposed eastern flank. When the *Survey* asserts that such operations would have been impossible because of disruption of the communications system, it acknowledges that the Iraqis were defeated *because* of the disruption of the communications system.

It must be understood that for the first time in history, airpower

was so effective against lines of communication that what was, in effect, an army group was rendered immobile. That is big news.

If one accepts the premise that the war was won by destroying the enemy commander's ability to command and control first his air defense system and then his ground forces, it follows that the systems that achieved this end were the strategic weapons of the war. In that case, the two strategic weapons were the F-117 stealth fighter and the Tomahawk.

It is an interesting argument as to which did more damage to key facilities, but the argument is really moot. Both had the ability to range freely about the battlefield and bomb with such precision that well over half of their missions—at a most conservative estimate—were successful. Moreover, they carried out missions that no other weapon system could have accomplished. F-15s, F-16s, F-111s, and the rest could not have shattered Iraq's center of gravity so completely and with so little cost. Without F-117s and Tomahawks, the war would have taken a very different shape.

An infrequently used weapon is worth noting in this context—the conventional air-launched cruise missile (CALCM). The Air Force fired fewer of these missiles (35 compared to the Navy's 298 Tomahawk Land-Attack Missiles [TLAM]),[50] but these were no less effective. Two things made the CALCMs interesting. First, they were fired from aircraft—B-52Gs, which left Barksdale Air Force Base in Louisiana nearly twelve hours before H hour, in an exercise more notable for its endurance than its necessity. Second, their guidance system was of a radically new design. In 1985, the Air Force replaced the nuclear warheads on about a hundred of its cruise missiles with conventional thousand-pound fragmentation warheads. At the same time, it replaced the old inertial guidance system with a radically new system, built around the global positioning system (GPS).

During the 1980s, the United States had orbited a constellation of satellites known as Navstar. At the time of the Gulf War, sixteen of these were in orbit. Each satellite emitted a steady signal, which allowed anyone with an appropriate receiver to locate himself in three dimensions (altitude was only available sixteen hours a day, while ground position was knowable twenty-two hours a day, due to the incomplete nature of the satellite constellation). GPS became the rage of everyone in Desert Storm, and for the first time in a war, a space-based system became universally indispensable for tactical operations. It also allowed the CALCM an extraordinary degree of precision. For the first time, guidance was provided by a space-based system.[51]

The CALCM, like the Tomahawk, causes us to turn our attention from the air war to the space war. There can be no mission unless the planner knows what he wants to hit and knows precisely where the target is located. There is no mission unless the terrain is precisely defined with the latest shifts detailed—including destroyed buildings, new structures, and camouflage. As important as it is to blind your enemy, it is that important to be able to see for yourself. And that is what Desert Storm introduced to us—not only a new way of sensing, but also a new platform from which to sense. In short, the cruise missile, and indeed the entire war, depended on the new center of gravity of American military power—space.

12

End Game

THE SENILITY OF THE MANNED AIRCRAFT

Until the Persian Gulf War, the technological foundations of airpower doctrine remained essentially unchallenged. The accepted weapons platform of the air war was the manned, air-breathing aircraft. Operating at speeds cycling around the speed of sound, it was able to carry about seven tons of ordnance for a distance of less than a thousand miles, without refueling. Arguments raged about whether every such platform should be able both to attack ground targets and to engage enemy aircraft or whether different sorts of platforms, each dedicated to a different mission, should be designed. But it was always assumed that the only way to attack a target was for it to be approached by a manned aircraft with the pilot or his crew releasing the bomb at the correct moment.

The emergence of precision munitions seemed to confirm the wisdom of this approach. Because of laser-guided bombs and similar devices, the historical defect of aerial bombardment—inaccuracy—seemed to be abolished. The manned aircraft was finally equipped to become a precise surgical tool, just as the advocates of daylight precision bombardment had dreamed during World War II.

But hidden in the success of the manned aircraft were the seeds of its own doom. The same precision technologies that permitted attack aircraft to increase accuracy from one in three hundred bombs in Vietnam to one in two in the Persian Gulf would inevitably increase the accuracy of antiaircraft systems. The same array of sensors—radar, infrared, laser, acoustical—that allowed fliers in Desert Storm to decimate Iraqi targets could be turned on the manned aircraft. Indeed, we saw a hint of that when the Allied Command banned all manned aircraft, save for the F-117A stealth fighter, from downtown Baghdad because it was too dangerous.

Obviously, manned aircraft have been able to stay viable by using ever more complex and costly techniques to carry out their mission: electronic countermeasures designed to confuse enemy radar systems; flares and other heat-emitting devices designed to confuse infrared sensors; missiles designed to destroy radar systems that seek targets; a host of low-observable technologies going under the rubric of "stealth"; and the standoff munitions that allow aircraft to attack enemy targets without having to enter the most well-defended areas.

The most important question facing airpower advocates—a question rarely confronted directly or honestly—is whether and for how long the manned aircraft can continue to function as the primary weapons-delivery platform on the battlefield. At what point will the resources necessary to keep an aircraft flying outstrip the value of its mission? At what point will the stress placed on the pilot—both by the maneuvers he will have to execute to survive and by the quantity of data he will have to absorb and make decisions on—make it impossible for the manned bomber to operate? Can the pilot remain in the cockpit? Must the pilot remain in the cockpit to remain in the loop?

Airpower planners have raised these questions very quietly because they understand that doing so could undermine public support for airpower in general. They also know that, for those men who have lived their entire lives in a world where the manned aircraft was seen as *the* decisive weapon, imagining its senility would be unbearable. Certainly, the demise of the manned aircraft—if this were to happen—would not mean the demise of airpower or aerospace power. Fear of this occurring has limited thinking about and deployment of cruise missiles and will, inevitably, continue to do so.[1] Debate over the manned bomber vs. the cruise missile will become the fundamental issue within the Air Force and between the Air Force and civilian planners over the next decade.

The outcome will be at least as fateful as the one between admirals of battleships and aircraft carriers prior to World War II.

The Air Force has already fired the first shot, arguing in a 1989 study conducted by the Air Force Studies Board:

> Sustained hypersonic flight in the atmosphere between the two extremes of [Mach 8 and near orbital speeds] presents major technical difficulties. Problems of surface heating, thrust, vehicle stability and control, infrared signature, aiming, and weapon release could make any potential military advantage in this speed range unlikely.[2]

These same problems are considered soluble at the sub–Mach 8 level—speeds that might well include pilots. As a result, one wonders about the extent to which they are technical and the extent to which they are psychological.

The rapid movement of munitions through the air and over extended distances remains both a technological possibility and a military necessity. The question is whether a manned aircraft has to be involved. We know from Desert Storm that it is now possible to select targets from space-based sensors and order projectiles to travel hundreds of miles and attack them with near-perfect precision. The Tomahawk is, one must remember, the Model T of cruise missiles. Its range, speed, and sophistication will only increase over the next generation. If history is any guide, the increase will be dramatic—indeed, exponential. In the end, weapons may exist that can guide themselves to targets in different continents, at hypersonic speeds—eight to thirty times the speed of sound—dispensing smaller submunitions on arrival, and they may be able to influence the tactical development of a battle from intercontinental distances. A weapon launched from an Air Force base in Colorado could travel to an armored brigade in Korea in a matter of minutes, locate each individual vehicle, and destroy it. Even if we find that technology cannot be extended into the intercontinental, hypersonic domain but only to intracontinental, supersonic ranges, will manned aircraft still be necessary?

Even if manned aircraft can take effective countermeasures against the threat of precision-guided munitions, **are manned aircraft the best way to drop explosives on enemy positions?** Certainly, prior to Desert Storm, this was the only way to bring power to bear beyond the range of artillery. After Desert Storm, the question is not only whether the manned aircraft can survive in an attack role, but whether it needs to.

Agility and Invisibility: Trying to Survive

The modern aircraft is constantly endangered by attackers that are faster than it is. Both air-to-air missiles and surface-to-air missiles can move at speeds far in excess of maximum aircraft speed. A supersonic fighter may be capable of speeds greater than Mach 2, but it is extremely rare that the pilot would push his aircraft into that region. During Vietnam, not a single second of flying time was recorded above Mach 1.8; a few seconds were recorded above 1.6; a few hours above 1.2. Almost all combat time was recorded at Mach 1.2 or below.[3]

There are two limiting factors on speed. The first is fuel. The rate of consumption above the speed of sound is so high as to make prolonged flight impractical. The second is maneuverability. The faster a plane is going, the more limited its maneuverability, as higher speeds place increased stress on the aircraft—and the pilot. High-speed maneuvers, such as rapid turns and changes in altitude, increase the number of g's. The higher the transient g's of a maneuver, the more rugged the plane has to be, the better trained the pilot needs to be, and the lower the margin of error. As a result, the overwhelming majority of combat flying is done at speeds from Mach .6 or .7 to just below Mach 1 at less than 9 g's.

The missiles that are dedicated to killing aircraft are much faster and more agile. For example, the Soviet SA-10 surface-to-air missile has a maximum range of about fifty miles, a maximum altitude of about one hundred thousand feet, moves at about six times the speed of sound, and can sustain 100 g's (one hundred times its weight)—all of which gives it an impressive maneuvering range.[4] The Soviet SA-14, a man-portable surface-to-air missile weighing 22.8 pounds, can generate speeds of Mach 1.75 against low-level attack aircraft flying well below the speed of sound.[5]

Air-to-air missiles fired by other aircraft also offer no comfort. For example, the new advanced medium-range air-to-air missile (AMRAAM), being developed by the United States, has a range of nearly fifty miles and a maximum speed of Mach 4. The HAVE DASH II, being developed as a future version of the AMRAAM, will increase the g's from its current 35 to 50. The older Sparrow missiles claimed to have a 73 percent kill probability on any launch.[6] And these are the older missiles!

There is clearly a mismatch between aircraft and the threatening technology. Indeed, given the difference in gross characteristics, it is surprising that aircraft can even now survive in any environment in which the enemy has reasonably modern antiair missiles. But, as we have seen in the Persian Gulf, gross characteristics are not the only factors that permit aircraft to survive.

The success of the air campaign during Desert Storm depended on the allies' ability to suppress the Iraqi air defense system; that is, to prevent the launching of both surface-to-air and air-to-air missiles. If that failed, the allies had to confuse the missiles' sensors; if that failed, then they had to use aircraft that could not be sensed; and if all else failed, they had to have aircraft sufficiently agile to evade the missiles.

The Iraqi air defenses consisted of radar installations linked into a command and control system capable of activating ground-based air defenses and aircraft. The United States had excellent intelligence on the system, from the French, who designed it, and from the Soviets, who supplied many of the surface-to-air missiles used. The United States also knew a great deal about it from signals and electronic intelligence.

The Iraqi system had two weaknesses. First, it was highly centralized. There is good reason to place real-time command of the battlefield in the hands of a single authority. This ensures that limited resources are allocated by a small group of individuals fully aware of all operational conditions. The defect of this organizational structure is that if the center is destroyed, the system cannot operate. The Iraqi system provided no way for devolution of authority without inefficient resource allocation. Soviet influence, which emphasized centralization in air defense operations, contributed to the problem. Thus, when the Iraqi center went down—along with a few key nodes—the entire system collapsed.

Second, the operation of the system depended on active sensors: radar. Both the system as a whole and individual launch sites were built around search radars that would locate the position of approaching aircraft. Once the planes entered the intercept envelope of a particular missile—which of course would depend on the range and altitude capabilities built into the system—the missile would be launched. From that point on, guidance of the missile to the target could proceed in several ways. Ground tracking radar would use radio commands to guide the missile to its target. This was the case with the older Soviet SA-3. In another method, semi-active homing, the ground radar would detect, track, and illuminate the aircraft, and the missile would home in on the point of reflection. In yet a third method, search radar would detect the aircraft and launch in its direction SAMs guided by an infrared homing device.[7]

The problem with the Iraqi radar system was not that it didn't work, but that it worked too well. Radar is active. The radiation that it emits can be traced to its origin. Thus, during Desert Storm, the United States turned radar into the Achilles' heel of the Iraqi air defense system and, thereby, the Achilles' heel of the Iraqi military.

During Vietnam, the United States had developed a class of aircraft known as the Wild Weasel, which fly into enemy airspace, hoping to trigger enemy radars.[8] Armed with sophisticated sensors and processors, the Wild Weasel locks onto the radar beam tracking it, then fires an antiradiation missile that follows the beam down to the radar transmitter and destroys it. During Desert Storm, the role of Wild Weasel was carried out for the Air Force by the F-4G, a Phantom II converted to the role, and for the Navy by the EA-6 Prowler. The main antiradiation missile was the high-speed antiradiation missile (HARM).

High speed is critical. It is imperative that the antiradiation missile hit the transmitter before an antiair missile is launched or, if that is impossible, before the latter can lock onto the Wild Weasel with its own sensors. If the HARM is able to carry out this mission effectively—and during the Persian Gulf War it was spectacularly successful—then the Wild Weasel's radiation sensors coupled with the missile's internal guidance system can disrupt or even destroy the enemy's air defense system.

This is certainly what happened in the Persian Gulf. It is not an exaggeration to argue that the Wild Weasel gave the allies control of the air above Iraq. Merely turning on a radar system invited a Wild Weasel to destroy the transmitter.

On the surface, it would seem that the ability of the Wild Weasel to cope with ground-based radar meant that we had entered an era in which aircraft would be much safer than before. During Vietnam and the 1973 Arab-Israel War, Soviet-style surface-to-air missiles had been extremely effective, imposing substantial losses on aircraft. The Wild Weasel would have appeared to have solved the problem by making it impossible to use radar for tracking aircraft. By suppressing radar, the foundation of air defense since the Battle of Britain, it would seem that airpower had entered a new era of preeminence.

In fact, all that this meant was that airpower was capable of coping well with an air defense system organized in the Soviet style—just as the Israelis could cope with Syria's Soviet-supplied air defense system in 1982. The key weakness in the antiradar strategy was, of course, that the HARM-style missile was only effective against a radar transmitter. That could be anywhere from feet to miles from the missiles, launchers, or command center—or even from the radar's data processors. The transmitter or antenna is the cheapest element of the system. The ability to destroy it does not have to take the entire system down.

Some simple innovations could have blocked allied success.

- Create a system of multiple sensors, networked to a wide range of launchers. A Weasel flies into an area and is illuminated not by one radar transmitter but by twenty or thirty, all locked on it, and some locked on from beyond the range of the HARM—said to be from ten to ninety miles, with the optimal range being forty to sixty-three miles.[9] Every launcher within range can be controlled from every radar unit.

- Create a network of decoy transmitters. Saturate the area with microwave emissions for all of the radar bands being used without hooking the transmitters into the weapons control system. Turn on the operational radar—and twenty other radiation sources at the same time. Let the Wild Weasel try to figure out which radiation source is hooked into the command and control system and which are merely decoys—and do this before the missile going after the Weasel locks on. If there is a problem of signals stepping on each other, develop programs that will allow semiactive homing systems to home on illumination from a source—such as laser-guided bombs.

- Undertake a crash miniaturization program so that a surface-to-air missile could carry its own active radar seeker without suffering from weight penalties. Shooting a HARM down the throat of a transmitter is easy, but try hitting a surface-to-air missile closing at Mach 6 and pulling 50–100 g's in maneuvers!

- Assuming that you have an American-type air force with plenty of onboard sensors and avionics, shut down your radars and let your fighter planes go Weasel hunting. Antiradiation operations assume that the Weasel will not be jumped by fighters or that the threat will always be managed. Imagine it was the U.S. Air Force against the Israeli Air Force instead of the Iraqi. It would be hard indeed for Weasels to survive in a world of hostile interceptors. And when F-15s and F-16s come to the rescue, then turn on all your radars for a turkey shoot. Weasels stop the missiles; air-superiority fighters destroy enemy interceptors. Which comes first—the chicken or the egg?

- If all else fails, abandon active sensors and go to passive ones. Advances in infrared detection and electro-optical detection, which we have discussed earlier, also apply to aircraft. If to these two obvious choices we added areas in which advances are now being made, such as ultraviolet detection and acoustical detection, a wide array of passive sensors might be available for sensing and vectoring surface-to-air missiles.

The Wild Weasel is a useful tool against a very specific sensor. It was the good fortune of the allies that the enemy in Desert Storm depended almost exclusively on that sensor to operate its air defense system and that the system's design made it extremely vulnerable to disruption on all levels. The Wild Weasel needs to be respected, but it doesn't make active-radiation air defense systems obsolete. Nor does it make a wide range of passive sensing alternatives obsolete. If Wild Weasels fail, it leaves fighters in a nasty fix, twisting and turning for their lives.

The Wild Weasel is a brute-force assault on radiation—it works by blowing up the transmitter. A more subtle, and even more traditional, approach to radiation is counterradiation: electronic warfare. This approach interferes with the sensors' ability to see the target by disrupting communication between various parts of the air defense network, between the ground sensor and the surface-to-air missile, or it hinders the ground sensors from seeing the aircraft in the first place. It must be remembered that, in principle, there is little difference between radio communication and radar reception. Both use the electromagnetic spectrum, albeit different parts, to function. And both are subject to disruption by jamming.

To jam a transmitter, its *precise* frequency must be known and then noise generated on the same frequency. One defensive approach is to move the signal about on a random, but prearranged, sequence, on the assumption that the enemy's jammers will lack sufficient agility to cope with the shifts. The counter to this might be to jam all signals simultaneously on a wide band of frequencies. Obviously, the problem with this is that in jamming your enemy, you will be making important parts of the communications and radar spectra unusable to your own side. During the Cold War there was real concern that, unless other NATO forces provided the precise frequencies on which they were operating, U.S. jamming would knock out their communications as well as the enemy's.[10] Thus wide-area, wide-band jamming is rarely used.

One can see from this why electronic warfare is shrouded by so much secrecy. The principles of electronic warfare are not complex, nor are the technologies. What are precious are the frequencies used by sensors in weapons systems and the frequency-hopping plans that are employed by communications systems. Just as precious is information about the capabilities of jamming units—their range, their agility at moving from frequency to frequency, their sensitivity to frequency.

The United States spends a substantial part of its intelligence budget collecting this data in large, secret databases and designing countermeasures to the specific frequencies and vulnerabilities of weapons and radios that American armed forces might encounter on a battlefield. Trying to bring some intellectual coherence to this area, the U.S. Army has defined this domain as EW/RSTA: electronic warfare/reconnaissance, surveillance, and target acquisition.[11] The United States does everything possible to keep secret the capabilities of American jammers and the frequency patterns of American equipment, always with the knowledge that the theft of a single specimen of a system, or its capture in combat, might compromise the electronic security of an entire class of weapons.

Almost every advanced attack aircraft carries equipment designed to foil enemy sensors. For example, the AN/ALE-45 electronic countermeasure ejection system used on the F-15 is nothing more than an updated version of the old World War II metallic chaff that used to be dumped out the door of B-17s. The metallic-coated Mylar forms a cloud of radar-reflecting material, making it impossible for any radar-guided system to see the aircraft. To counter infrared seekers—particularly common in the low-altitude, man-portable missiles—an aircraft drops magnesium flares, supported by parachutes so that they remain airborne but at an increasing distance from the aircraft. The flares are even hotter than the jet exhaust and engines. Hopefully, a missile will divert from the airplane, attacking the decoys instead.

The difficulty with chaff and flares is in knowing exactly when to use them. If they are deployed too early and they dissipate, it can be disastrous since fighters carry few loads. If they are deployed too late, they will fail and the plane will be shot down. Against a missile approaching at Mach 6, release must be timed to the second—a problem that is even more daunting than it appears because of the need to sense the precise point not only of radar lock-on, which can go on indefinitely, but also of missile launch.

Apart from chaff and flares, another approach aircraft can use is to attack the sensor directly. In the case of infrared sensors, countermeasures—such as the AN/AAQ-4 and 8—can be used. These focus an intense laser beam on the missile's sensor, blinding it or even causing it to detonate by making it assume that it has reached the target. However, the more general problem, particularly for long-range and high-altitude surface-to-air missiles, is radar and the radio system that controls the flight of the missiles.

A wide range of American fighter and attack aircraft now carry systems, such as the ALQ-184 jamming pods, against radar-guided systems. These fighter-based jamming systems must, of necessity, be light and compact. This severely limits their range and power as well as the complexity and sophistication of the jamming signals they emit. Therefore, the United States uses a class of dedicated electronic-warfare aircraft to provide broader suppression of electronic signals, both radar and communications. During Desert Storm the United States used several different jamming aircraft: the Air Force's EF-111 Raven, a variation of the F-111 attack aircraft; the EH-130H Compass Call, a variation of the C-130 Hercules transport plane; and the Navy's carrier-based EA-6, a variation of the A-6 attack plane, and the only one of the three also able to launch HARMs as does the F-4G.

An airborne jamming system is capable of three different actions. The first is simply to generate enough noise to disrupt enemy communications and radar. The problem with such jamming is that it is obvious; it is hard to miss the fact that your communications have just turned to static or that your radar screen has nothing but snow on it. The second action is to create pulses rather than continual waves. During the 1973 Arab-Israeli War, Israeli electronic-warfare specialists managed to disrupt communications between Syria and Egypt by precisely overlaying the Arab Morse code with their own pulses.[12] In choosing frequencies and methods, most advanced electronic-warfare systems also contain a computerized library of signals against which radio and radar signals can be compared. The computer can then apply the appropriate protocol to disrupt it.

The third ability of an airborne jamming system—particularly of an attack aircraft's or a fighter's—is to use the transmitter not merely to mask the aircraft's position but to actually relocate it or create a false, second aircraft. Remember, all that a radar system is going on is the signal it receives back. It assumes that the signal returned because it hit an object. Imagine, however, that instead of bouncing off the aircraft the signal bounced off a beam of electrons—another radar beam—configured in such a way as to change the shape and angle of the return. Instead of a return showing an attack plane coming straight ahead, it might seem as if an attack plane is approaching two hundred yards off to the side—and the missile would react accordingly.

Disrupting radar emissions is the most important function of an electronic-warfare system on board a tactical fighter or attack aircraft. Prior to the Gulf War, U.S. Air Force and Navy estimates were that in the event of a war with a sophisticated enemy, all aircraft without jammers would be destroyed within the first twenty days of battle. With the jammers, 44 percent would survive for thirty days.[13] These dire predictions did not come to pass in Desert Storm, in large part because the Wild Weasels and other aircraft with precision-guided munitions collapsed the

Iraqi air defense system in the first hours of the war. Had the system been robust enough to survive the assault, or had Iraqi aircraft been able to threaten Weasel sorties, then the allies would have been forced to fall back on airborne jamming and interference alone.

The problem with all of these jamming techniques is, of course, that they expose the jammer to the same weapons he is trying to suppress. An aircraft must emit tremendous energy to suppress entire bands of communication. The jammer assumes that his emissions will be so overwhelming or so confusing that no seeker will be able to home in on the source. But in the end, the emission does have a source, and however chaotic the emerging pattern, it does have a built-in order—and therefore is vulnerable.

More important, the need to disrupt radar signals is not the only need; there is also the need to detect the launch. While the precise means for doing this are a tightly held secret, survival dictates that new systems known as missile-approach warning systems (MAWS) be developed. MAWS under development are said to be able to detect an incoming missile, determine its guidance type, and automatically respond to it with chaff, flares, or jamming if available.[14] Certainly, the aircraft can use passive means to detect launches, including electro-optical and infrared. Indeed, there are plans for introducing ultraviolet sensors to spot launches.[15] Each attempt to protect the aircraft, therefore, poses new problems. Dispensing chaff makes it difficult for the aircraft to use its tracking radar, and dispensing flares makes it difficult to use infrared sensors. Turning on the radar opens the aircraft to jamming or antiradiation missiles. And so on.

Multispectral seekers could make jamming a nightmare. Consider advances in U.S. surface-to-air missiles. The Patriot missile is being given a radar seeker able to hop from frequency to frequency as it tracks an enemy aircraft, making jamming difficult. All U.S. radar-based surface-to-air missiles are being shifted into the highest frequency bands available because they are more sensitive to irregularities in aircraft hulls, such as air intakes or radar apertures. It also appears that the Russian SA-10 surface-to-air missile operates in these high-frequency bands, which are beyond the jamming capability of the EF-111 and many other electronic-warfare aircraft.[16]

Radars will do what soldiers do—spread out and take cover, moving swiftly from position to position. Radars will emit on several frequencies in several bands, simultaneously or in sequence. In principle, the sensors will always be faster than the jammers. Just as the aircraft can hit the target with sensor-guided munitions, so too sensor-guided munitions can hit the aircraft. Two parts of the equation make this inevitable:

- Missiles are much faster and more agile than manned aircraft can ever be.
- Aircraft are several tons of hard material moving around in the sky. That is hard to hide from all of the spectra that might see it.

Aircraft can run, but they cannot hide. However, they might try, which is the point of the stealth aircraft.

From Invisibility to Unconsciousness: Stealth Warfare and G-LOC

The extreme complexity of suppressing sensors leads to an apparently simpler solution—an aircraft that is invisible or barely visible to those same sensors. As we have described, one specimen of stealth was the F-117A. But the mother of all stealth aircraft was the B-2 bomber. Original plans, in 1981, were to purchase 132 B-2s for a total cost of $36.6 billion—a per-unit cost of about $277 million. By 1989, total costs had risen to $70.2 billion, over half a billion dollars per plane. By 1991, the plan was to purchase 75 planes for $61 billion—$800 million per plane. Flyaway cost—the price you pay to take the plane with you, which does not include the research and development and engineering costs—was lower, $434.7 million a plane.[17] By 1993, only 20 B-2s were to be built, at a total cost of $44.2 billion—over $2.2 billion per plane.

What was the Air Force going to get for its money? The B-2 has a maximum bomb load of 50,000 pounds, compared to 60,000 for the B-52, the older bomber. Its maximum speed is Mach 0.85; the B-52's is Mach 0.9. With a 37,000-pound load, the B-2 has a high-altitude range of 7,255 miles; the B-52 can carry 30,000 pounds over a 7,139-mile range. On the surface, the United States was not getting much more for its billions. Though the two aircraft have very similar performance characteristics, there is one primary difference.[18] The B-2 is designed to survive against a sophisticated enemy; the B-52 was not. Both the B-2 and the B-52 were designed to carry out the same mission—enter Soviet airspace, evade Soviet air defenses, deliver nuclear weapons on their assigned target. By the mid-1970s, however, it was becoming increasingly apparent that the B-52 could no longer carry out its mission in the face of increasingly sophisticated Soviet air defenses. Although the B-52 was modernized to include an array of electronic-warfare systems, including sophisticated jamming, chaff and infrared dispensers, and warning radars, most Air Force officers believed that the B-52 could no longer carry out its mission.

If we assume that the manned bombers must be retained in our strategic force and that enemy air defenses cannot with certainty be suppressed, then the B-2 is the only rational solution. It was designed to be invisible to Soviet radar.[19] Unlike the F-117, which achieved a low radar cross section by using angles to deflect or focus search radars, the B-2 attempts to evade radar by providing only curved surfaces, which were lined with radar-absorbent material. In effect, the B-2 is a flying wing, without fuselage or distinct tail section. It was stealthy by a somewhat different, more radical means. The question, of course, is whether it works and, even if it works, is it worth it?[20]

Col. John Warden, one of the most thoughtful analysts in the military, defended the B-2 in a White Paper he wrote for the Air Force arguing that while the B-2 might be detected, the time of detection would be too short for an air defense system to react. Moreover, to track a stealth aircraft for fifteen minutes, the Soviets would have to construct a system of acoustic sensors 150 miles deep. Assuming that acoustical sensors had a range of five miles, 27,000 sensors would be needed.[21]

The problem with Warden's analysis is the assumption that an air defense system would require fifteen minutes reaction time. Now, it is certainly true that fifteen minutes are desirable for theater-defense systems such as IHAWK or Patriot, but in the world of antiship missiles, such as the Navy's new RAM missile, which travels in excess of Mach 3, fifteen minutes is an eternity. Constructing an air-crust defense system that could cycle from initial contact to intercept in less than five minutes seems in the realm of possibility; and one would be taking an extremely conservative view of the future of land-based sensor/missile systems to believe otherwise.

Warden is quite right in warning about the proliferation of stealth technology—particularly to missiles. The proliferation of stealth technology almost inevitably includes the proliferation of antistealth technology. Indeed, counterstealth research has been going on for years in the United States and at any number of research centers from Moscow to Tel Aviv to Paris.

Several respectable sources are already claiming that stealth technology is not only not low-observable but fully observable. For example, in February 1994, the Russian first deputy defense minister, Andrei Kokoshin, was quoted as saying, "Incidentally, our newest radar stations and other detection systems have no trouble detecting American 'invisibles' at quite considerable distances."[22]

This claim came as no surprise to the United States, since Gen. Merrill McPeak, Air Force chief of staff, said of the Russians, "I expect that certain parts of their air-defense setup would be able to detect the B-2 today. Nobody's ever argued that the B-2 is invisible or immortal. What we've argued is that it is a very hard target to shoot down. I expect that will be true in ten years."[23]

Well, actually, some people did claim that the B-2 was invisible, and for what it cost, immortality would seem more in order than a ten-year guarantee. But all General McPeak was claiming was that while sensors will be able to *see* the B-2, they will not be able to *shoot it down*. To his credit, McPeak played a key role in reducing the number of B-2s bought.

The Russians and the Air Force are not the only ones who agree that the B-2 can be seen. An official with British Aerospace pointed out that "if an aircraft shows a contrast with the surrounding temperatures of more than one degree centigrade, its electro-optic signature will show up."[24]

And the deadliest enemy of the Air Force, the U.S. Navy, issued a study from its Center for Naval Analysis that stated:

> The Navy believes that stealth is not absolute, that the advantage of stealth may erode in the future as air defenses are reconfigured in response to stealth. . . . [The B-2] could be at risk if it flew very near a high-performance surface-to-air missile such as the Soviet SA-10, or if somehow a fighter aircraft flew close enough to a B-2 to pick it up with optical or infrared sensors, or perhaps even radar.[25]

Of course, everyone has his own reasons for debunking stealth. The Russians do not want to admit that they are vulnerable. The Europeans do not want to be forced to build their own stealth aircraft and so are comforted by the thought that they are not particularly effective. The Navy loves to pick at Air Force wounds under any circumstances. But it is hard to dismiss the Air Force chief of staff's admission or to give much credence to his argument that seeing the B-2 does not mean that it is vulnerable. Certainly, the evolution of surface-to-air missiles and radars is much less costly than the development of stealth technology. By 1993, the Air Force had already been forced to update the fuselage to make it more stealthy—an upgrade costing a mere "couple hundred million."[26]

The B-2 is the perfect case of the senile weapon system. It is not a question of whether it works. Conceding every claim made for it, it works at an enormous price, at best over $400 million a plane (flyaway cost), at worst, over $2 billion (actual cost). In return, you get a plane that can carry out exactly the same mission as the B-52 bomber thirty-five years before—delivering nuclear weapons to targets in the Soviet Union or about twenty-five tons of conventional ordnance elsewhere. **Defenses against the manned bomber have grown so sophisticated that the bomber can only survive by acquiring defensive characteristics costing more— per plane—than the total defense budgets of Sri Lanka or Ecuador (1989 flyaway cost) or of Indonesia or Brazil (actual 1994 cost).**

Proponents of the B-2 argue that it would do an excellent job. At those prices it should. Certainly, if what the United States wants to do is drop twenty-five tons of bombs on the enemy using a manned bomber, the B-2 is a fine choice. But first, proponents should explain their reason—not the reason for dropping the bombs but for insisting that they be delivered by men in a very expensive plane.

Imagine what else the United States could buy with $44 billion. Think of spending $44 billion on maintaining our carrier battle groups at current numbers, or maintaining manpower over the next few years so that key specialties are not lost, or developing precision-guided munitions with long ranges and high speeds so that we don't have to spend billions of dollars on bombers to drop a few bombs.

A study conducted by MIT summed it up:

> Stealth technologies would only marginally improve the B-2's ability to penetrate Soviet air defenses, and would certainly not justify the system's tremendous cost. More importantly, even if stealth technologies *guaranteed* that every B-2 would penetrate, it would still be a less cost-effective weapon than either the B-1B or the cruise missile.[27]

Even if the B-2, F-117, and F-22 are currently capable of evading sensors, there is a near certainty, historically, that this advantage will be eroded in due course. As it is eroded, the manned aircraft will be forced to fall back on the virtue that has kept it in action since World War I— agility. The best defense of the manned aircraft on the battlefield is not electronics or stealth but the ability to maneuver, to evade, to ambush, to hide, to attack.

Unfortunately, the manned aircraft has run into an insurmountable obstacle there too—the human body. Until recently, aircraft were more vulnerable to gravitational forces than human beings were. Pilots had to take care in their maneuvers not to place excessive stress on the aircraft's structure lest its wings come off or other disaster strike. In the latest vintage of fighters, however, this is no longer the case. F-15s, F-16s, and F-18s are all capable of surviving both rapid acceleration and turns that render pilots unconscious and give rise to a condition known as gravitational loss of consciousness—or G-LOC.

G-LOC strikes without warning in high-performance aircraft. As someone said of the F-16:

> The airplane was too good. In fact, it was better than its pilots in one crucial way: it could maneuver so fast and hard that its pilots blacked out. It wasn't only that the pilot confronted more total g-forces than in any other combat airplane; the F-16 was so responsive, so tough and so maneuverable that everything was happening too fast. Instead of building gradually, suddenly spikes nine or ten times the force of gravity were slamming the pilot.[28]

G-LOC results in decreasing blood pressure in the head, resulting in blackout. Following this, complete muscular collapse occurs. Basically, at 9 g's, the maximum an F-16 can pull, a 170-pound man can suddenly be hit by 1,360 pounds—in seconds, inside and out. Some things can be done to protect the pilot. He can perform exercises to stabilize his blood pressure; he can wear a suit that inflates and constricts the lower part of the body, forcing blood to his heart and head; he can try special diets.[29]

He can do all of this and more, but the human body is not built to suddenly accelerate to 9 g's and function properly. And it must be remembered that a plane being maneuvered at 9 g's is facing missiles, unencumbered by human protoplasm, that can pull ten times that many g's.

One solution is to computerize some aspects of flight. There are plans, for example, to develop sensors that will report when the pilot is unconscious and allow a computer to cut in to continue flying—and perhaps even fighting—the plane. The Navy has a modest project under way to introduce sensors that will warn the pilot if loss of consciousness is on the way.[30] The Air Force also has in mind some innovations, such as one that will prevent the plane from flying into the ground, no matter what the pilot's condition.[31]

But, in general, the Air Force has opposed computerized pilot surrogates.[32] The reason is simple. If the Air Force concedes that the pilot is no longer able to maintain consciousness under modern flying conditions and that a computer is an adequate substitute for him, then an obvious question arises: Isn't it time to take the man out of the cockpit and let him control the aircraft in other ways?

Even the Air Force's attitude appears to be changing. In a document released in early 1996 called *New World Vistas*, intended to be the Air Force's road map for the next thirty years, the Air Force expressed interest in an Advanced Research Projects Agency idea called the Unmanned Tactical Aircraft—UTA. UTA is intended to be unmanned in the sense that the pilot is left behind on the ground, but it will still be managed by a controller on the ground; intelligence will be provided by biology rather than microchip. With the pilot out of the aircraft, the size of the aircraft will be reduced by 40 percent and it can be highly maneuverable and capable of pulling up to 20 g's.[33] The seriousness of the proposal can be gauged by the fact that Boeing has already proposed to deploy its new Joint Strike Fighter in an unmanned form, ready for deployment by 2020.[34]

Obviously, there is opposition to the idea from the manned aircraft community in the Air Force—in other words, from most serving Air Force personnel. However, the underlying problem is not that the man is out of the cockpit, but that by still being in control, the aircraft suffers from potential time lags in data uplink and downlink, as well as from the problem of the reliability of the data link in an electronic warfare environment. Ultimately, the unmanned but piloted aircraft is an intermediate step, extremely useful in that it allows high g maneuver in the face of evolving missiles, while we await the maturation of machine intelligence over the next century.

The fundamental and unavoidable reality in aerial warfare is high-speed, high-agility missiles going after much lower speed, much less agile manned aircraft. And the fact is, a human being simple cannot survive the kind of stresses that ultra-high-speed maneuvers will place on him. In the F-16, pilots are already at their physiological limits. Electronic warfare, antiradiation missiles, the entire range of systems designed to permit the manned aircraft to survive in an increasingly dangerous world, are merely stopgap measures. Worse, they are astoundingly expensive pseudo solutions that divert scarce resources away from the revolution in warfare that is currently under way—a revolution in which the manned aircraft, an increasingly decadent, complex, and costly technology, moves to senility and is overthrown by a technology that is infinitely superior—the hypersonic, intercontinental projectile.

PART THREE

SPACE AND PRECISION

Introduction

THE LOGIC OF SPACE WARFARE

The United States won the war against Iraq because of the ability of its mechanized forces to flawlessly execute a classic encirclement of Iraqi forces in the Kuwaiti salient. The Iraqi Army was unable to challenge the maneuver for two reasons: first, their intelligence was so poor that they were not aware of a multidivisional force maneuvering within miles of their front lines; second, both their strategic and tactical communications had been so badly damaged that the chain of command no longer functioned well enough to assure control during a counterattack. Because of this, the Iraqis lost their key advantage in the war: the ability to prolong the war and impose casualties on the allies in excess of the allies' political ability to absorb them.

The strategic imperative of Desert Storm was to force the war to a rapid conclusion, avoiding at all costs a war of attrition of the sort that Saddam threatened and that was his most rational move. The coalition leaders sensed that they lacked the political base for a multiyear conflict. They also knew that logistically, supplying a high-intensity, long-term

301

conflict in the Persian Gulf might prove impossible. Therefore, allied strategy had to be the rapid envelopment of enemy forces rather than direct assault. Fortuitously, their political needs coincided with military geography. Kuwait's long western flank was an open invitation to an envelopment that would bring a rapid conclusion to the war.

Schwarzkopf and his staff also understood that for the strategy to work, Iraqi forces would have to be immobilized, otherwise the enveloping force—and particularly its lengthening supply line—would be exposed to Iraqi counterattack. One prong of the strategy was to develop a force capable of moving rapidly northward, then wheeling to the east along the Iraq-Kuwait border so as to surround and destroy the bulk of the Iraqi Army. The other prong of the strategy was to eliminate the Iraqis' ability to maneuver their forces by attacking their command, control, communications, and intelligence (C^3I) facilities; in other words, by making Baghdad blind and dumb.

Strategic weapons are not the ones that make the loudest bang but, rather, the ones that contribute most to a particular strategy. The assault on Iraq's communications and intelligence was carried out by two such weapon systems: the F-117A using Paveway laser-guided bombs, and the Tomahawk cruise missile. Without these two weapons, the assault on the key air-defense-control facilities and the communications facilities in downtown Baghdad would have been impossible or would have entailed tremendous casualties.

But it is important to remember that the effectiveness of the F-117 and cruise missile depended on superb intelligence. Before either weapon could take off or be launched, it was essential to know, with utter precision, which targets were to be hit, and where they were located. In the famous picture of a bomb heading down a ventilation shaft in downtown Baghdad, it seemed to be forgotten that before the miracle of precision could take place, a miracle of intelligence had to occur. Since the target was important and therefore protected by enemy troops, it had to be seen with enough clarity by U.S. intelligence sources so that the entire building could be carefully examined, and a small vulnerability could be located—a ventilation shaft. Only then could the aircraft be sent to attack it. All of this is doubly true for cruise missiles, which must know before they are launched where they are going, how they are getting there, and what they will do once they arrive. All of the maps and pictures contained in the missiles' onboard computer memories had to come from somewhere. And the place they came from, to a great extent, was space.

This suggests that if the strategic weapons of the Gulf War were the

F-117A and the Tomahawk, the center of gravity of the allied war effort was the reconnaissance and intelligence satellites the United States had in orbit during the war.

Satellites in space have both altitude and safety. They can see far more than aircraft, and unlike aircraft, they are exceedingly difficult to shoot down—at least for the time being. They can also carry out a wide range of functions, from image reconnaissance to intercepting enemy signals and radar emissions to detecting rocket launches to relaying communications. Had these satellites not existed during the Gulf War or had they been destroyed, it is unlikely that war planners would have known the structure and weaknesses of the Iraqi air defense system and, therefore, could not have planned the campaign to destroy it. Without electronic ferret satellites to locate communications and intelligence centers, for example, the United States would not have been able to map the Iraqi radar-warning system. Without infrared satellites, they could not have detected SCUD launches. Without weather satellites, they could not have planned the air and ground campaign. And without communications satellites they could not have controlled and commanded a multidivisional force.

Space is not only about reconnaissance. It is about communication as well. It is an extraordinary fact that virtually all communications within the theater of operations and between the theater and the United States were carried out via satellites. The U.S. Defense Satellite Communications System (DSCS), operating at superhigh frequencies, had two satellites in place at the beginning of the war. One, in geostationary orbit over the Indian Ocean, was ready to go. Another, over the eastern Atlantic, had to be tilted to access the Persian Gulf region. Another DSCS satellite being held in reserve at 180 degrees over the Pacific was maneuvered into position to 65 degrees east over the Indian Ocean, and a reserve Indian Ocean satellite was activated. These satellites, along with British systems, some leased commercial systems, and a complex of ground stations created the first space-based communications network in the history of warfare.[1]

If the Iraqis could have destroyed that system, most of the allied advantage would have dissipated. **For the first time in history, the center of gravity of a military operation was located outside the earth's atmosphere.**

Victory depends on being able to paralyze your enemy's power source, the capability that makes his war-fighting possible. For the United States, the most important war-fighting capability has become the ability to *see* the enemy—the intelligence platforms that allow it to gather information on a global scale and in all the spectra. Every enemy of the United States that observed the defeat of Iraq noted the combination of precision-guided munitions and superb space-based intelligence that lay behind it. Thus, to prevail against the United States, a future enemy must be able to ground U.S. precision weapons. Grounding these weapons means blinding them—which means, in turn, destroying or blocking America's space satellites. To protect these valuable assets against assault, the United States will need to defend by attacking.

Let us now consider the inevitable outcome of this challenge and counterchallenge: the conduct of war in space.

13

A New Foundation

SPACE AND CONTEMPORARY AMERICAN STRATEGY

During the Cold War, the United States and the Soviet Union pursued a similar strategy. Offensively, the goal was to be able to destroy enemy missiles, bombers, and submarines before they could launch. Defensively, it was to deter enemy attack by assuring that enough missiles would be left after a first strike to devastate the enemy. This meant that each side needed to know a great deal about the precise location—both above and below ground—of enemy missiles, radar installations, air bases, command and control facilities, and communications nets and frequencies, which were hidden thousands of miles from home, in the interior of vast continents. Discovering their whereabouts would require an elaborate espionage network that could canvass the country and penetrate the national command structure of the enemy. Alternatively it would mean sending aircraft over the enemy country and taking photographs that could be interpreted if the plane returned home.

Some of this reconnaissance could be carried out without crossing into enemy territory. Cameras

could take pictures deep into the coastal regions from high-altitude aircraft loitering just outside the three-mile limit. Aircraft could probe the
enemy's borders, triggering enemy radar and mapping their defensive
coverage, so that war plans could be developed to neutralize those
defenses. To this end, the United States converted long-range bombers,
such as the B-47, into electronic-intelligence platforms, probing enemy
radar and communications capabilities, locating radar sites and listening
in on communications. They probed for targets and for weak points in
Soviet defenses.

The United States had tremendous geographical advantages in this
task. U.S. aircraft based in Europe, Turkey, Iran, Pakistan, Korea, Japan,
and the United States could probe virtually any portion of the Soviet
periphery, from the North Cape of Norway to the Bering Strait. The
Soviets were unable to reach either coast of the United States (only
Alaska was within reach) because their aircraft lacked the range to travel
that distance and return.[1] During the 1950s, the Soviets were almost
completely blind. They had no choice but to fall back on ground-based
human intelligence.

During the late 1950s, when missile installations stood above
ground, a moderately good intelligence agent with a road map could
locate U.S. missiles in Germany, Turkey, or the United States. And a
large nuclear blast anywhere within miles of the exposed launch gantries
would destroy them. Therefore, the Soviets concentrated on human
espionage to discover the location of large multimegaton nuclear warheads and large rockets or heavy medium-range bombers, such as the
TU-16A.

For the United States, acquiring information on Soviet missile
installations was a chancy business. The Soviet Union was vast and movement was slow, difficult, and dangerous. Casual tourists in the vicinity of
Soviet missile bases did not have extended life expectancies. Whereas the
Soviets had been developing espionage networks in the United States
since the 1920s, the United States was completely dependent on British
sources and on sources the United States has inherited from the
Germans. So, while the United States had excellent technical data about
the periphery of the Soviet Union, the Soviets probably had better
human intelligence on the American and European interior. But in the
final analysis, neither side could base their nuclear war plans on the quality of intelligence available—it was all too uncertain. In the same way that
long-range indirect-fire artillery had posed a tremendous challenge to
intelligence capabilities of the nineteenth century, intercontinental
weaponry was posing a challenge to postwar intelligence.

If ground reconnaissance was impractical, the only alternative was aerial reconnaissance. But conventional aircraft, such as the B-47 Stratojet flying at a maximum altitude of forty thousand feet, could easily be shot down by highly competent Soviet air defenses. Additionally, the number of transits at forty thousand feet that would be required to cover the Soviet Union would be prohibitive. Finally, there was the minor matter that the presence of such an aircraft over Soviet territory would be regarded as an act of war.

The United States had to develop an aircraft that could photograph the Soviet Union with sufficient precision to locate missile launchers, air bases, radars, and so on. It also had to have the ability to do this without being caught red-handed in an act of international outlawry. The plane had to be able to fly at such an extraordinarily high altitude that it was invulnerable to enemy countermeasures and would have a panoramic view of the Soviet Union; it had to be so fast that it could complete continental reconnaissance missions in hours; and it had to be equipped with such advanced cameras that it could photograph vast swatches of enemy territory in a single sweep.

The Soviets had not developed long-range bombers during World War II, concentrating instead on more urgently needed battlefield support.[2] In the aftermath of the war they generally focused on developing aircraft able to target Western Europe. As a result, they had little knowledge of the esoteric sphere of long-range, high-altitude aircraft. Building a plane that could be based in the Soviet Union, travel to and cross the United States, and return to the Soviet Union at altitudes in excess of seventy thousand feet was beyond Soviet design and engineering capabilities. On the other hand, designing a plane that could fly from Pakistan or Turkey to Norway was not at all beyond American capabilities. Such a plane was, in fact, built and was called the U-2.

U-2 and *Sputnik:* The Origins of Space-Based Reconnaissance

The U-2 was manufactured by Lockheed's Skunk Works, later to produce the Stealth aircraft. It was one of the first aircraft produced by the United States whose very existence was both classified and actually kept secret, and it demonstrated a tremendous American technical capability. To develop the U-2, the United States had to be able to sequester scientists and engineers, commandeer materials, and build giant facilities, while leaving civilian aircraft production—and the development of conventional aircraft such as the B-52—completely unaffected. Whereas the Manhattan Project had a significant impact on the U.S. war effort, creating critical shortages because of its demands on manpower, brainpower, and material, the United States managed to build the U-2 without causing any great pressure to be felt. Starting with the U-2, the United States created a vast, secret aircraft industry operating alongside an even vaster public industry. This is a capability that the United States retains to this day and is of enormous importance.

The U-2 was a strange aircraft. It looked more like a glider than an aircraft, with eighty-foot, thin, flexible wings with wheels on their end for use during landing. It could fly as far as a B-52 without refueling, at an altitude of about seventy thousand feet and a speed of about five hundred miles per hour. Its stall speed was five knots less than its do-not-exceed speed, making it brutally difficult to fly. It was completely unarmed, its only protection being speed and altitude. It had onboard cameras and electronic sensors for signals intelligence.[3]

The camera (the Hycon-B) was the true marvel. Designed by Harvard astronomer James Baker, it could cover an area 745 miles wide, with the central 150-mile strip photographed in stereo.[4] The U-2 systematically mapped the interior of the Soviet Union, locating Soviet missile sites as they were being constructed, as targets for American bombers and intermediate-range missiles located in Turkey and Germany. With a range of 2,600 miles, the U-2 provided complete and overlapping coverage from bases in Norway, Pakistan, Japan, and Alaska. Though the U-2 did not provide real-time coverage, the turnaround time for processing was brief. As a result, the United States was no longer strategically blind; that is, not until Francis Gary Powers's U-2 was shot down over central Russia. The Soviets, by contrast, were always blind from the air. Lacking any aircraft capable of carrying out such a mission, they turned to a radical new solution—satellite reconnaissance.[5]

When the U-2 was shot down by an SA-2 over Sverdlovsk in 1960, the United States was made to look foolish, dishonest, and vulnerable. It also looked as if the United States had been blinded by the Soviets. There are many mysteries about the U-2 shootdown. What is not in doubt is that U-2 flights could not continue and that unless alternatives were found, the Soviets could construct their network of bases for intercontinental ballistic missiles without the Americans knowing where they were located. In short, it seemed that U.S. intelligence about Soviet capabilities and targets was about to disappear—unquestionably, a massive, strategic failure. Along with this, *Sputnik* made it appear as if the United States had been caught napping and the Soviets had a strategic reconnaissance capability far greater than that of the Americans. In fact, neither perception was true. The Americans were not blinded and the Soviets did not have a superior capability.

The United States had begun thinking about space-based reconnaissance soon after World War II. In 1947, the Rand Corporation published a study on satellites, which stated that "by installing television equipment combined with one or more Schmidt-type telescopes in a satellite, an observation and reconnaissance tool without parallel could be established."[6] Three problems had to be solved in developing a reconnaissance satellite. First was the mechanical question of boosting a satellite—a problem essentially solved with the German invention of the V-2 rocket. The second problem was photography. Extremely high resolution optics had been worked on prior to World War II by the Germans and was perfected during World War II by the Americans, in their strategic bombing campaign. Ultra-fine-grain photographic films, extremely high and wide-angle lenses, had also been developed, as had a solid body of knowledge on how to interpret photographs. The issue that had not been resolved was how to get the pictures back to earth from a satellite. A reconnaissance aircraft could take off and land, but a satellite could not.

The first idea, television transmission, proved unworkable because the quality of the signal was so poor that the picture was degraded beyond recognition. The second idea, de-orbiting the satellite and having it return to earth, was tried by both sides, but its weaknesses were that it took so long to get the pictures back, if one or two satellites were being relied on, and it was prohibitively costly if a lot of satellites were used. The third idea, which proved practical, was implemented in 1960 when a Discoverer satellite armed with a camera was boosted into polar orbit and snapped pictures of the Soviet Union. A capsule containing the film was returned to earth via parachute, to be caught by a waiting aircraft.

Reconnaissance photography has an inherent paradox. It is impossible to achieve the twin goals of reconnaissance intelligence—covering the largest area of a target possible while also seeing the smallest objects—with the same optical system. A wide-angle lens reduces a camera's resolution. A telephoto lens reduces a camera's field of vision. Moreover, for high-resolution work it is preferable to have a low orbit, so that the subject is as close as possible. For wide-area coverage, the obvious choice is a higher orbit. The Discoverer series solved the problem by supplementing the KH-4A wide-angle system, with a resolution of ten feet, with the KH-7 close-up system, with a resolution of eighteen inches.[7] Thus, reconnaissance now required two perishable platforms, which still did not give real-time intelligence and warnings.

The importance of the Discoverer satellites, the public name of the secret Corona project, showed itself quickly, in a fairly disconcerting way. During the 1960 presidential campaign, Kennedy claimed that the Eisenhower administration had allowed a missile gap to develop—that the Soviets had surged ahead in the development of intercontinental ballistic missiles (ICBMs) and were about to deploy hundreds of city killers. Once in office, Secretary of Defense Robert McNamara personally took charge of examining Discoverer pictures, looking for evidence of massive Soviet missiles. He knew roughly where the massive Soviet missiles would have to be located—near major highways and railroad tracks—and after carefully examining all possible deployment sites, he discovered that the Soviets were not deploying hundreds of missiles, but, at most, had deployed only ten to twenty-five ICBMs in September 1961.[8]

Therefore, information supplied by the Corona satellites had an enormous political payoff. It showed that the Soviet nuclear force was in no way capable of challenging American power in the immediate future. The American B-52 force remained secure; it could not be taken out by a Soviet first strike. On July 20, 1960, the USS *George Washington,* a nuclear submarine, successfully test-fired a Polaris missile.[9] By October 1, 1961, the United States had deployed 187 Atlas and Titan I ICBMs throughout the country.[10] So, on the eve of the Cuban missile crisis, in October 1962, Discoverer had taught the new administration what it had not known during the Berlin crisis or the Bay of Pigs: that the United States had hundreds of missiles able to hit the Soviet Union, while the Soviets had only a handful that could respond. As a result of the Discoverer photos, Kennedy could take actions that would have been impossible otherwise.

In retrospect, Kennedy's brinkmanship had more to do with the quality of his intelligence than with any sort of risk-taking personality. He knew that whatever other options Khrushchev might have had, launching a nuclear first strike against the United States was not one of them. Kennedy also knew that Khrushchev, given his satellites and intelligence sources, probably had a pretty good idea just how massive the American arsenal actually was. Khrushchev had to act with extreme caution not to provoke the Americans. Once he understood that the United States might well choose a nuclear strike at the Soviet Union rather than permit the placement of missiles in Cuba—and he desperately needed them there *because* he could not hit the United States from the Soviet Union—he had no choice but to capitulate.

Corona proved its worth as a strategic weapon barely two years after being launched by providing data that stabilized a near-war situation and prevented the United States from stumbling into war through miscalculation. The president was able to act in ways in which he could not have acted otherwise: he could press Khrushchev to the wall because he *knew* the cards Khrushchev was playing. And Khrushchev knew he had to fold. Because of the Discoverer, the United States appeared to win a spectacular war of nerves. The intelligence communities in both countries fell in love with the reconnaissance satellite, and the space agencies, civilian and military, were given blank checks on the national treasury, for better and better intelligence satellites.

Both the United States and Soviet Union continued to use the launch/return satellite technology throughout the 1960s. In the early 1970s, the United States began to depart from this pattern in two important ways. The first was the launching of the first Defense Support Program (DSP) in 1970. Its neutral-sounding name notwithstanding, the DSP was America's first line of defense against nuclear attack. Until 1970, the first warning of Soviet attack would come from radar sensors stretched across northern Canada—the Distant Early Warning System (the DEW Line)—which would report incoming missiles to the joint American-Canadian North American Air Defense System (NORAD). Precious minutes were lost between the time the missile was launched and the time it was detected, minutes that could mean the difference between the United States launching its own missiles and scrambling its bombers, and losing its ability to retaliate.

DSP was the world's first *real-time* satellite reconnaissance program. It could detect the launch of missiles and report the launch almost simultaneously.[11] DSP did not have visible-light cameras on board; it had only infrared sensors. And rather than transmit crisp images, its task was to detect the exhaust discharges of intercontinental ballistic missiles as they were being launched. DSPs were placed into geostationary orbit, about 22,300 miles above the earth, the altitude at which the velocity of the satellite matches the rotation of the earth. From that altitude, a satellite is able to observe a little over 40 percent of the earth's surface, so that, theoretically, three satellites should provide global coverage. In fact, DSP's sensors were not then able to sense that much of the surface efficiently, so several were placed in geostationary orbit over the Soviet Union. Information on launches was downlinked to several earth receiving stations, including Alice Springs in Australia, and from there to NORAD in Colorado Springs, where it was analyzed and acted upon.

DSPs served an unanticipated purpose during Desert Storm. They had, of course, been much improved since 1970. There were more sensitive sensors, three times as many infrared detectors, a much more powerful telescope, and superior data-transmission capabilities. This allowed DSPs to detect the plumes of fighter planes and to provide enough warning of SCUD launches for Patriot missile crews in Israel and Saudi Arabia to set themselves for intercepts.[12,13]

Extending DSP's relatively simple breakthrough into real-time reporting was the KH-9, the first full-service reconnaissance platform designed to remain in orbit for extended periods. Powered by huge, ungainly solar panels, the giant, which was dubbed Big Bird, also did double duty photographically. It employed a special Kodak camera for wide-area pictures, and mirror-type telescope for close-ups. Wide-angle shots were developed on board, and the images, captured by onboard television cameras, were transmitted to earth receiving stations. For finer, close-up shots, exposed film was returned to earth by the old Discoverer method—in pods, which deployed parachutes that were snagged by waiting aircraft.[14]

Big Bird was a great advance over the Discoverer series, but it also had many defects. The number of shots that could be taken was limited by the amount of film on board. The TV shots were still quite poor, while the close-up shots returned to earth via pod had all the drawbacks of the Discoverer satellites—extensive delays in receiving data. Real-time, high-resolution reconnaissance was still not at hand.

Nevertheless, the KH-9 was a tremendous advance. It had two cameras, allowing for stereo photographs, making precise measurements and superior photo interpretation possible. As important, it carried a type of infrared film that allowed it to take pictures on hazy days. And most important, the camera's field of vision was greater than with the Discoverer series. But for all of these improvements, it had the chief defect of all satellite reconnaissance: it was still dangerously slow for the needs of national security. It took too much time to drop and intercept the film, ship it to an analysis lab, and disseminate data. The time between the ordering of pictures and the availability of results ran to days. As a result, U.S. satellite reconnaissance virtually missed the 1973 Arab-Israeli War. Films of the beginning of the war were available about the time the war ended. The gap had to be filled by SR-71 Blackbirds, successors to the U-2.

Somehow, it was necessary to find a way to transmit clear images quickly and easily, in as close to real time as possible. And try as they might, scientists could not solve the problem until they achieved a fundamental breakthrough—which was both practical and theoretical. **They needed to rethink what was meant by pictures, and what was meant by data.**

The Digital Revolution Comes to Space and War

Photographic film is based on the reaction between certain mirror-type telescope chemicals and different wavelengths of light. It produces an image that is *analogous* to reality: an analogue. The analogue is the whole picture—it has no discrete parts. There is no intermediate domain between the visible picture and the unprocessed film that could be extracted for data management and transmission. It was a matter of "what you see is what you get." Since the picture could not be reduced to its constituent parts, the entire film had to be returned to earth for viewing.

An alternative means of organizing a picture had to be found, one that allowed the image to be expressed in an efficiently transmissible form. The solution was found in digitalization, a process that overturned our entire understanding of reality, how it ought to be thought of and represented.

Imagine a simple black-and-white drawing. Imagine the drawing fragmented into about 100,000 tiny points, arrayed 316 by 316, each so small as to be invisible to the naked eye. Now, assume that each of these points, called picture elements or pixels, was assigned a value, 0 for white and 1 for black. Assume that we set a rule saying that the first number would correspond to the first pixel in the upper-left-hand corner, and the numbers would progress left to right, as if we were reading. Then, you could represent a black-and-white picture as a string of 100,000 numbers, each either 0 or 1. In viewing the number array, the picture would make no sense. But a computer program and a computer could easily read each number and order each corresponding pixel on a computer screen to be either black or white. The original drawing would then be retrieved.

Digitization does not have to be confined to black and white. Suppose that, instead of using one digit, we used a string eight digits long, each having a value of 0 or 1. Each pixel could then have one of 256 values, because using 0 or 1 in eight places allows for 256 possible combinations, 256 different meanings, or 256 different colors or shades of gray. (The now primitive chip on which the first IBM personal computer ran had an eight-bit processor.) A complex picture could then be represented by a long string of numbers. A 316-by-316 pixel display would contain 99,856 pixels, with colors being represented in a string of just under 800,000 digits.

Imagine a camera, therefore, that registered light as a string of numbers, rather than as a chemical interaction with film. Unlike film, that data would not have to be returned physically to earth; instead, the strings of data, 0 or 1 in blocks of eight, could be transmitted back to earth via radio—an easy task for a medium that began with Morse code, a form of digital communication.

The KH-11, Big Bird's successor, carries just such a camera. The KH-11 used lenses and mirrors to focus light on an array of sensors known as a charge-coupled device, or CCD, instead of on film. The sensors turned the light into electrical impulses of varying intensity, corresponding to the value of the light striking it. Each electrical output had a value between 0 and 256, representing a palette of 256 colors or shades of gray. Each sensor corresponded to a pixel in the picture being captured by the camera's lens. The resulting string of data was stored in an onboard computer, then relayed to earth as a string of radio-transmitted digits, then reconstituted as a picture.[15]

All of the data about the KH-11—and, indeed, about all reconnaissance satellites—is classified. The very existence of the KH-11 is not acknowledged by the Department of Defense. In fact, KH-11 may or may not be the real name for the electro-optical reconnaissance satellite that we are discussing. However, the public literature is full of references to a KH-11 and advanced KH-11 satellite and is quite consistent in describing its characteristics. Moreover, we can infer the existence of the equivalent of such a satellite from public-policy decisions made by the government. Put differently, if the United States didn't have the technical capabilities attributed to KH-11, then some of the policy decisions reached by presidents ought to be considered impeachable offenses, and some of the military decisions of ranking military men ought to result in courts-martial.

For example, during the 1970s and 1980s, the United States entered into strategic arms agreements with the Soviet Union that required high levels of verification. For the United States to identify Soviet strategic weapons systems it needed the ability to scan large portions of the USSR and return data that was sufficiently refined to be able to locate and identify relatively small objects, such as ABM launchers, construction sites for ICBMs, and so on. Assuming that the United States was not continuing high-altitude reconnaissance flights of the Soviet Union, one can only posit the existence of a high-resolution, electro-optical, digitized-data satellite reconnaissance system—a rose by any other name is still a KH-11. Indeed, one of the reasons that information about KH-11 was leaked by successive administrations was to reassure a wary public and Congress that the United States had the ability to verify the agreements.[16]

How well can a KH-11 see? Reconnaissance is measured by five levels of precision: the ability to detect a target, to recognize what it is, to identify it precisely, to describe its characteristics, and to provide technical information about it. For a satellite merely to detect a bridge, it would need a resolution of about 18 feet; to know what it is would require about 13.5 feet; to identify it precisely, 4.5 feet; to describe it in detail, 3 feet; and to be able to see it well enough to know how to blow it up, about 1 foot. To locate rockets, however, which is what our "national technical means of verification" is supposed to be able to do, a resolution of 3 feet is needed just to detect it, 1.5 feet to recognize what it is, about 6 inches to identify it, 2 inches to describe its features in detail, and perhaps 1 inch to describe it technically—and rockets are harder to detect than planes or tanks.[17]

Knowing this, we can safely guess the *minimum* photographic capability possessed by U.S. satellites at the time of the Gulf War. At the very least, they could identify Soviet rockets, which would mean that the KH-11 had a resolution of about six inches. However, since we know that the KH-11 had reached a six-inch resolution in 1984 when it first went into orbit, it is difficult to believe that later models had not improved on this.

We know that during the Gulf War, cruise missiles were able to attack buildings through their ventilation shafts or were able to hit a particular corner on a particular floor of a building. Aside from being a testament to the cruise missile's precision, it is a remarkable testimony to a reconnaissance that was able to read technical details on buildings so precisely.[18] At least some of these remarkable feats of intelligence had to have been carried out by satellite. On the other hand, we know that bomb-damage assessment, particularly on tanks hit by the A-10's antitank rounds, was weak. Reconnaissance could not detect openings measuring less than an inch or two. A reasonable assessment is that the electro-optical, clear-day, optimal-lighting capability of the KH-11 was probably down to nearly, but not quite, two inches. Not bad for government work.

At the time of the Persian Gulf War, by most accounts there were three KH-11 satellites in orbit, one launched in 1984, another in 1987, and a third in 1988.[19] Each had been placed into an elliptical (186 by 621 miles), near-polar orbit, at a ninety-eight-degree angle. The orbit is synchronized with the sun, which means that it returns to each earth point at the same time of the day. This provides an important advantage in that shifts in shading always represent changes on earth rather than changes in the angle of the sun. This is particularly important in high-speed, computerized analysis of photo data.[20]

The fact that the orbit is elliptical adds to its reconnaissance capabilities. As we have discussed, the lower the satellite's orbit, the better the picture. In an elliptical orbit, it is possible to photograph from the lowest point of the orbit. For American reconnaissance satellites, maximizing the amount of low-altitude time over the Soviet Union was the object, the reverse was true for the Soviets. Satellite orbits are a political statement about friends and enemies.

Whatever the virtues of the original KH-11, and they are great, it is an optical system. Although optical systems intrinsically have better resolutions than other spectra, since shorter wavelength increases resolution, this advantage frequently pales in the real world. In clear weather optical systems can tell which missiles the enemy has deployed and count the number of his tanks. But the atmosphere is rarely perfectly transparent and sometimes it is completely opaque—with clouds and at night. Clearly, the KH-11 was not enough.

The advanced KH-11 was introduced to begin solving some of these problems. New sensing capabilities were introduced—for example, an infrared capability called false color, which gave it the ability to see through haze and darkness, substantially increasing its effectiveness. Just as important was the development of a fully multispectral sensing system. The optical and infrared systems could see an object but could not tell if it was a mock-up or the real thing. By adding another analytic spectrum, such as radar, an object of interest could be analyzed for composition—did it consist of metal or was it merely a wooden model? This was indispensable in analyzing Soviet missile deployments during the Cold War, or in the Persian Gulf War for distinguishing actual tanks from decoys.[21]

Perhaps the most interesting innovation reported on the later KH-11 was the presence of flexing mirrors.[22] Under the best of conditions, the earth's atmosphere distorts light passing through it. With smoke, fog, and other haze, the distortions can become substantial. If the light captured by a camera lens could be systematically reshaped to compensate for the distortion, the picture could be significantly enhanced. If the newest KH-11s do, in fact, have such flexing mirrors, this would indicate that they also have sophisticated computers on board, which can run complex programs to manipulate the mirrors as images are being captured. If such computers are present, the level of onboard image enhancement must obviously be soaring. If the mirrors can be manipulated, then probably images can be integrated with the advanced KH-11's signals intelligence (SIGINT) capability—locating command vehicles in armored formations and so on.

The first KH-11s were placed in a near-polar orbit out of Vandenberg to provide full coverage of the Soviet Union. The newer versions were placed in a radically different orbit, reflecting changes in geopolitical reality. Launched from Canaveral during the slow collapse of the Soviet Union—in August 1989 and March 1990—the advanced KH-11s were intended to provide detailed coverage of other areas, particularly the third world, where challenges were considered likely to arise.

The placement of these satellites—one is in a fifty-seven-degree orbit, the other in a sixty-five-degree orbit—represents a gamble. A fifty-seven-degree orbit cannot see anything north of St. Petersburg; a sixty-five-degree orbit is blind north of the arctic circle. Expensive resources are being committed beyond recall to the notion that northern Russia is no longer worthy of serious interest.

The advanced KH-11 seemed to have another vital capacity that it may have shared with earlier versions of the KH-11—maneuverability. Not only can it change its attitude—the direction that it and its cameras face—it can actually change orbits. It can shift the shape of its orbits, thereby increasing coverage of certain areas, or the angle of its orbit in relationship to the equator. Perhaps most important, the advanced KH-11 can maneuver to avoid antisatellite weapons. To do this, the KH-11 has to have a fairly large engine. Indeed, it is likely that much of its bulk is taken up by engines and fuel, rather than by cameras and other sensors.[23] Some accounts claim that the advanced KH-11 is powered by something called the Bus-1, produced by Lockheed—a nuclear-powered maneuvering rocket.

The KH-11 even in the advanced version had several limitations. For one thing, as impressive as its resolution was, its area of coverage was limited. At maximum resolution, one KH-11 image covers about four square kilometers. This is useful for close examination of particular installations or deployments of troops but not very useful in generating large-scale maps for use by ground forces. U.S. forces turned to commercial sources, including the American Landsat and, more importantly, the French SPOT satellites. The SPOT has a resolution of only about ten yards, but that is sufficient for locating bridges, buildings, and even concentrations of tanks. Best of all, a single SPOT image covers 3,600 square kilometers, allowing maps made to a 1:50,000 scale.[24]

The usefulness of the SPOT surprised American commanders.[25] They had concentrated so heavily on locating Soviet missiles—for which the KH-11 was superbly suited—that they had neglected to recall basic rules of warfare: the importance of geography. Ironically, the thoroughly unclassified world of satellite imaging for commercial and academic uses had produced the needed mapping capability. As a response to this discovery, more recent versions of the KH-11 are equipped with lower resolution, wider angle lenses that emulate SPOT's broader sweep with higher resolution.[26] In another ironic note, Saddam Hussein may have purchased SPOT data of Kuwait just prior to his invasion in August 1990.[27] The growing importance of commercial space-based imaging in areas such as agriculture, mining, and water management raises an interesting question of how long military reconnaissance will be able to retain its secrets, particularly if it is forced to share technology with the private sector.

Playing LACROSSE: Radar Imaging

Darkness covers the earth about half the time. During the winter, the globe north of the forty-fifth parallel is covered by clouds 70 percent of the time. Weeks can go by without the ground's being visible, weeks in which a new missile silo could be constructed, manned, armed, sealed, and camouflaged. Visible-spectrum imaging systems have an important defect—they work only during the day and in relatively clear weather. The infrared sensors on later KH-11s can work at night or in haze, but its primary function is to penetrate camouflage—camouflage, such as netting and cut vegetation, has a different temperature than live vegetation. Although there are reported to be night-vision systems on board KH-11s, they would work only if there are no clouds, and then with a tremendous loss of resolution.[28] Solid cloud cover blocks and distorts infrared radiation, just as it blocks visible light. Much of the time, therefore, the KH-11 is useless. More precisely, while some of the KH-11's pictures are invaluable, they need to be supplemented by other reconnaissance means.

One solution is to utilize other parts of the electromagnetic spectrum—which is what is done by another variety of imaging satellite, codenamed LACROSSE. The eye can see in a narrow band at about 10^9 megahertz. This means that the energy wave of light oscillates about 10^9 million times a second. (Oscillation is the *frequency* of waves, that is, how often they occur. The higher the frequency, the shorter each wave, since greater frequency requires more compact waves. Microwaves, for example, have extremely short waves and extremely high frequencies.) The human eye can see subtle variations in wavelength, seen as color, within a very narrow band. Infrared sensors, which "see" what we experience as heat, operate in a broader band, from 10^5 megahertz to 10^8 megahertz, a lower-frequency, higher-wave area of the spectrum. This area is close to human vision. Indeed, cats and other nocturnal animals see in the infrared range.

From about 100 megahertz to just short of infrared is a range of radiation that is used in commercial radio, television, short-wave radio, and microwaves. Radio and television are at the lower frequencies. At the higher frequencies, from 300 megahertz to 300,000 megahertz, are the various radar bands. The higher the frequency of a wave, the easier it is to focus it into a narrow beam. So the lower-frequency bands are perfect for *broad*casting. Radar, which requires as narrow and focused a beam as possible, is therefore to be found in the higher-frequency ranges. Very high frequency, very low wavelength *microwaves* are the most useful, because they can be more narrowly and precisely focused than any other radar frequency. Thus microwave radiation is an essential part of our normal communications net. Microwave towers can be seen on the roofs of city buildings or on hilltops in the country. Microwave ovens can also be seen on kitchen counters, where their high-frequency, high-energy beams excite water molecules in foods and cook them with what used to be considered miraculous speed.

In addition to being capable of precision focusing, microwaves, unlike higher-frequency radiation, such as visible light and infrared, can also see through clouds, haze, and smoke—and through solid obscurants such as vegetation. However, since they have a lower frequency than visible light, they cannot, all other things being equal, have a high resolution.

The LACROSSE was a fundamentally different satellite from the KH-11, but was on a very similar mission taking pictures of the earth below. KH-11, like any optical system, was passive; it collected light that came to it. LACROSSE, on the other hand, was active and transmitted energy from its antennae. Therefore, it needed a long-term energy source, which it found in large solar panels that converted sunlight to electricity. Interestingly, Soviet radar reconnaissance craft use small nuclear reactors on board to produce power.[29] Unlike the KH-11, LACROSSE could be used day and night, through clouds as well as in clear weather. LACROSSE satellites were probably built with additional capabilities, including infrared sensing and some sort of signals intelligence.

This was extremely important, as one of the ways to fool planners of air strikes was to construct a large number of decoy planes or tanks. Radar could distinguish between steel and wood, but if the decoy was covered in a metallic foil, it could be confused. Infrared, however, which sensed ambient temperature, could detect if an engine had recently been used. Finally, signals intelligence could see if normal communication patterns were occurring among the tanks. All this would make decoying more and more difficult.[30]

It is not certain what resolution LACROSSE had. It is generally assumed that, at best, it was about three feet, as such resolution would have been necessary merely to recognize the presence of some sort of vehicles and perhaps distinguish between tanks and cars. This would also have allowed LACROSSE to locate ships at sea or aircraft on the ground, but it would not have been sufficient to provide much information about what they were, or even how many there were.

If its resolution was worse than three feet, LACROSSE might still have recognized aggregate formations of armor—for example, the presence of an armored brigade, sensed as a mass of metal—but not individual tanks. This might well have been LACROSSE's primary purpose—either to locate previously unknown concentrations of armor so that air reconnaissance or KH-11 pictures could then identify them, or to track already identified formations at night, during bad weather, and between more detailed aerial reconnaissance by KH-11 sweeps.

The problem with radar has always been that the farther from the target and the greater the desired resolution, the larger the radar antenna had to be. In 1978, the United States launched a satellite known as Seasat, which had a resolution of about eighty-two feet. The satellite was in an orbit about a hundred miles above the earth. A traditional radar would have required an antenna about 1.2 miles wide, obviously impractical.[31] What made Seasat possible were two new innovations: phased-array and synthetic aperture radar—both combined into a single system on reconnaissance satellites.

Phased-array radar does away with the mechanical antennae with which we are all familiar. Instead of an antenna sending out a beam and receiving a reflection, small transmitters and receivers, arrayed in columns and rows, emit energy and receive echoes. This allows for much finer manipulation of beams, as each array can be focused on a different target. Additionally, since phased-array radars have no moving parts, they are much more reliable—and reliability is nice to have in a space satellite.[32] Phased-array radars are also a critical component of synthetic aperture radar (SAR). This focuses extremely narrow beams of energy that encounter the object and return a series of signals while the beam is passing over the object, producing a series of electronic snapshots.[33]

These signals are also analyzed by a method called Doppler radar. We are all familiar with the Doppler effect from the sound of trucks or trains approaching and leaving us. As a truck approaches, its sound becomes louder and louder and higher and higher pitched. As it leaves, the sound quickly declines in volume and frequency. Sounds propagate as waves, at frequencies below two hundred kilohertz. As the truck rushes toward you, the source of the waves gets closer and closer. New waves close in on waves released earlier, crowding them, increasing their frequency. Sounds get louder and higher pitched. As the truck passes you and moves farther away, the waves get more and more spaced out. Pitch and volume quickly drop off. To put it differently, they doppler off.

By measuring shifts in the frequency of waves, it is possible to measure motion—how fast something is moving, and whether it is moving toward you or away from you. This can be done in any spectrum. Astronomers, for example, can examine the wavelength of light coming from stars to determine if the shift is toward the red (movement away) or toward the blue (movement toward you). In the radar spectra, Doppler radar can measure changes in frequency to determine speed—something that many of us have discovered, to our sorrow, in speed traps.

By using the return data and the Doppler data, and processing it digitally, the system can be fooled into thinking that it is using a very large antenna. Since the satellite is moving, there is a small Doppler effect. By comparing Doppler effects of successive shots, along with the strength of the signal return, a clear picture of the surface emerges. Much of this blending of data takes place on earth-based computers, rather than on board the spacecraft. In this way, an extremely refined picture can be returned at substantial distances, regardless of cloud cover, obscurants, or night.[34]

ELINT and SIGINT: Listening In

Ever since Mordecai overheard Haman plotting to kill the Jews, eavesdropping on the enemy has been the center of spy craft. Being able to insert an agent into the inner circles of a foreign government is invaluable precisely because of what spies can overhear. Such human intelligence is always difficult and dangerous. But in the era following the invention of the telegraph, telephone, and wireless, another means has become available for hearing what others say—intercepting electronic emissions.

There are many ways to intercept conversations. During World War I and World War II, patrols routinely found and tapped into enemy telephone lines. The U.S. Marines used Navajo Indians in the Pacific as communicators because English-speaking Japanese soldiers would not be able to understand them. As wireless communications grew, tapping into them became critical. In fact, the entire area of the electromagnetic spectrum, from radio to microwave radiation, became a key sphere for intelligence gathering. Penetrate that sphere, and you could know everything from the type of radar guarding the approaches to your enemy to cabinet changes being planned in a friendly country.

There are many ways to collect intelligence—aircraft, ships, submarines, ground stations, all can position themselves to overhear conversations. During the 1967 Arab-Israeli War, the USS *Liberty,* which was attacked by the Israelis, had positioned itself to overhear Arab and Israeli communications. The USS *Pueblo,* captured off Korea in 1967, was on a similar mission. Embassy roofs bristle with antennae designed to pick up signals intelligence (what people are saying) in the host city. The United States did quite well at this. In an operation known as Gamma Guppy during the late 1960s, American intelligence managed to listen in on the car phones of the Soviet leadership.[35] We cannot wait for the day when we find out what Brezhnev had to say on his car phone.

During the mid-1960s, the United States launched a class of satellites commonly referred to as signals and electronic intelligence satellites—ELINT and SIGINT—whose task was to wait and listen. These ferrets used the radio-radar region of the electromagnetic spectrum to collect virtually any signal stronger than the background radiation of the universe.

In 1972, about the time the DSP and KH-9 were first going up, the United States launched its first geostationary signals intelligence satellite, code-named RHYOLITE. We know something about RHYOLITE because of the arrest and trial of Christopher Boyce—made famous in the movie *The Falcon and the Snowman*—who sold information on RHYOLITE to the Soviets while working for TRW, RHYOLITE's manufacturer and operator. In another case a few years earlier, a British spy was arrested for selling RHYOLITE data to the Soviets.[36] During the Boyce trial, it was revealed that the Soviets had re-encoded their missile telemetry after receiving Boyce's information. So we know one of the functions of RHYOLITE—and we know that it was good at it.

The contemporary version of RHYOLITE is code-named MAGNUM. The first was launched on January 24, 1985, the second on November 23, 1989. Both were launched by the Shuttle because their enormous antennae were too large to be contained in a missile. A third ferret, a smaller variety code-named CHALET, was launched on May 10, 1989, by a Titan 34D from Canaveral. Most reports describe MAGNUM as large, weighing 3.6 tons.[37] MAGNUM's antenna, which resembles an umbrella, is said to be twice the size of a football field, so that it can pick up the faintest radio signal from 22,300 miles up in the sky.[38]

Nothing is certain about U.S. signals intelligence. Unlike the KH-11 program, which leaks like a sieve, SIGINT is kept under tight control. We do not even know whether the different code names represent different sorts of satellites or merely new names meant to confuse. Those who talk, don't know—those who know, don't talk. But we do know, from various espionage cases that have become public, that the United States has signals intelligence satellites and, from observing Shuttle launches, that two large loads were put into geostationary orbit, amid consistent reports that they had enormous antennae and were intended for SIGINT. We also know that the National Security Agency (NSA) is located in Laurel, Maryland, and that it employs thousands of workers, and that they must be doing something. And we know that billions of dollars are spent on signals and electronic intelligence by the NSA, which means either that SIGINT is capable of strategically significant performance—or that a large number of federal officials are committing felonies.

What is the American SIGINT capability? Oliver North gave us a sense of NSA's capabilities by reporting that, on November 22, 1985, as a plane left Tel Aviv carrying missiles for Iran, NSA instantly notified the CIA and placed in motion plans for monitoring activities moment by moment. According to North, NSA set up "some very specific, targeted intelligence collection that would give us, almost instantly, exactly what was happening very, very accurately. . . . Within hours, we would have detailed information on what these people were saying to each other, and the plans they were making. It is probably the most reliable form of intelligence there is."[39]

North's comments were less than discreet, but probably accurate. As far back as 1977, when Stansfield Turner, then director of the CIA, decided to dial back on human intelligence in favor of electronic intelligence, it has been obvious that the United States has the ability to ferret out and analyze the smallest electronic emissions. These things can only be inferred. For example, there were undenied reports that the United States provided both Gorbachev and Yeltsin with information about possible coups. It is difficult to imagine that the United States would have better human intelligence (HUMINT) in Russia than two old hands like these. Therefore, we must surmise that the source was satellite SIGINT. SIGINT is obviously imperfect. When Noriega fled, the United States could not find him for days; and it did not know about Iraq's invasion of Kuwait. Nor did it know about terrorist activity such as the bombing of the U.S. embassy in Beirut. We assume, therefore, that SIGINT can only know about *electronic* transmissions.

There is, of course, another explanation for apparent failures—the

Coventry dilemma. During World War II, the British, using the Enigma machine, had almost perfect access to German operational codes. They therefore knew, at the same time that the German field commander did, that the Luftwaffe had been ordered to destroy the city of Coventry in central England. Churchill faced a classic dilemma. If he ordered the evacuation of Coventry and the Germans found out, they would know that their code had been penetrated. The Germans would change the code, and the British would lose an invaluable tool, perhaps costing thousands of lives or even the war. On the other hand, what was the use of intelligence if Churchill allowed an English city to be devastated? Churchill properly allowed Coventry to be destroyed without evacuation, reasoning that, if he did not do so, far more lives would be lost in the future.

U.S. intelligence might well have faced equally heartbreaking choices, from Vietnam to Beirut. Any intelligence asset has a value, but using the asset carries with it a risk. The benefit of use must outweigh the risks of giving away the capability. The more valuable the capability, the less likely that the risk is worth taking. There is no science to this. It merely shows that intelligence work, regardless of technique, remains practically and morally difficult and murky.

If SIGINT is to be used prudently, our policymakers must frequently act as if they know less than they do, failing to respond to situations in a timely fashion in order to preserve the mystery of technological capabilities. Unfortunately, from the outside it is sometimes difficult to distinguish prudent employment of precious intelligence from incompetence. And this is the citizen's dilemma in passing judgment on the performance of public officials.

Reading Tea Leaves: The Data Crisis

Imaging, the Defense Support Program, signals intelligence, electronic intelligence, and all other types of reconnaissance platforms, in space and elsewhere, collect vast amounts of data—all of it, in its raw form, useless. The endless stream of digital material is incomprehensible unless some system turns the data into information, analyzes the information, and then distributes that information to people who are making decisions or fighting wars.

During Desert Storm, the sheer amount of data being downloaded from satellites became daunting, and the lag in getting the results to the battlefield frequently meant that the information was irrelevant by the time it arrived. A report of the staff of the House Armed Services

Committee released in August 1993 found that the sheer volume of data and information generated, and the appetite of tactical commanders for more and more images, swamped a system not designed to handle it. (You can get a measure of just how vast the data could have been when you consider that a *single* weather photo runs about forty megabytes in size—almost forty times larger than this book on disk.)

This distinction between tactical and strategic intelligence is at the root of the problem. Satellite reconnaissance was developed for strategic purposes—to aid in a nuclear war. Imaging satellites had, as their first mission, monitoring the enemy's nuclear forces—they observed silo construction and counted missiles. The most important function of SIGINT and ELINT satellites was monitoring telemetry from missile tests and emissions from radars and other electronic systems associated with the operation of nuclear warfare systems.

Most reconnaissance missions had two characteristics in common. First, the data were not time sensitive. Locating missile silos, maintaining counts of enemy missiles, mapping out radar systems—all of these were critically important, but the translation of data into useful information could take days or even weeks without a problem. It takes a long time to build a missile silo or deploy a new type of missile. Seconds do not count. Second, access to data was severely restricted because the data was highly sensitive. It went to very few places and was seen by very few people. Data flowed into collection and interpretation centers, which dispensed it with utmost care, so that intelligence capabilities would not be compromised.

Desert Storm posed a completely unexpected challenge to the space reconnaissance cultures of the National Reconnaissance Office and National Security Agency. First, conventional warfare is far more fluid than the nuclear stalemate with which the U.S. space intelligence program was designed to cope. For decades, the nuclear ballet went on at a snail's pace in anticipation of a conflict lasting a few hours. Conventional war, on the other hand, is fluid. It shifts minute to minute, hour to hour. This is not merely because it is tactical. Air strikes at strategic targets require battle-damage assessments in minutes or, at worst, hours, so that additional strikes can be called in as needed. With the advent of precision munitions, it follows that precision information is essential to plan and order air strikes. Moreover, turning the data into information and getting the information widely distributed to the troops in the field is an absolute necessity. This necessity flies in the face of the system that sees the wide dissemination of information as a threat to the integrity and security of the entire system.

During Desert Storm, the processing and distribution of information showed itself to be the system's weak link. The system tended to centralize data collection, which allowed the collection point to determine what would be distributed to whom and when. Obviously this maintained system security—which it was designed to do. It also transferred tremendous power from the hands of field commanders to officials at the National Reconnaissance Office and the National Security Agency, who, in interpreting and analyzing data, could build policy judgments into distributed intelligence. In retaining control over who would receive what intelligence, these same officials could control operations. In a short, successful war, this is not likely to be a problem. In a long, divisive war, such as Vietnam, intelligence analysis could and did become an acrimonious issue.

Centralization was built into the technology. For example, it took less than ten minutes for a SCUD to hit targets in Israel and Saudi Arabia, while information from the Defense Support Program had to travel the strategic nuclear route to Colorado Springs, then be bounced by satellite to Central Command and the Patriot units on watch. It got there, using advanced terminals that had been rushed into the field, but the warning system had to be jury-rigged around the centralized structure.

There were even more serious problems with KH-11 imaging data. KH-11s, after taking their pictures, encrypt the data so it cannot be intercepted, then download it to receivers in Greenland and Nurrungar in Australia.[40] There is a choke point right here. Downloading these pictures, even with advanced data-compression programs, takes a long time, given the number of pictures and the level of resolution they contain. Once received at the ground stations, the data is retransmitted via Satellite Data System (SDS) to Fort Belvoir, Virginia, where the CIA's National Photographic Interpretation Center (NPIC) computers transform the vast data sets into information—pictures.

At this point an even more time-consuming process begins—interpreting the pictures. During Desert Storm, demand was so great that NPIC was at its limits and falling behind.[41] It is not clear how much interpretation is being done by computers and how much by human eyes. There is little evidence in the commercial market that artificial-intelligence programs have been developed with the ability to "read" complex graphics consistently. On the other hand, we know that Tomahawks have some ability to compare pictures prior to making a final attack—which requires some sophisticated imaging capability. We must assume that NPIC has developed some capabilities, above the

Tomahawk but below humans. We might assume further that routine images are first processed through some sort of screen that alerts humans to possibly interesting targets, while data from critical areas is simultaneously screened by humans and computers. Either way, interpreting the photographs takes time.

Another problem has been what to do with the photos once they are interpreted. During Desert Storm, KH-11 and LACROSSE satellites undoubtedly located and mapped Iraqi armored positions. If those pictures were sent into the field, where they could be seen by thousands of field-grade officers and noncoms, one of the most precious intelligence assets of the United States would be compromised. The world would know the capabilities of U.S. imaging satellites and would also learn how to defeat those capabilities. The creators of the system always saw it used to develop tightly held intelligence within the highest echelons of the intelligence community. For it to be used otherwise meant that it could no longer be considered secure.

This problem was compounded by the introduction of a secret terminal code-named CONSTANT SOURCE. CONSTANT SOURCE computers were capable of receiving KH-11 pictures directly from NPIC via the classified Tactical Event Reporting System (TERS), which is operated by Space Command in Colorado. Apparently, SCUD launch data was transmitted by this means as well.[42] The natural reluctance of the intelligence community to compromise reconnaissance assets by wide distribution was simply brushed aside. Certainly one wonders if the data received in the field during Desert Storm was the best available or simply what intelligence officials thought would do. One suspects that a painful compromise between the reconnaissance community on one side and senior military officers on the other had to be worked out in determining what level of resolution would be made available.[43]

With thousands of servicemen returning from the Gulf after having seen KH-11 images via CONSTANT SOURCE, intelligence officials must assume that imaging capabilities are now fully known to foreign officials, although American policymakers have been showing the pictures around for years to prove points and get cooperation.

Obviously, if there were problems with disseminating image intelligence, one can imagine the problems with distributing signals intelligence from MAGNUM and CHALET. Downloaded to general headquarters in Cheltenham in Britain, Pine Gap in Australia, and Bad Ablein in Germany, SIGINT data is then bounced to National Security Agency headquarters in Laurel, Maryland, where it is processed by high-speed computers.

This data proved indispensable in allowing the allies to map Iraq's air defense and communications apparatus. Electronic emissions from Iraqi radar were noted, along with their characteristics, permitting aircraft to jam them or put them out of action on the first day of the war. Intercepts of a wide range of communications emissions permitted planners to pinpoint key command facilities, as well as to track the deployment of units—down to the smallest fragment—so long as they were emitting an electromagnetic signal.

The ability of SIGINT and ELINT to track electromagnetic emissions is well known. The ability to identify individual units speaking in the clear follows. What is not known, and what is the deepest secret, is the extent to which the National Security Agency can read encoded emissions. Reading a normal phone conversation is easy, if ferrets' antennae pick it up and computers select it for analysis. But most nations now encode their national security transmissions.

If NSA is able to draw up an order of battle of Iraqi forces, it follows that field-level encoding can be broken. But what level of encoding can NSA break? Computer scientists, mathematicians, and intelligence analysts work to establish key words and phrases, key frequencies or telephone numbers, which are used as screens to extract data most likely to contain useful information. If the code has already been broken, material is decoded. But were Iraq's most secret codes broken? Was the United States able to not only locate transmitters and overhear open phone lines but also to intercept deeply encoded transmissions? Was it able to literally eavesdrop on Saddam's meetings with his senior commanders? To what extent can NSA eavesdrop on any conversation—even if it is not being electronically transmitted? And if such capabilities do exist, how could the information be used without destroying the security of the space-based intercept system?

The Coventry dilemma plays itself out daily for the National Security Agency and the National Reconnaissance Office. Every use of intelligence threatens to compromise the method that accumulated it. The best strategy is to deny that the asset exists, to falsify its capabilities and so on. But if soldiers in the field are constantly being ordered to act on mysterious information that invariably turns out to be correct, it will eventually be clear that we have some means of knowing what the enemy is saying or doing.

Space-based technology and the culture of intelligence have collided. Now that the technological means are available to get the information to the field quickly, the political pressure to get it there fast is close behind. As more and more data becomes available, the power and mystique of the intelligence community will tend to dissolve. It was no accident that Norman Schwarzkopf, during his report to Congress, lashed out at only one group—intelligence officials in Washington who failed to provide him with useful intelligence. Schwarzkopf told Sen. Sam Nunn:

> Again, I think that we have wonderful systems out there, and I hesitate to talk about a specific system of sorts. But I think that the intelligence community should be asked to come up with a system that will, in fact, be capable of delivering a real-time product to a theater commander when he requests it. . . . I just think that that's a void and if we—we have now a system because we focus too much on what might be called national systems, which respond more to national directive out of Washington.[44]

Lt. Gen. Thomas S. Mooreman, former commander of the USAF's Space Command, drew the battle line clearly during Desert Storm:

> I am committed to maximizing the use of space systems for space applications, particularly tactical applications. I continue to believe we are not taking full advantage of the tactical utility of space systems. One of my priorities is to ensure that the people who ultimately use space systems to maximize war-fighting capabilities—such as the aviators, ground forces, and sailors in Desert Storm—know what space capabilities are available to them, how to get the data, and how best to exploit it.[45]

But Air Force Space Command does not control KH-11, LACROSSE, MAGNUM, or whatever they are being called today. Indeed, even the new unified, interservice Space Command does not control them. This makes Mooreman's manifesto dependent on a redefinition of intelligence principles—and a redefinition of the chain of command when it comes to space-based intelligence. The Department of Defense can fund undertakings, such as the Tactical Exploitation of National Capabilities (TENCAP) project, designed to use "strategic" means for "tactical" ends—or in our terms to use reconnaissance assets for actual warfare—and formally work with the intelligence community on this.[46] But the Department of Defense's initiative will not come to pass until the distinction between intelligence and warfare is abolished, or at least until the distinction between strategic and tactical is not translated to mean important and trivial. This will, of course, become more and more important as war spreads into space—and space intelligence will become less a matter of managing satellites and more a matter of defending them.

The ongoing warfare between the intelligence and defense community over control of these space based assets reflects the increased practicality of space based reconnaissance. It is a symptom of a deep evolution in warfare.

Superb intelligence is now available. For all of its weaknesses, the space-based system has permitted us to see deeply and strike precisely. As the pretenses of nuclear warfare dissolve, and conventional wars become more common, space assets will cease to occupy a metaphysical realm controlled by mysterious bureaucrats holding jobs in agencies that do not exist. The creation of a united U.S. Space Command is, ultimately, incompatible with the control asserted by such agencies as the National Security Agency and the National Reconnaissance Office. This is true not because secrecy is unimportant. *Secrecy is essential.* The problem is that, for these agencies, space-based intelligence is too precious to waste on war. As Space Command matures into an essential part of conventional warfare, that tendency will be checked.

The creation of a Space Command will, of course, be seen as a vital event in the history of warfare. For the first time the leading military power in the world will have treated space as a clear-cut sphere of warfare. Just as the creation of the various air corps and forces in the interwar period signaled the emergence of aerial warfare, so the emergence of SpaceCom signals the emergence of space. Clearly, space was the center of gravity of the U.S. effort in the Persian Gulf. Now the question is, how secure is that center of gravity, what will threaten it, and what will war in space look like?

14

Space and the Future of American Strategy

The use of space parallels the use of air. The first use of the air was to provide reconnaissance information for ground-based weapons. The second purpose was to destroy enemy reconnaissance aircraft. The third was the defense of reconnaissance aircraft, by arming them, increasing their agility, and creating a class of aircraft dedicated to fighting other aircraft. The fourth and most important use was to merge the reconnaissance platform with the weapons system by placing explosives on the aircraft. From that point on, control of the air was understood to be a means to dominate the land and sea, both militarily and politically. This same process is now unfolding in space. Space is already being extensively used for reconnaissance. It follows from this that the next phase of warfare will be an attack on space-based reconnaissance systems, along with attempts to protect these platforms from destruction. Finally, just as the use of aircraft as reconnaissance platforms for artillery gave way to the bomber, space satellites will evolve from platforms passing targeting information to earth-based weapons into weapons platforms in their own right.

This evolution is well under way. During Desert Storm we saw the first comprehensive use of space-based reconnaissance to support ground-based weapons. The architecture of this system was still incomplete. Neither sensors nor information-delivery systems were fully evolved, nor was satellite imagery delivered as data that weapons could use without human intervention.[1]

The evolution of space-based reconnaissance will involve more efficient sensors and computers, able to fuse data gleaned from throughout the spectrum—including wavelengths shorter than visible light, such as ultraviolet and X-ray. In April 1994, the Air Force opened a Space Warfare Center as part of the Air Force Space Command. The motto of the Center—In Your Face From Outer Space—captured a new self-confidence and aggressiveness for space warriors. The purpose of the Center is to develop procedures for using satellite data to enhance conventional operations. One project, code-named TALON SCENE, is designed to demonstrate how imagery from a wide range of sources can be integrated and used in advanced precision weapons. Another, TALON ZEBRA, combines global-position systems data with imagery.[2] In a project carried out by the Department of Defense, the Joint Targeting Network demonstrated, during an exercise in Korea, that a wide variety of data could be merged and provided to weapons systems in real time.[3] Thus, the U.S. defense community is working to convert assets created for nuclear warfare into systems for conventional war-fighting—the purpose of a project called Tactical Exploitation of National Capabilities, or TENCAP.[4] Other sensor platforms will continue to be used. Even as subsonic manned aircraft, such as JSTARS, become increasingly vulnerable to smart antiaircraft weaponry, an entire generation of unmanned aerial vehicles (UAVs)—sometimes little more than radio-controlled model airplanes, sometimes as sophisticated as flying saucers—stand ready to assume the tactical reconnaissance role.[5] Powered by the sun, lithium batteries, or microwave radiation beamed from the ground, the vehicles will be able to hover thousands of feet above the battlefield, while using the same sort of sensors used by satellites.[6] They will be useful supplements to the broad and deep coverage provided by satellites. Indeed, they will become the poor man's satellite system—low cost, but able to provide comprehensive intelligence over a limited area.

Unmanned aerial vehicles can never replace satellites because of strict limits on their field of vision. Even from high altitudes, they can observe no more than a radius of a couple of hundred miles. Where the battleground is clearly defined, as in Kuwait, these systems are useful. When targets are at intercontinental distances and commanders need to track enemies with centimeter precision, they will be insufficient. So, one can expect UAVs to serve as tactical and operational reconnaissance platforms for ground forces already committed to combat in a specific theater. For the general deployment of global firepower, space-based systems alone will provide adequate coverage. As space-based reconnaissance is perfected, and the flow of data is integrated with weapons systems able to strike at targets on land and sea, command of space will come to mean command of the earth. Any nation wishing to defend itself against a powerful military opponent will have to try to deny the enemy the use of space—it will have to destroy or paralyze the enemy's satellites and, with them, the ability to see globally and to use intercontinental weapons. It follows from this that as antisatellite (ASATs) weapons are developed, anti-ASAT weapons will also be developed. As was the case with aircraft, satellites will have to become increasingly agile to evade threats, while other types of spacecraft—space fighters—will also evolve to protect reconnaissance craft from predators. As the long data link between earth-based weapons and space-based surveillance becomes more and more vulnerable, and as the experience of combat in space increases, explosive missiles will move from earth into space. There, they will take advantage of gravity as well as of shortened lines of communication. Once long-range hypersonic cruise missiles are perfected, their use will depend on space reconnaissance. Indeed, the perfection of these missiles will set the stage for space warfare.

Why Space? Intercontinental Hypersonics

The unmanned, intercontinental missile had already entered history, in the form of the ICBM. The ICBM could travel from North America to Eurasia in under thirty minutes. But it suffered from several defects that made it impossible to use on a conventional battlefield. First, before launching it, its precise destination had to be known. It could not be fitted with sensors and allowed to seek out and destroy its target at the end of its flight. Second, even if the target's precise location was known, the probability of error was greater than the blast radius of high explosives—which meant that to be effective it needed to carry enormous, nuclear explosives. Finally, each ICBM was extremely expensive and required costly launch facilities.

The cruise missile has none of the ICBM's defects. As we saw in the Persian Gulf, it is nonballistic. Its engines continue to provide sustained power until the end of the flight. Moreover, even its target is not *necessarily* determined at the moment of launch; it is capable of searching for its target on arriving in the target zone, recognizing it, and hitting it. Theoretically, it can be maneuvered without limit on the basis of information generated by both its own sensors and information—and commands—received from external ones. Its terminal maneuverability does not detract from its accuracy. In fact, it appears to have a much smaller circular error of probability (CEP) than do ICBMs, certainly superior to submarine-launched ICBMs. Moreover, because it can adopt any sort of mission profile, rising to eighty thousand feet, falling to treetop level, or anything in between, it can take maximum advantage of weaknesses in enemy detection systems. Of course, it must be remembered that currently available cruise missiles such as the Tomahawk are not nearly as flexible as cruise missiles will become. The Tomahawk is assigned both target and course at launch—in fact, several hours prior to launch to prepare its computers. It can maneuver impressively, but not autonomously; therefore it cannot take full advantage of external data. Nonetheless, cruise missiles will have to take on several characteristics before substituting for airpower—and programs are under way in all of these areas:

- Their range will have to increase from the current 1,000 to 1,500 miles, to at least 5,000 miles.
- They will have to be given improved warheads, including warheads capable of dispersing intelligent submunitions.
- Either their cost will have to decline or they will have to be able to return to base after dropping submunitions on the target. In effect, they will have to become unmanned bombers.
- Their ability to sense, suppress, and evade threats in tactical situations will have to improve dramatically.
- During flight, they will need to accept override commands for strategic purposes, including aborting missions, changing targets, changing course.
- They will need to do their own bomb-damage assessment and be available for second strikes with available munitions.
- Above all else, cruise missiles will have to become *fast*—fast enough to influence the course of a battle an ocean or continent away. They will have to sustain speeds like Mach 20—making their time to target substantially less than that of a manned attack aircraft stationed within the traditional boundaries of a theater of operations.

This is no fantasy. On March 9, 1992, the Air Force's Wright Laboratory at Wright Patterson Air Force Base in Ohio issued a request for proposals for a project entitled Hypersonic Aerodynamic Weapon Demonstration Program Definition.[7] The weapon was to be an extension of the precision technologies seen in the Gulf. Performance guidelines were set precisely, including the ability to hit a target three thousand miles away in about one hour. Of particular interest was the call for an endo-atmospheric weapon—one that would remain within the atmosphere at all times and not be boosted into space where it would need to rely on rockets. This would indicate that it would differ from the Tomahawk by coming in fast and high. In this sense it would not be a true cruise missile but, rather, a glide weapon that would, after initial acceleration, continue to target using its own aerodynamic lift, much the same way as the Space Shuttle lands.

This first approximation of a hypersonic, long-range, smart weapon attracted a great deal of industry interest. McDonnell Douglas quickly expressed interest.[8] General Dynamics created a Hypersonic Technology Center in Fort Worth (later purchased by Lockheed—owner of the famous Skunk Works). According to a General Dynamics spokesman, the center would perform work on a wide range of secret programs, explicitly including the Hypersonic Aerodynamic Weapon (HAW).[9]

The logic of long-range hypersonics is obvious; the question is how we can achieve a system that is intelligent, maneuverable—and cost-effective. An ICBM, which costs tens of millions, is too costly, too inaccurate, and too unintelligent to be used in surgical attacks on key communications and intelligence installations.

If chemical rockets, as we have known them, are out, two very different technologies remain:

- Single-impulse technology, in which a projectile is hurled a tremendous distance by some force outside of it.
- Cruise technology, in which a scramjet provides continual power to the projectile.

A baseball has no internal power source, but it can travel for several hundred feet when powered by a pitcher's arm. Theoretically, something powerful enough could hurl it hundreds or even thousands of miles. Under the aegis of the Strategic Defense Initiative, widely known as Star Wars, which involved searching for a way to knock down long-range missiles, several means for propelling solid objects at extremely high speeds were investigated under the general rubric of the Exo-atmospheric

Reentry Interceptor System (ERIS).[10] ERIS was intended to be a hypersonic, intelligent system that would intercept incoming missiles outside the atmosphere. Finding a low-cost boost system was ERIS's goal—and stumbling block.[11]

In the wake of the collapse of Star Wars, the exo-atmospheric project has gone in two directions. One direction—by far the dominant one—has been to develop hypersonic interceptors out of the basic design of the chemical-rocket Patriot missile—only smarter, faster, and more agile. The joint U.S.-Israeli project to design the Arrow missile is the result of this initiative. A smaller, less well funded, and much more interesting project is under way at Lawrence Livermore Labs. Under the leadership of Lowell Wood, a group of scientists who had been trained by Edward Teller, father of the H-bomb, had been working on the problem of hypersonics for missile defense.[12] John Hunter, another Livermore scientist, came up with the most intriguing, low-cost solution to hypervelocity—the Super High Altitude Research Project gun, the SHARP gun. What had made the ICBM so costly—and inefficient—was that it had to carry its energy with it. It had to boost itself and its fuel, along with its payload. As a result, most of the rocket was made up of fuel, with the payload constituting less than 10 percent of total weight. Barring a radical improvement in fuel efficiency, such as a new high-thrust, low-weight, low-cost fuel, it seemed unlikely that rocket engines could economically serve this mission. The Livermore solution was to separate propulsion from the projectile. The SHARP gun works very much like a traditional gun. An explosion takes place at the bottom of a long tube, and a projectile is fired out, albeit at phenomenally high speeds and distances. A mixture of air and methane is ignited, driving a piston down a tube and hypercompressing hydrogen. The compressed hydrogen breaks a dam containing it, and it rushes into a second tube, containing a projectile, which is forced out of the tube at about nine thousand miles per hour.[13]

Fully developed, the gun would cost about $7 billion. However, payloads such as food, water, and construction supplies needed for space stations or moon stations can be launched economically, along with low-earth-orbit satellites at a cost of about $225 a pound, compared to about $9,100 a pound in the Space Shuttle.[14] The weakness of the gun is that it operates from a single initial impulse, which means that initial acceleration will be at about 1,500 g's. These enormous accelerations and pressures mean that special materials, such as high-cobalt, high-nickel steel—the strongest steel alloys available—need to be used in the gun.[15] It also means that projectiles and payloads have to be designed that can survive these accelerations. Most especially, this means humans can never be shot out of the gun.

All discussion of SHARP gun technology has been focused on its ability to place objects into orbit—or, if it is large enough, to send them to the moon. Another application has not been discussed—firing projectiles at distant points on the earth. For example, let us assume that a SHARP gun in the hills behind Livermore Labs fires a suborbital shot at a target on China's coast—a distance of about eight thousand miles. If the initial speed of nine thousand miles an hour could be sustained, impact would be in about fifty-three minutes. The most obvious question is this: Why bother? We can already achieve these kinds of speeds and distances with ICBMs. However, there are two advantages: first, at $225 a pound, SHARP could deliver a thousand pounds of explosives for about $225,000, a quarter the price of a Desert Storm Tomahawk, which was subsonic and limited in range.

The second advantage is terminal maneuverability. An ICBM is designed to minimize initial drag in order to get the most efficient use out of its engine. As it reenters the atmosphere, it cannot maneuver toward a precise impact with its target because it does not have a lifting body—an aerodynamically configured shape, such as the Space Shuttle—that can control its descent without using engines. Without that sort of terminal control, an ICBM is fairly inaccurate. With the SHARP gun, however, a thousand pounds of aerodynamically shaped explosives—in a lifting body—could be launched toward a target in China. On reentering the atmosphere, sensors could pinpoint the target and onboard computers could control the descent with the precision of laser, infrared, or radar.

The Hypersonic Aerodynamic Weapon was intended to be just such a lifting body. Wrapped in a sabot, a shoe that would give it an appropriate shape to be fired from the SHARP gun, the projectile could enter space and then discard the sabot or use it as a heat shield on reentry.

But the HAW was not designed to leave the atmosphere—and for good reason. The high-arc trajectory of even a suborbital shot is easily detected from space by systems like the Defense Support Program (DSP). In this essentially ballistic model, the general area of impact can be defined, permitting theater defense systems like the Patriot and its successors to intercept. Keeping the weapon in the atmosphere—indeed, as close to the earth as possible—makes its detection more difficult and, more important, makes predicting the point of impact impossible. But endo-atmospheric flight, as it is called, is extremely difficult to achieve at very high speeds, since the energy required to punch through the atmosphere and resist gravity at hypersonic speeds is beyond the reach of conventional weapons. Indeed, it is beyond the reach of even the SHARP gun, since, after the initial explosive impulse, friction and gravity cause energy to bleed away quickly. SHARP technology, therefore, must be supplemented by another technology that will *sustain* speeds already attained within the atmosphere and, just as important, permit the aircraft to maneuver on its way to the target.

Thus, 1993 tests of the SHARP gun did not fire a dead weight, but a *scramjet*—an engine capable of sustaining flight at speeds greater than Mach 6. In one test, the eleven-pound, nineteen-inch-long Rockwell International scramjet reached Mach 8.2, flying an experimentally tremendous distance of seventy-five feet into a wall of sandbags.[16]

The scramjet idea was originally intended for the National Aerospace Plane (NASP), or X-30. NASP was first proposed during the early Reagan years, as the next generation of commercial aircraft. Intended to fly at over Mach 6, it would cut travel time between New York and London to under an hour. It was to be about the size of a 727, traveling at the edge of space, with passengers experiencing a normal flight—and never pulling more than about 2 g's. Apparent engineering difficulties led the Bush administration to cancel the project.

At about the same time that the NASP was canceled, however, rumors began to fly that a hypersonic plane was already operational, and that the NASP had been canceled because it had become irrelevant. In December 1992, *Jane's Defence Weekly* ran a story describing, and providing sketches of, a new hypersonic reconnaissance plane called the Aurora, which had been spotted over the North Sea,[17] the Republic of Georgia, and the Mojave Desert.[18] The engines were said to thrum, emitting a beating sound. A United 747 reported a near collision with a supersonic aircraft, and there were several reports of an aircraft that was emitting rings of smoke, like doughnuts on a rope.[19]

All of this may sound like the search for a UFO, but it overlays an important question. According to *Jane's,* the Aurora has a maximum speed of about Mach 8, which, it might be added, was the precise target speed recommended for an SR-71 successor by the Air Force Studies Board in 1989.[20] If, as *Jane's* says, the Aurora exists, then scramjet propulsion is already a reality and a host of hypersonic weapons are possible in the coming years.[21] A cruise missile might be launched by a SHARP gun, sustain speed with a scramjet, decelerate and reaccelerate, all without worrying about the effect on protoplasm.

It is important to understand how a scramjet would work, for just as the piston engine was the foundation of airpower until 1945, and the jet engine throughout the Cold War, the scramjet will power aircraft in the next era. A jet engine draws air into a chamber where it is compressed by turbines. Fuel is injected into the chamber, where it is ignited. The sudden heating of the compressed air causes it to expand explosively and surge out of the engine's rear, creating thrust. It follows that the greater the compression of air, the greater the potential thrust, and the greater the aircraft's speed. In jet aircraft, air compression, which is achieved through the turbines, is never so great as to produce heating. This, together with the fact that jet engines cannot cope with the extreme heat at speeds much greater than Mach 2, places a limit on jet speed.

At Mach 2—with a ramjet engine—something interesting begins to happen. As the plane moves forward, it encounters air trying to enter its jet intakes at twice the speed of sound. At these speeds, if properly managed, air compresses itself—without turbines—many more times than at slower speeds, using the shock at the inlet to provide energy. This allows the aircraft to travel at speeds up to six times the speed of sound. Above that, the shock of air flowing into the intakes begins to stress the engine, creating inefficiencies in compression.

The scramjet is merely the logical extension of the ramjet. By taking air in at oblique angles to minimize shock, the scramjet takes advantage of the supersonic flow of air to achieve hypercompression in the mixing chamber. By using hydrogen in the airflow to reduce temperatures, it allows even higher levels of compression. When fuel is injected into this hypercompressed mass—first at the mouth of the compression chamber to increase compression still more and finally in the chamber itself to induce expansion through the nozzles—the scramjet can reach speeds of up to Mach 25.[22]

All of this is very fine in theory. In practice, the problem is to find materials that can survive the kind of stresses—including friction heat and pressures—that will occur within the engines and in the aircraft at eight times the speed of sound. This includes air intakes and nozzles that could handle such airflows, combustors that could survive extreme temperatures without rupturing, and so on. The existence of Aurora is so important because it indicates that the materials problems have been solved.[23]

Two problems are particularly bothersome. One is called the shock boundary problem. Scramjets were predicated on minimizing the shock that occurs when the air enters the inlet. The problem has been that because engineers know very little about the behavior of airflows at hypersonic speeds, designing inlets and nozzles was virtually impossible.

Until recently, with the creation of large-scale hypersonic wind tunnels, there was no way to generate winds of that magnitude.

Another problem was the temperature—both the extreme heat of hypercompressed combustors and of surfaces passing through the air at hypersonic speeds, as well as the extreme cold of hydrogen—so cold that it stored at the consistency of sherbet.[24] At Mach 25, temperatures as high as 3,000 degrees Fahrenheit are possible. Commercially available titanium-aluminum alloys can only cope with temperatures up to 1,100 degrees Fahrenheit. Therefore, a wide range of alternative materials and processes must be developed. Some under consideration are boron-titanium metals, beryllium-based materials, as well as composites, ceramics, and carbon-carbon products. In addition, it is almost inevitable that the most exposed areas of the aircraft—the nose and the leading edges of the wing—will be cooled using super-low-temperature hydrogen, which will of course cause serious stresses in and of itself.[25] One should add that, in spite of ultrahigh temperatures, nose-cone sensors will have to be able to look out and find targets.[26] Clearly, scramjet technology is attainable, but per unit costs must be reduced. Reusable cruise missiles—missiles able to deliver submunitions to a target and then return to base—will surely be necessary.[27]

The U.S. Air Force recently announced that hypersonic technology ranked among the four key technologies to be developed over the next generation, along with unmanned combat aerial vehicles, laser and microwave weaponry, and new generations of weaponry.[28] The report stated that "Hypersonic air breathing flight is as natural as supersonic flight. Advanced cycle, dual-mode ramjet-scramjet engines and high temperature, high-altitude supercruise are the enabling technologies." The hypersonic concept, combined with new weaponry—including speed of light weapons—and the unmanned aircraft concept actually are different aspects of the same project, creating a global, tactical capability.

The Air Force currently has two projects under way in cooperation with NASA. The Air Force part of the project, known as Hypersonic Technology, or HyTech, concentrates on developing a hydrocarbon-powered scramjet-based hypersonic missile in the Mach 4 to 8 range that could be deployed after 2002. The NASA version of the project is even broader. Its Hyper-X project is designed to reach Mach 10 using hydrogen fuel rather than hydrocarbons. Both will draw on capabilities developed by the ill-fated National Space Plane project.[29]

Within the next generation, it will be possible to use an unmanned aircraft to deliver munitions on enemy targets at hypersonic speeds and at intercontinental distances—and that is important geopolitical news.

Imagine a cruise missile that combines the handling characteristics of hypersonic, short-range antitank and antiair projectiles and the range and endurance of the Space Plane. It would have an intercontinental range of 10,000 miles and a sustained speed of Mach 30 or about 30,000 miles an hour, or 8.5 miles per second. The missile would have onboard sensors as well as receivers for instructions from space-based craft, aircraft, or ground stations. It could be preprogrammed to attack a particular target. Alternatively, it could be sent to a general area, where it would loiter, use its own sensors to acquire the target, and then strike it. Finally, it could be fired in a general direction and, from satellites or other reconnaissance platforms, be guided to its final targets. In short, it would be an entirely flexible weapon.

Consider the effects of this system on warfare. A projectile fired from New York would reach Los Angeles in a little over six minutes; from Tokyo to Seoul it would take two minutes; from Berlin to Warsaw about forty seconds. At these speeds, it becomes possible to affect the tactical situation on a battlefield from strategic distances. A projectile launched from Fort Bragg, North Carolina, would impact on the Golan Heights—about six thousand miles away—in thirteen minutes. From the time of launch to impact, an armored company would be able to travel at most five miles. With proper guidance and munitions, such a projectile could destroy the armored company without American forces ever touching foot in the Middle East.

The hypersonic missile must be brilliant. Upon reaching the area of its engagement, it, or its submunitions, must be able to search the area, locate and identify the target, evade defenses, and strike with a hit/kill probability in excess of 50 percent. As we saw with Desert Storm, a wide range of sensors already exist. As we shall see, they will become a lot smarter. In the end, they will threaten the continued existence of the weapons systems of the great European epoch: tanks and ships. They will determine the course of land battles and determine control of the sea.

Long-range, hypersonic cruise missiles make the entire world vulnerable to attack from any point and completely reshape the geography of surface warfare. The compact battlefield, tied together by the weapons platforms that brought guns into proximity with each other, will be abolished. But expanding the geography of war does not mean expanding the intensity of war. Quite the contrary. War becomes more widespread but much less devastating. In a sense, the traditional distinction between military and civilian, obliterated in the inaccuracy of bombing, will reassert itself.

During the European epoch, the point of war was destroying

enemy weapons platforms. The hypersonic cruise missile will ultimately not require a highly visible launch platform. And it is so small and requires so little in terms of basing facilities that it can be located in any ship, plane, or backyard shed on the globe. This means that it can't be taken out in the usual way.

There are two ways to kill a cruise missile. One is to go in and occupy the ground it is stored on—or destroy the factories that produce it. The other solution is to detect the missile launch and destroy it in flight. Detecting a hypersonic launch and movement through the atmosphere is not difficult—friction will inevitably give the hypersonic missile a bright infrared signal. But intercepting a missile traveling at over Mach 25 and capable of maneuvering en route is not easy. Speed and unpredictability can help it defeat detection for a while. Nevertheless, in the long run, no matter how fast or how hard to find it is, interceptor missiles will be as fast and resourceful. One must remember that senility starts at birth.

Defending against the hypersonic cruise missile will require seeing it and having interceptors placed strategically. Ground observation, obviously, is inadequate. Given the global scope of cruise missiles, global observation is the only rational choice. Ground-based interceptor missiles will have limited zones of engagement—and they will be fighting gravity. Space-based systems, either in low earth orbit or geostationary orbit, will have much wider zones of engagement—and will have gravity as an aid.

One must move into space to stop hypersonic cruise missiles. One must also move into space to use them. The weakness of global warfare systems is intelligence—knowing what the enemy is doing as he is doing it. This leads us to the most important point. The most efficient way to disable a cruise missile is by blinding it before it is launched. Blinding a cruise missile requires destroying or interfering with the global sensing system that identifies its target and guides it there.

In a hypersonic age, the best defense is an attack on the enemy's space sensors, while protecting one's own sensors and interceptors from counterattack. Thus, the inevitable move from the manned aircraft to the hypersonic cruise missile carries with it an unintended consequence—war in space. It will also give war a new end—space control.

Invisible Order: The Topography of Space

On the surface, space appears to be a domain of freedom where weightlessness gives all objects the power to roam at will. Nothing could be further from the truth. Weightlessness gives us the illusion of freedom. In reality, space is a realm in which gravity and the laws of motion rule with an iron hand. Every object in space is trapped into a predetermined motion. It is possible to change this motion, but only with great care and tremendous energy. In a sense, aircraft are much freer than spacecraft. The physics of air flight provide airplanes with much greater energy relative to gravity than a spacecraft has. An airplane pilot can compensate for errors with amply available energy—both from his engines, partly powered by air, and from the lift provided by his wings. There is no air in space. Engine-produced energy is therefore scarce, and there is no aerodynamic lift to forgive error. Thus, every change of motion must be perfectly calculated. And every change of motion must be utterly necessary.

But out of this harshness arises a very real orderliness that is not readily apparent—certainly not to the naked eye. Space is a domain of strategic positions, choke points, wastelands. The space around the earth has as rich and powerful a topography as does the earth itself. This topography gives war in space its structure, much as geography gives war on the earth's surface a structure.

Earth is part of a two-planet system, with the system revolving around the sun. The earth is by far the larger planet, with eighty times more mass than its moon, and its gravity is six times that of the moon's. The moon rotates exactly once each time it orbits the earth, so that the same side always faces the earth, while the earth rotates once every twenty-four hours. This means that one side of the moon is always invisible from the earth, while one-half of the earth faces the moon at any moment, and all of the earth faces the moon each day. This places the earth at a distinct military disadvantage. Military forces on the moon could monitor activities on the earth continually using more advanced optics than currently exist (during dark phases, infrared and other means would have to be used), while weapons, logistics, and command facilities could be hidden out of sight of earth forces, on the moon's far side.

Moreover, much less energy is required to fire a projectile at the earth from the moon than at the moon from the earth. On the other side of the ledger, the moon lacks an atmosphere; orbital altitude is not limited by atmosphere, thus low-orbital approaches would always threaten lunar facilities with attacks with little warning. In spite of this, control of the moon would place all of the earth under observation and at risk, limiting an earth power's ability to mount a surprise attack on the moon.

No clear line marks the beginning of space. Atmosphere and gravity decline gradually, but both persist well out into what we would consider space. A useful rule of thumb is that space begins at that point where the orbital velocity necessary to keep an object aloft does not encounter so much atmospheric drag that the object is immediately brought down by friction. For practical purposes, this means that a satellite must maintain an orbit at least sixty miles above the surface of the earth. However, to survive atmospheric drag for at least a full day, an altitude of about one hundred twenty miles is required. For extended life—orbits lasting a year—orbits *at least* three hundred miles are necessary. Now, there are noncircular, elliptical, orbits in which the low point—perigee—can dip down quite low, while the high point—apogee—is many times the perigee's altitude. By minimizing the amount of time at low orbit, the atmospheric drag is minimized and orbital life extended. Orbits can be constructed that both extend life and permit low flight over certain parts of the globe—which is what is done with photo reconnaissance satellites.

Operational space begins at about sixty miles. Conventional fighter aircraft have a maximum ceiling of about ten miles; fifteen miles is the practical limit for air-breathing systems. This leaves a substantial gap between operational airspace and outer space. This forty-five-mile gap—referred to as the mesosphere—is a strange region. It is a transit area, where non-air-breathing rockets power through to reach orbital space and beyond—or where unmanned, hypervelocity scramjets will be able to operate in the future. It is also an area of extreme vulnerability.

Persistent gravity makes it extremely costly and risky to maneuver in this region—both on the ascent and the descent. Air is too thin to provide lift, but thick enough to increase friction, so that any fast-moving craft is in danger of burning up—and even if it survives, its heat signature is so strong that it can be seen by infrared sensors thousands of miles away. Mesospace will inevitably be a no-man's-land, where defenders on the ground take their best, last shot at projectiles heading down at them and through which attackers from space will peer to find their targets and fire their weapons. It will be a deadly battleground in coming generations, where no one can win a permanent victory.

Orbital space is a vast area, ranging from about 60 miles to 22,300 miles, or even out to about 60,000 miles—an arbitrary point where gravity is reduced to 0.05 percent of what it is on the surface.[30] Its vastness can barely be imagined, encompassing about 900 trillion cubic miles. To the eye, it is an empty cavern, governed only by gravity. But even this is incorrect. Other forces are present. Atmosphere persists as far out as 1,200 miles as a drag on spacecraft. More important, at the upper regions are the solar winds and flares, cosmic radiation, and the influence of solar and lunar gravitation, all of which have important, and not fully predictable, influences on the movement of spacecraft.

At the lower altitudes, other forces affect spacecraft. The earth has a powerful magnetic field that is particularly energetic between forty-five degrees north and forty-five degrees south latitude—from about the northern United States to the southernmost part of Australia. This field traps energized particles, such as electrons and protons, creating a strong electrical field that necessarily affects electrically operated equipment—which means all spacecraft. In addition, there are two Van Allen radiation belts, whose precise location and size varies with the season and solar activity. The lower belt starts at between 250 and 750 miles and goes out to about 6,200 miles. After a gap, the second belt resumes at about 37,000 miles, stretching to 52,000 miles.[31]

Spacecraft traveling extensively in these belts must be heavily shielded against radiation, which raises their weight dramatically and makes it more difficult and costly to launch them into orbit. The alternative is to minimize the amount of time spacecraft spend within the belts, by keeping them in low earth orbit or out beyond nine thousand miles or in elliptical orbits that bypass the dangerous altitudes between forty-five degrees north and forty-five degrees south. Under any circumstances, the amount of time spacecraft spend in maximum radiation zones has to be kept to a bare minimum.

There are clearly a number of critical points in space topography:

- The equatorial belt at an altitude of 22,300 miles. Critical areas are from 90° to 120° west, where U.S. communication satellites abound; 15° to 25° west, where satellites linking transatlantic communications are; 60° to 105° east, where Eurasian communication systems operate; and the area roughly around 180° where transpacific links operate.
- The area from 60 to 180 miles above the earth, in the Northern Hemisphere, particularly between 30° and 45° above the United States, and between 15° and 60° north and above Eurasia (15° west and 35° east).
- Sixty degrees south and about 75° west—the area where the Russian Molniya communications satellites hit a perigee of about 12,000 miles above the earth in the Southern Hemisphere.
- Each of the polar areas, between 90° and about 98°, where polar-orbiting and sun-synchronous satellites must pass.
- Atmospheric exit points easterly downrange from established launch facilities.

■ There are five points in space, invisible to the naked eye, where the gravity of the earth and the moon balance each other in such a way that an object placed at that point will remain there, without falling toward either of the planets. Libration point 1 (L1) is about 200,000 miles from the earth, while L3 is 240,000 miles away. L2 is behind the moon. These three are considered weak libration points, as objects there might be able to drift out and would require constant station-keeping. L4 and L5, however, are strong libration points, which would be able to hold objects readily. They are in the moon's orbit, but bracket the moon sixty degrees on either side. These are important points to bear in mind, for when luna becomes an important military base, L4 and L5 will become critical offensive and defensive positions. And holding L2 would permit a power to dominate the back of the moon—a precious asset.

As we can see, trackless space has very clear tracks. Our ability to utilize space around the earth is characterized by three variables—residual atmosphere, concentrated radiation, and gravity. This third variable is the most important in determining where things can go and what they can do. Until spacecraft have enough energy to freely counter gravity's force, they will be satellites, orbiting in a predetermined path around the earth. **This means that the efficient use of reconnaissance satellites requires careful choice of orbits.**

All objects propelled into space have a trajectory determined by speed. Some, going fast enough, head straight out from earth. Others, going slowly, fall back to earth. Still others, traveling at certain precise speeds, commence free fall. But instead of falling *toward* earth, they fall *around* the earth. They go into orbit. This orbit is determined by the speed and direction imparted them by the rocket that boosted them into space.

There are severe constraints on orbits, constraints defined by the astronomer Kepler in 1609. His first assertion was that the plane of every orbit must pass through the center of the earth. Every orbit has an inclination relative to the equator. An orbit directly over the equator would be said to have an inclination of zero degrees. An orbit that would pass directly over the north and south poles—a polar orbit—would have an inclination of ninety degrees. Inclination determines how much of the earth's surface will be traversed by a satellite. A satellite with an inclination of forty-five degrees never goes north of 45°N or south of 45°S. But the smaller the angle of inclination, the smaller the area covered and the more frequent the coverage. The greater the angle of inclination, the larger the area covered, and the lower the frequency of coverage.

Inclination is one variable determining coverage. Altitude is another. A satellite's field of vision is determined by how high it is.[32] At 22,300 miles above the equator—the point at which the speed of an orbiting satellite matches the earth's rotation—a satellite can see about 42.3 percent of the earth's surface. At 100 miles, just above the atmosphere, a satellite can see 6.7 percent of the earth at any one time. A tenfold increase in orbit to 1,000 miles would only triple the field of vision. Increasing altitude 126 times increases the visible area about 6 times. The higher you are the more you see—but you have to go *much* higher to see much more. The higher you go, the harder and more expensive it is to get there—and the harder it is to make out the details on the earth below. Thus, the decision on how high to put a satellite is complex.

A circular orbit is possible, but so is an elliptical orbit—one in which the satellite is at times closer to the earth and at times farther. Kepler's second law states that the rate of rotation around the center of the planet by a satellite will—measured against the planet's center—always be fixed. Imagine a satellite in an elliptical orbit, sweeping as low as one hundred miles then hurtling out to five thousand miles. Imagine a line drawn from the satellite to the center of the earth. At one hundred miles, the satellite would have to travel much more quickly to keep the line rotating at a steady rate than it would have to travel at one thousand miles, while at five thousand miles, it would have to be traveling at the slowest rate of all.[33] This is important. By setting an elliptical orbit properly, you can gain a great deal of control over how much time the satellite spends over any given place, and how much time it spends at any given altitude.

For example, reconnaissance satellites could be designed to have a high of five hundred miles and a low point of seventy miles, where it could quickly dip into thicker air to take close-ups. The perigee would be maintained at the same latitude, so that, as the earth rotates, countries of particular interest at similar latitudes could be monitored. If you were monitoring the central United States, for example, where many missiles are based, you could program a reconnaissance satellite to orbit at an inclination of 45 degrees and with a perigee of about 40 degrees north, to take photographs from less than a hundred miles over the target area, which would extend between 35 and 45 degrees north and between longitudes 90 and 110 west.

Note that each altitude requires a specific velocity to keep a satellite in orbit. If the velocity of a satellite is too low, the satellite will be pulled back to earth. If it is too high, the satellite will break out of its orbit on a straight line and head out into space. At the lowest possible altitude, just outside the earth's atmosphere, orbital speed must be about 4.8 miles per second, which means that a satellite will orbit the earth once in about 89 minutes. As altitude increases, earth's gravitational force decreases and so does the speed needed to keep a satellite in orbit. By the time you reach 22,300 miles, orbital velocity decreases to match the velocity of the earth's rotation—about 0.3 miles per second. From one orbit every 89 minutes, you've gone to one orbit every 1,436 minutes—23 hours, 55 minutes, 48 seconds, almost, but not quite, a day.

Every orbit, including a geostationary orbit, has a degree of uncertainty, deriving from the imperfect curvature of the earth, the difficulty of precisely maintaining the satellite's position, solar wind, and so on.[34] All orbits shift, and geosynchronous satellites are no exception. It is necessary to anticipate this shift in planning ground stations and missions, or to eliminate the shift through station-keeping—using thrusters to periodically correct for satellite drifts.[35]

Geostationary orbit is an ideal spot for reconnaissance. The difficulty is that 22,300 miles is so far away from the earth's surface that no available optics could provide the kind of information that is available at low earth orbit. From a station on the equator over the Indian Ocean, just south of Sri Lanka, for example, complete coverage of the extreme north is difficult because of the earth's curvature. This is compounded by the fact that, because of the increased angle, energy—broadcasting, microwaves, and so on—must pass through far more atmosphere on its way to the poles than to the equator, increasing distortion substantially. Seeing or transmitting to areas beyond forty-five degrees latitude becomes first difficult and then impossible. This has been a problem for Russia and Canada, which have had to devise more creative orbits for communications satellites.[36] Because of these drawbacks, geostationary orbit has remained the domain of communications, signals intelligence, and very low resolution reconnaissance satellites—and will remain so until much better electro-optical and radar-sensing equipment is developed.[37]

Another problem with putting satellites into high orbit was the need for powerful boosters. It has generally been too difficult to launch a geostationary satellite on a single boost from earth. The normal procedure has been to place high-orbit satellites into a low parking orbit and then have a booster rocket—small because it is accelerating a mass in low gravity—give the satellite the final push into geostationary orbit. This multistage process was inevitably expensive.

Apart from this, placing satellites into orbit over the equator posed a special difficulty for both the Americans and the Soviets. It is impossible to put a satellite into equatorial orbit directly from the Soviet Union or the United States. The only way to do this is to launch a satellite into an inclined orbit and then, as it crosses the equator, fire rockets to change its orbit into an equatorial one. The final step is to fire an additional rocket to boost it into geostationary orbit. Assume that a satellite is in polar orbit. It would have to execute a right turn, a full ninety-degree shift, at the equator to go into geosynchronous orbit. To overcome the satellite's inertia, as it hurtled along at four to five miles a second, would require tremendous energy, and an impossibly large load of fuel on board.

The smaller the inclination of the orbit, the smaller the correction necessary. This is one reason why Canaveral, and not a site in Maine, was chosen as home to the American space program. In addition, because Florida is much closer to the equator than is the southernmost point in the Soviet Union, the Soviets had to boost a lot more fuel to attain an equatorial orbit than the Americans. It followed that Soviet rockets were always massive compared to American rockets, something that concerned American defense planners during the Cold War.[38]

Since Canaveral is located at 28.5 degrees north, and since, according to Kepler, the orbit must have its plane pass through the center of the earth, it is impossible to launch a satellite from that location at an inclination less than 28.5 degrees. It is also impossible to launch a satellite at an inclination of more than 35 degrees relative to Canaveral. Rockets have a nasty tendency to explode, sometimes at great altitudes, strewing debris for hundreds of miles along the flight path. A launch due north from Canaveral would pass over Savannah, Georgia, and so on. A shot due south would come close to passing directly over Havana, Cuba. To clear Cape Hatteras, no launch can be inclined more than 35 degrees from Canaveral. This means that nothing can be fired into polar orbit from Canaveral without an adjustment in orbit—a costly operation in terms of fuel. For this reason, the United States opened a second launch facility, run by the Air Force—at Vandenberg Air Force Base—in California. Vandenberg is not far from Santa Barbara and has a clear shot due south.

Geography gives order and logic to the sea. Physical characteristics—currents, straits, water depth—coupled with geopolitical considerations—major trade routes, naval expenditures, location of naval bases and developed harbors—combine to distinguish between strategic sea-lanes and points and mere ocean.

This same is true in space. A space-control strategy, therefore, does not require that control be exercised over all 900 trillion cubic miles of space, any more than sea control requires domination of every inch of ocean surface. For example, a power seeking to shut down American space launches could concentrate on two points—Canaveral and Vandenberg. By placing interceptor forces in an area stretching from due east of Canaveral to about thirty degrees northwest, from one hundred to three hundred miles downrange, at an altitude of about ninety miles, it would be possible to intercept launches as they emerge from the atmosphere.

American planners would be forced to switch to dangerous overland launch trajectories or to trajectories that make equatorial insertion much more costly, or to shift launches to Vandenberg, whose facilities are much more limited and which can also be targeted for interception. For example, there is a tremendous advantage in launching due east, as this takes advantage of the earth's rotation. While a due south launch would still be possible from Vandenberg, providing a polar orbit, an easterly launch could be impossible, as it would travel over land. Thus, controlling the launch path to low earth orbit from launch facilities determines whether a nation will be able to project power into space.

Every nation's space mission has, built into it, certain necessary constraints—and therefore some clear predictability. The Americans need to monitor Eurasia with satellites launched from Canaveral and Vandenberg. This constrains American behavior and makes it predictable. Israel's interest in a very small number of countries constrains their launch patterns. Japan's program, as it develops, will be conditioned by its political interests, its northerly position, and the availability of launch sites. Europe's northern location and lack of clear launch sites has resulted in launches from French Guiana—commercially intelligent but militarily dangerous, as Guiana is a long way to transport launchers and satellites in time of war.

Moreover, the technical needs of missions constrain topography. Most reconnaissance missions require low earth orbit with low perigees at the important latitudes. Communications satellites require geosynchronous orbit to provide continuous coverage at low cost, as is also the case with meteorological satellites, and infrared missile warning systems. Satellites are not likely to be found in circular orbit in the Van Allen belts, nor in wide elliptical orbit without a very good reason. As on earth, military topography constrains and renders predictable the movements of friend and foe. And this predictability means that it is possible to make plans to thwart the enemy.

From Star Wars to Our Wars: First Thoughts on Space War

It is one thing to speak about the geography of space control, yet another to be able to control space. Denying access to space to enemy forces requires weaponry. That weaponry does not yet exist. Indeed, at this moment, *relatively* little thought is being given, at least publicly, to weaponry for space control. There are several reasons for this. First, no nation has both an immediate political interest in challenging American reconnaissance platforms in space and the ability to construct antisatellite systems. Nations that can build such systems, such as Russia, Japan, the Europeans, are not politically motivated. Nations that are politically motivated, such as Iraq, Iran, or North Korea, are incapable of mounting a challenge.

During a time of declining defense budgets, when U.S. space platforms also appear to be secure, it is not likely that scarce resources will be diverted for what appears to be a distant threat. In addition, the first great American drive for the command of space, the Strategic Defense Initiative (SDI) of the Reagan years, has been shattered by political forces, leaving behind a pale ghost in the Ballistic Missile Defense Organization, which concentrates on defending military theaters from incoming missiles. Indeed, from the beginning the great defect of SDI was that it sold itself as a defense against ICBM attack, rather than as an attempt to seize control of space before anyone else did. Throughout the Cold War, strategic relationships were constructed around the idea that, for the first time in history, a weapon existed—the intercontinental ballistic missile—against which there was no defense. They were launched on the other side of the earth, traveled at speeds approaching Mach 30, and could arrive at their target before anyone knew they had been launched. Since it was unlikely that these missiles could be intercepted before they struck, it appeared that the only way to prevent a nuclear war was to be able to threaten a devastating counterattack. This doctrine became known as MAD: mutual assured destruction.[39] It was assumed that only MAD allowed each side to deter the other from initiating a nuclear war.

The Strategic Defense Initiative, or Star Wars, challenged MAD by arguing that ICBMs were no longer invulnerable—that technology had evolved to the point that they could be destroyed before hitting their target. Ronald Reagan, in a speech on March 23, 1983, declared that, with Star Wars, "we embark on a program to counter the awesome Soviet missile threat with measures that are defensive. Let us turn to the very strengths in technology that spawned our great industrial base and that have given us the quality of life we enjoy today."[40] Defense Support Program satellites could already detect the heat of a launch, and with more advanced sensors, the path of missiles could be tracked even after the initial boost phase was ended.

If space-based sensors were to be the foundation of Star Wars, its edifice was to be the weapons that could knock out ICBMs. The most attractive options were speed-of-light weapons. Knocking out an ICBM in flight was frequently compared to hitting a bullet with a bullet. Both were small and fast, and hitting one of them required eyesight and reflexes beyond imagination. But with a weapon that traveled at the speed of light, the ICBM—moving at less than 1/25,000th the speed of light—was virtually standing still.

Reagan and his supporters saw this as a way out of the nuclear nightmare. The United States and the Soviet Union would no longer have to rely on their ability to obliterate the other side to prevent nuclear war. Rather, nuclear war would become impossible because speed-of-light weapons would make long-range missiles helpless. Ironically, some of the most severe critics of MAD became the harshest critics of Star Wars. At times, their criticism did not make a great deal of sense.[41]

In a way, proponents of Star Wars oversold the project, leaving the impression that what they were proposing was a leak-proof shield against long-range missiles. Later, as critics were able to demonstrate that Star Wars was porous, defenders tried to make the case that they had never promised absolute protection.[42] Of course they were right, as no sane person ever argued that any system, let alone a weapon system, would be foolproof.

Proponents of Star Wars were actually arguing that the number of missiles that would leak through the screen would be so small that they could not inflict decisive damage on the United States, unless the other side went to the astronomical expense of increasing the number of ICBMs exponentially. The machine gun didn't actually make cavalry impossible. It merely increased the number of horses and riders required for a single pair to get through. But here, the defenders of Star Wars were put on the spot. If a horse got through, a few men might die. If an ICBM got through, an entire city would die. Enemies of the program managed to make the argument that it had to be fail-safe, because a single mistake imposed an unacceptable cost. In the end, the complexity of the case for Star Wars gave its opponents a clear advantage—as well as a good argument rarely made explicitly. Star Wars was an untested theory. MAD had worked for decades.

Star Wars began to dribble away in fruitless political argument. But the real problem was that the speed-of-light weapons that made it so intriguing were not anywhere near ready for deployment. The plan originally envisioned an interlocking space-based system of sensors and weapons.

Sensors might be able to track ICBMs and computers might be able to draw a bead on them, but what was going to intercept and destroy them? Star Wars considered a cluster of speed-of-light weapons:

- Long-wavelength lasers: These would be lasers using energy in the infrared band. Energy is chemically generated, and the beam can be redirected by mirrors positioned at strategic points. The infrared lasers heat the surface of an ICBM, causing internal damage.
- Short-wavelength lasers: These generate free-electron beams, generated by various power sources, including possibly nuclear reactors. Focused by an array of magnets, powerful, coherent electrical energy is projected. Wavelengths can be tuned to penetrate the atmosphere or for other purposes. On the whole, these lasers are more powerful than long-wavelength lasers.
- X-ray lasers: These utilize small nuclear explosions to generate X rays. Up to fifty surrounding laser rods would focus the energy from the bomb at targets. A fraction of a second later, the rods themselves would be destroyed by the nuclear explosion.
- Microwave weapons: These use high-powered microwave radiation to disrupt the guidance system of missiles. They would be powered either chemically or by space-based nuclear reactors.
- Neutral-particle beams: A beam of neutrons is charged and directed at ICBMs. They would penetrate the target and destroy critical systems internally.[43] They would require large chemical chambers or a nuclear reactor.

Any one of these weapons, based in space, controlled by computers hooked to sensors, could take out a large number of enemy ICBMs—*assuming* that a power source was available. The power source would have to generate a great deal of energy very quickly, then repeat the surge of energy as the weapon retargeted a few seconds later. During the debate over Star Wars, a great argument raged over whether software could be designed to manage the system. But the real problem was the generation of energy in space.[44] The problem of high-energy sources has still not been solved, and therefore, any dream of a speed-of-light weapon has faded.

Until recently it appeared that speed of light weaponry would be delayed for decades because of the unsolved energy problem. A joint U.S.-Israeli project called Nautilus has created a laser that destroyed a 120-mm rocket at the White Sands High Energy Laser Test Facility on February 9, 1996. Nautilus, run by TRW and Israel's RAFAEL, is the demonstration project for a technology called the Mid-Infrared Advanced Chemical Laser (MIRACL). MIRACL is a deuterium-fluoride laser that was powerful enough to melt the surface of the missile. Small enough to be transported in a C-130 aircraft, MIRACL is now able to engage a target out to about six miles.[45] Obviously, the range of the laser is insufficient to affect space control. However, it does indicate that the problem of power is being managed and that the development of speed-of-light weapons is possible during the next generation.

As an alternative to speed-of-light weapons as well as to traditional chemically powered rockets, Star Wars proposed the kinetic-energy weapon (KEW)—solid projectiles that would travel extremely fast, although not at the speed of light—to intercept slower intercontinental ballistic missiles. These space-based interceptors (SBI), parked in low earth orbit but high enough outside the atmosphere that they could survive for at least seven years, would be able to intercept enemy long-range missiles outside the atmosphere. They would be supplemented by ground-based systems designed to intercept missiles inside and outside the atmosphere.[46]

When the Reagan administration was succeeded by the Bush administration, interest in space-based interceptors declined. The technology appeared too exotic and there were treaty considerations as well. Perhaps most important, the military men who had done so splendidly in the Persian Gulf were generally uneasy with Star Wars. They were concerned that increasingly scarce funds were being diverted from conventional needs to an exotic technology that, with the collapse of the Soviet Union, seemed increasingly unnecessary.

The Iraqi SCUDs were the death knell to Star Wars. The inability of American forces to suppress SCUDs at the launch site meant that U.S. forces had to defend themselves against incoming missiles. The problem was that the Patriot missile intercepted SCUDs inside the atmosphere, so that as the SCUD disintegrated, it still rained debris around the countryside—doing as much if not more damage than if it had not been intercepted.[47]

The fault was not the Patriot's. It was designed to intercept planes, destroying them before they could release their weapons. A SCUD's only mission was to hit the ground, destroying itself in the process. It didn't matter if it did this in one piece or in forty pieces. A Patriot could fragment missiles or change their trajectory—sometimes toward more populous areas. But it could not destroy them outside the atmosphere, nor could it vaporize them.

Thus, from one side, the pressure grew to develop a system that could protect a theater several hundred miles on a side from tactical and intermediate-range ballistic missiles. On the other side, it was argued that Star Wars had become archaic. Its advocates capitulated to the inevitable and attempted to save what they could by focusing on a ground-based system designed to intercept relatively short-range missiles. In the end, all that remained of the bold dreams of the Strategic Defense Initiative was the Ballistic Missile Defense Organization, high endo-atmospheric defense interceptors (HEDI), and some dreams about Brilliant Pebbles—small kinetic-energy weapons to attack missiles in space.

Star Wars was written off as a failure, but if it was a failure, it was a brilliant one. It will be remembered by history as the first time that men thought seriously about war in space. That many of their concepts were premature, and the political basis for their project dissolved both domestically and internationally, does not detract from the fact that it was the seedbed for both a wide range of technologies and for critical operational principles. Perhaps most important, the rarely discussed stepchild of the program, antisatellite (ASAT) systems, becomes the inevitable next step of warfare. In this sense, military planners and thinkers at all levels and in all services will be drawing on Star Wars throughout the coming generation.

Tracking and Killing: Space as a War Zone

Let us return to the main thread of this chapter. Long-range, hypersonic missiles require global surveillance and targeting to be effective. If that surveillance is unavailable, the employment of long-range systems is impossible. Without real-time information about the battlefield, it is impossible to influence tactical situations using intercontinental systems. Thus, if a nation like Iraq wanted to defend itself against the type of air and ground assault mounted by the United States, its logical counter would be against American reconnaissance satellites.

Control of space, and particularly control of strategic sectors of space, is becoming the foundation of military operations of the post-European epoch. In the first phase, space control required merely a presence. A satellite launched into orbit asserted control simply by being there—nothing could threaten it. The next phase will go beyond assertion of control, to the denial of control. Reconnaissance satellites will be threatened by antisatellite systems, ranging from space destroyers of various sorts to systems for disrupting space-earth communications. Space control will then depend on the ability to defend one's own assets and destroy the enemy's.

The systems envisioned by the Strategic Defense Initiative for interdicting missiles can also be used against satellites. Indeed, satellites are generally more vulnerable to interception because of their predictability. The precise trajectory of an ICBM is unknown until launch. We know a great deal about where satellites are going to be by monitoring their initial launch and insertion into orbit. In addition, satellites are noisy things, periodically emitting bursts of electromagnetic radiation as they transmit signals to earth and other satellites. Thus, signals intelligence can tell us a great deal about where satellites are—as well as about what they do.

The problem with both signals intelligence and extrapolating from initial launch data is that it isn't precise enough for interception and kill by antisatellite missiles. Assume that we could know, from a distance of several hundred miles, where a satellite is within one cubic mile of its exact position. Hitting a Soviet Cosmos series satellite that is about eighteen feet long and six feet in diameter in a cubic mile of space is like finding a needle in a haystack.[48] And the assumption that orbit projection and signals triangulation could provide accuracy within a mile is extremely optimistic. Even if we were to monitor a launch, note the orbit, and load the data in a powerful computer, we would not know the satellite's precise location. All sorts of forces, such as the irregular curvature of the earth, variations in atmospheric height and density, solar activity, and so on, cause orbits to shift enough that interception would be impossible. More important, satellites can change their orbits on their own, using onboard thrusters. Thus, the first element of a successful space-warfare system is a comprehensive, real-time tracking system. Controlling space begins with seeing space.

Currently, satellite tracking takes place from the earth's surface rather than from space and relies primarily on optical systems, including some infrared tracking, rather than other spectra, which lack the range necessary to track and identify space objects. This means that space objects have to come to the tracking station—and that what happens outside the field of vision of the tracking station remains invisible. A space-craft that initiated a maneuver on the other side of the earth would have completely shifted orbit by the time it reached the tracking station's area. Moreover, the optical scanning systems used have excellent imaging capabilities—they can see very small objects—but are much weaker in scanning huge volumes of space. It might take days or even weeks to find a spacecraft that had unexpectedly shifted orbit.

In spite of the weaknesses, ground optical tracking has some real advantages. First, it uses technologies that are familiar. We know a lot about optical systems. Second, it is earth-based—men and equipment are in the same place. Thus, the costs of launching and maintaining space platforms are not incurred. The United States, Britain, and Russia have all built their space tracking systems around optical scanning, with the assistance of radar systems primarily to track ICBMs in low earth orbit.

The key U.S. space-tracking facility is located on Mount Haleakala volcano on the island of Maui in Hawaii. There are several reasons for this location. Maui is blessed with dust-free, low-humidity air. Its location—on an inactive volcano—gives it a 250-mile horizontal visibility on a clear day. Perhaps most important, Maui is located due east of Soviet launch facilities in Central Asia, which means that it can track Soviet satellites as they are being launched into orbit. Mount Haleakala is home to three separate tracking operations:

- Air Force Maui Optical Station (AMOS): uses both visible-wave-length and infrared sensors to identify objects as small as three inches. AMOS is one of the finest telescopes in the world.
- Maui Optical Tracking and Identification Facility (MOTIF): a lower-resolution telescope used to scan larger areas of space to track and catalog man-made objects.
- Ground-Based/Electro-optical Deep-Space Surveillance System (GEODSS): has the lowest-resolution optical system but is able to see deep into space and make observations in relatively low light. GEODSS can track objects in geostationary orbit.[49]

These facilities, used in conjunction with various radar systems that constitute the Space Surveillance and Tracking System (SSTS), provide us with a pretty clear picture of what is in space at any given time. Certainly, SSTS is more than adequate for tracking spacecraft during a period when antisatellite missiles pose little threat.

The key weakness of SSTS is that its optical sensors are at one location and not tied together efficiently enough to manage a war in space. Data takes too long to get to Space Command, too long to be integrated, and much too long to be analyzed. An antisatellite missile attack on a KH-11 might well be completed before CINCSPACE had all the necessary data in hand.

Thus, the Defense Department is pressing for improvements in the management of data by ground-based space scanners. On March 30, 1994, *Aerospace Daily* reported that PRC, Incorporated, had been awarded a $97.5 million contract to support

> the Integrated Tactical Weapons Attack Assessment Sensors System (ITWAASS), which includes ground-based electro-optical deep space surveillance, ballistic missile early warning system, Have Stare radar, Cobra Dane radar, Pirinclik [Turkey] radar site, PARCS radar, Eglin radar, Pave Paws radar, the Defense Support Program, and the associated system program agencies. Contract is expected to be completed March, 2000.[50]

Very little can be found in the public literature on ITWAASS, and no mention of it is contained in *Commerce Business Daily*, which contains all federal contract solicitations and awards. Obviously, it represents an attempt by the Air Force to link a wide range of surveillance systems to monitor tactical weapons—nonnuclear, long-range projectiles. What is interesting about this project is that, if successful, it would tie together systems that would provide near-global, twenty-four-hour coverage of everything from the low mesosphere to geostationary orbit—but not with perfect, multispectral coverage. Things could still slip through. But it is obvious that the Air Force and the Space Command are concerned with developing global, integrated tactical surveillance systems.

Even more intriguing is work going on at the Phillips Laboratory, an Air Force research center at Kirtland Air Force Base in New Mexico. Since 1992 the Starfire Optical Range, located at Kirtland, has been soliciting bids in an intensive effort to develop an advanced, ground-based deep-space surveillance system. On June 8, 1993, a call was issued for bids for "nonconventional imaging techniques and device development, risk reduction, and prototyping for high resolution imaging of space objects from LEO to GEO (low earth orbit to geosynchronous earth orbit)."[51] The key concept here is "nonconventional imaging." There is a grave deficiency in all ground-based optical tracking—the atmosphere. Light traveling through the atmosphere is distorted by dust, humidity, and atmospheric turbulence. It is difficult enough to see stars on a clear night. Clearly seeing objects a few meters long hundreds and thousands of miles above the atmosphere is nearly impossible. Since the radar bands have great difficulty projecting thousands of miles out, any earth-based system has to figure out a way to compensate for atmosphere. A wide range of technologies is under way, including computer programs designed to enhance distorted images, the creation of nonlinear optical systems for passing lasers through the atmosphere, and enhanced optical systems for imaging. By managing the relationship between light and the atmosphere, the project might well make it possible for lasers to retain power as they move through the atmosphere— and into space to kill satellites. As a consequence, much lower powered laser systems could be used against satellites, solving the power generation and storage problems discussed earlier. As a sidelight to all this, it is amusing to note that the earliest publicly announced contract in this area, on April 1, 1992, was listed in *Commerce Business Daily* as a call for agricultural supplies. Undoubtedly a clerical error.[52]

The work at Starfire has dual significance. It will allow ground-based reconnaissance to see objects in space much more clearly. It will also allow space-based reconnaissance to see the ground much more clearly. We have discussed reports that the advanced KH-11 was using flexing mirrors to solve the problem of atmospheric distortion. Adaptive and nonlinear optics would make that solution obsolete and represent a radical breakthrough in space-based reconnaissance as well.

Clearly the Air Force is extremely dissatisfied with conventional approaches to space imaging—both the current Space Surveillance and Tracking System and the system integration enhancements proposed under the Integrated Tactical Weapons Attack Assessment Sensors System initiative. Any power seeking control of space needs a sensor system that will see with great precision and report in real time everything happening from low earth orbit all the way to geosynchronous orbit, globally, on a twenty-four-hour basis.

Starfire is an antisatellite project, combining space surveillance with a speed-of-light capability. But it raises fundamental strategic and operational questions. Is the most efficient means for space control a ground-based or a space-based system? Should surveillance systems be located on earth or in space? Should antisatellite systems concentrate on destroying satellites and interdicting data transmissions, or on destroying ground stations? Is a missile-based antisatellite missile system better or worse than a directed-energy system? These questions, beginning with the location of reconnaissance facilities, must be considered.

In the United States, the nation currently most active in space warfare, there is a great deal of interest in ground-based surveillance. But it is not clear that this is the most efficient way to monitor space. Until recently, space was a slow-moving game, and the gradually accumulated and perpetually dated information gathered at Maui and by earth-based radar systems sufficed. However, as it becomes imperative for nations to be able to shut down enemy reconnaissance satellites, more up-to-the-minute information is necessary. As a satellite orbits, it passes over the earth at a fairly high speed—once every ninety minutes in low earth orbit—its range of vision limited only by its altitude and the quality of its sensors. Alternatively, a satellite at 22,300 miles remains stationary over one spot but can cover one-third of the earth. Thus, a relatively few satellites can provide coverage of the entire globe—and of a vast volume of space surrounding it.

An earthbound station is far more limited in what it can observe in the sky. First of all, it needs particular conditions to work—dark skies, but with the sun still reflecting off the satellite's surface; low humidity; no cloud cover or pollution and so on. In addition, an installation like Maui takes snapshots of space rather than a moving picture. This means that real-time ground surveillance of satellites is impossible. As work at Starfire progresses, earth-based sensors might be able to operate under much less than optimal conditions, making a network of earth-based sensors possible. But the number of such sensors and the complexity of their technology must always be greater on earth than in space.

Star Wars planners understood that ground-based sensors could not provide the coverage—or accuracy—of a space-based system, but as the program became increasingly circumscribed, they had no other choice. By placing sensors in space, a much broader view of the battlefield becomes possible, providing earlier warnings and allowing interception at much higher altitudes by land-based missiles. By one estimate, a space-based sensor system would increase the effectiveness of ground-based interceptors by a factor of three.[53]

SDI's search for space-based sensors went under the name Brilliant Eyes. Brilliant Eyes was a plan to launch a constellation of about fifty satellites equipped with optical and infrared sensors. A wide-area, short-wave infrared sensor would scan the surface, looking for signs of a launch, while a longer-wave infrared sensor would lock onto the rocket, tracking it during the midcourse portion. The optical sensor would be used during peacetime for monitoring space objects.[54] Brilliant Eyes, for all of its focus on the missile problem, also provides what is in some ways a more important capability—tracking satellites. All satellites emit heat from their electrical systems. A space-based system that can detect heat, point its sensors toward the source, and then locate the satellite optically using other satellites to triangulate the precise location of the target can vector any sort of antisatellite missile to the target. Indeed, one suspects that the inclusion of an optical system on Brilliant Eyes was an attempt by planners to quietly give the system an antisatellite support capability in the face of congressional opposition to any such program.

Whether Brilliant Eyes is ever built or not, the concept of a constellation of sensors focused on space is a natural evolution. Though it is extremely unlikely that any nation will have a comprehensive constellation of reconnaissance satellites in place in the near future, it is obvious that some sort of low-error space surveillance system, either space-based or ground-based, will be created over the next decade. The purpose will be to devise a way to kill satellites.

During the 1970s, the Soviets were quite active in this area and developed an antisatellite system designed to take advantage of certain peculiarities in the American reconnaissance satellite program. Unlike the Soviets, who orbit and de-orbit satellites with great frequency, the United States orbits very few reconnaissance satellites, leaving them in relatively stable orbits for several years.

The Soviet Interceptor System (SIS) was not fast. It required a rocket launch, which required preparation, preset guidance, and the ability to put the antisatellite missile into an orbit where its own thrusters would be in range of the enemy satellite. The Soviets had to have a high degree of confidence that they knew what satellites they were going after and where they were going to be.

The Soviet Interceptor System could be put into a coplanar orbit (on the same orbital inclination as an American satellite) and then maneuvered to a rendezvous with the American satellite, using its own terminal guidance or infrared system, at a speed of about 1,200 meters/second.[55] Theoretically, each American satellite could have a Soviet SIS assigned to it, tracking it, waiting for the command to close and kill. All this is theoretical, of course, since it appears that the Soviet program didn't work out.

The Soviet pattern, as we have said, was to launch and recover dozens of satellites each year. The United States could not afford to boost an antisatellite missile with each new Soviet launch, nor could it afford a leisurely coplanar orbit, since Soviet satellites could be de-orbited rather suddenly. Thus, the American plan was to go for speedy interception using rockets. Indeed, the first American attempt at a satellite killer was an Army Nike-Zeus using a nuclear warhead—quick, but very messy.

The most sophisticated attempt by the United States was the miniature homing vehicle (MHV).[56] This program was canceled during the late 1980s. In part this occurred because of technical limitations. The missile could only reach low earth orbit, and since it required an aircraft—an F-15—for launch, it was limited by the location of F-15 bases. Most important, the program was not in keeping with Congress's desire to keep war out of space. Nevertheless, the MHV lives on, at least on the drawing board, in U.S. Army antisatellite plans.[57]

The Army has an obvious interest in ground-based interceptors. These not only give it a role in space warfare but can give it control in allocating resources for an antisatellite system in the event of war. The same is true for the Navy. The Army also favors rocket-based systems, since building rockets and integrating rocket systems with ground-based tracking systems is something the Army has been comfortable with since the end of World War II. No breakthrough technologies are needed here. Moreover, there is an institutional bias. Integrating antisatellite systems with F-15 squadrons appeals to fighter-pilot generals. Including antisatellite missile batteries in the Army Air Defense Command maintains institutional equilibrium in the Army. And adding these weapons to the submarine fleet or to surface vessels simply enhances naval warfare capabilities.[58] These are not trivial points. The military is vast and complex, and a weapons system that minimizes social and institutional disruption is, all other things being equal, preferable to one that does not.

But ground-based missiles have an obvious problem. They are intrinsically inefficient. Being both solid and based on the ground, they must fight gravity to get at satellites. In addition, to intercept the satellite before it can maneuver or take other countermeasures, they need to get into orbit quickly. Because the amount of maneuvering fuel they carry is limited, the payload must be fairly small—too small, in fact, to do the job properly. What is needed are radically more efficient boosters that can accelerate large payloads to six to eight miles per second before reaching low earth orbit.

There are two alternatives to ground-based, rocket-powered antisatellite missiles. One is speed-of-light weaponry, whether space-based or ground-based. The other is space-based kinetic weapons. As Star Wars planners knew from the first, speed-of-light weapons are by far the most attractive—theoretically. They make interception relatively easy, and they do not shatter themselves and the target in a high-speed collision, leaving debris scattered in and around strategic orbits.

There is little question but that a fleet of space-based and ground-based speed-of-light weapons would easily deny access to space to enemy satellites. There is also little doubt that we know how to generate energy beams. The problem is that it is not clear that we know how to generate beams that are sufficiently powerful to destroy enemy satellites at great distances. And just as important, it is not clear that, having generated power once, we have the means to continue to generate powerful pulses over and over again. Over the next generation, it is likely that both of the problems will be dealt with and that a kinetic-energy antisatellite system will also be developed. More precisely, a space-based kinetic-energy system will be deployed first, as it takes advantage of space's characteristics while being a less exotic technology. As relations between the leading international powers shift and deteriorate, more and more nations will have the ability to blind, mute, or destroy reconnaissance, navigation, and communications satellites. At that point, the problem will be to protect those satellites—and war will have come to space.

Space Wars

Assume that a power such as North Korea faced war with the United States. It would start with substantial advantages. Its armed forces would be larger than those deployed by the United States. It would have more armor and artillery than the United States, shorter supply lines, and more familiarity with the climate and terrain. The major U.S. advantages—overwhelming ones—would be the quality of its satellite and airborne intelligence, the efficiency of its communications network, and the ability of its navigational and sensor systems to guide munitions to their targets.

Still, it is not inconceivable that a tertiary power such as North Korea could mount an assault on American satellite assets and thereby disrupt American communications and intelligence to such an extent that the war would be fought on much more equal terms. In the event of such a conflict, three sets of satellites would be strategic targets:

- Optical and radar reconnaissance satellites in polar and fifty-eight-degree, elliptical, low earth orbit.
- Navstar satellites in semisynchronous (11,000 mile) circular orbit.
- Electronics and signals intelligence satellites in both geostationary and medium earth orbit.

There are four basic strategies for rendering satellite systems inoperative:

- Attack ground stations directly, destroying the ability to receive and distribute information.
- Attack satellite systems physically, using solid projectiles.
- Attack satellite systems using high-energy beams.
- Disrupt, corrupt, or supplant earth-space data flows, using electronic-warfare techniques.

Threatening satellites in low orbit is easier than threatening them in high orbit. G. Harry Stine suggested that "any sounding rocket capable of lifting a payload of 100 kilograms to an altitude of 500 kilometers makes an amazingly effective ASAT."[59] The rocket would be fired straight up toward the orbital path of a satellite, with enough fuel to achieve maximum altitude at that point. The rocket would then release sixty pounds of sixpenny iron nails. A low-altitude satellite traveling at over four miles per second slams into the nails—and finis.

This sounds simpler than it is. It would take a lot of luck for an unguided rocket to deploy its load at the right altitude and orbital inclination. But Stine's point is still well taken. Guidance technology is now over thirty years old, and guidance systems are readily available on the official and unofficial markets. This dramatically increases the possibility of taking out a satellite through a relatively simple antisatellite system. When multiple rockets are launched into a satellite's track, the probability of a hit increases exponentially.

For the United States, or another advanced technical power, threatening the survival of enemy satellites is less a matter of ability than a matter of will. As the United States increases its ground-based space surveillance system and ties real-time data from this system together with an antisatellite weapon, the ability of the United States to control space—and therefore dominate ground operations—increases to a point where it is politically decisive. The United States is not pressing this technology only because no immediate enemy has a space-based capability.

As we have discussed, the ground-based intercept system will come first, not because it is best or even cheapest, but because it appeals to military planners and commanders who are familiar with ground-based missile technology and uneasy with systems operating outside the atmosphere. The limits of ground-based systems are clear. Direct intercept—launch to satellite without insertion into orbit—requires that the enemy's satellite's track pass within range of the ground-based system, whether boosted by fighter plane or a single-stage rocket such as Israel's proposed Arrow. If this isn't the case, then a coplanar orbit is necessary—a time-consuming maneuver when time may be the one commodity missing.

A ground-based antisatellite system operates at the bottom of a gravity well shaped like a cone imposed by the earth's curvature and the density of atmosphere. Space-based systems suffer from none of these problems. If placed in a higher orbit than the target, gravity actually favors the ability of the interceptor to sense and strike its targets. Thus, a space-based antisatellite system such as Brilliant Pebbles is a far more logical foundation for space control than any earth-based system.

In the long run, of course, once the problem of storing large quantities of energy is solved, we will undoubtedly see a shift to speed-of-light weaponry, which will have many advantages, both in hitting the target and in not leaving debris in space. In the immediate future, the threat to satellites will come from kinetic weapons—solid masses designed to smash into satellites or swarms of matter deployed so that satellites smash into them. Certainly, the United States in particular must now operate with the assumption that many nations have developed ground-based systems for destroying low-orbit satellites.

It is important to understand that any nation able to produce a rocket that can go into low orbit and a guidance system capable of some degree of precision could have developed an antisatellite capability unknown to the United States. Indeed, nations able to boost satellites into geostationary orbit must also be assumed to be able to threaten satellites in geostationary orbits. Indeed, any satellite currently in orbit with thrusters attached must be assumed to be capable of being used as an antisatellite.

It follows, therefore, that any nation basing a substantial part of its military operations on space-based intelligence—which is certainly the case with the United States—must also be prepared to defend its satellites. Satellites facing attack from kinetic weapons have three possible tactical responses:

- To maneuver in patterns so radical that either the sensors of the pursuit craft cannot monitor the maneuver or the pursuer's thrusters lack sufficient energy to match the maneuvers.
- To create a system of passive defenses. For example, the satellite might be hardened to absorb a certain amount of punishment. The satellite might carry certain antisensor devices, such as deceptive heat sources or electronic-warfare systems designed to

spoof infrared and radar seekers. Finally, the satellite constellation might carry a large number of decoys intended to divert scarce antisatellite resources from the real target.

- ▪ To create a system of active defenses designed to destroy antisatellite weapons. Satellites themselves might carry anti-antisatellite modules to be fired at pursuers. Alternatively, satellite constellations might include weapons platforms designed to attack enemy antisatellite systems—or other enemy satellites in nearby orbits. Finally, constellations of weapons platforms, from geostationary down to low earth orbit, might be constructed to engage and destroy enemy spacecraft.

Obviously, all three strategies will be used. But the most important defense will be the ability to maneuver while under attack, along with the ability to confuse tracking by unpredictable relocation. The latter is also the most difficult to achieve.

To maneuver, a satellite must have power. Until now, the primary source of propulsive power has been chemical. And one problem with chemical propellants is that they are inefficient; they must generate all of their energy from the reaction of the chemicals themselves, without using exogenous energy.

If, however, energy could be used directly on some inert material, the amount of propellant needed would be reduced dramatically. This is the point of a new technology called electric propulsion. With electric propulsion, electrical energy, in the form of heat or electromagnetic energy converted from solar energy, is used to excite a propellant—ammonia, hydrogen, or xenon—so that it provides thrust.[60] The principle is essentially that of boiling a cup of tea by heating it with an electric element.

For example, in one technology developed by the Japanese, an electrical chamber is used to ionize xenon, which is then used as a propellant. According to the researchers, this will efficiently produce enough energy to maintain a two-ton satellite in geostationary orbit.[61]

The U.S. Air Force's Phillips Lab and TRW have moved along with the Electric Insertion Transfer Experiment (ELITE), intended as a precursor to an electrical orbit-transfer vehicle. Solar power will be used to generate an arcjet that will stimulate an ammonia thrust. The new system would be powerful enough to boost a satellite into geostationary orbit from a low parking orbit—and maintain it on station for its lifetime.[62]

Electric thrust is attractive in that it provides a substantial amount of thrust at relatively low weight. Its weakness is that the absolute levels of energy produced are fairly low, which means that it can take weeks or months to move from low earth orbit to geostationary. While this amount of thrust might well be sufficient to maneuver out of the way of antisatellites that are equally equipped, these weapons, like fighter planes, have a distinct advantage. Since their only mission is to destroy satellites, they can carry more chemical propellant with higher thrust than can satellites weighted down with equipment for other missions.

Another solution is nuclear power, used not only to provide internal power, as is the case with several satellites, but also to provide propulsion power. Lockheed has already produced a nuclear propulsion system called Bus-1, which is reportedly the propulsion system used by the advanced KH-11 reconnaissance satellite.[63] The Soviets have also developed nuclear reactors and are now using their metallurgical skills to develop space systems able to withstand the superhigh temperatures of nuclear-heated propellants.[64] As with electric engines, propellants are heated by the direct application of energy rather than by using chemicals. But unlike electric engines, nuclear power applies much higher temperatures—between 2,400 and 2,700 degrees Kelvin.[65]

The problem with nuclear reactors is that they are extremely heavy and costly to launch. Thus, they are unlikely to be used by the commercial users looking for a low-cost way to change orbits. But nuclear power is already being used in the American reconnaissance program to shift satellite orbits and, one can assume, to allow KH-11s to maneuver out of the path of enemy antisatellites. The optimal military solution might well be to use nuclear reactors for sudden, radical orbit shifts and electric thrusters for more leisurely maneuvers.

The key to survival is unpredictability—and the ability to remain unpredictable for extended periods. Thus, it follows that more and more reconnaissance satellites will go into orbit with maneuvering thrusters—able to shift orbit to evade and hide and return to station when appropriate. And of course, antisatellite systems will become increasingly sophisticated in tracking these shifts and pursuing. This, in turn, will create anti-antisatellite systems, designed for maximum agility.

The struggle for space supremacy will, like the struggle for air superiority before it, spawn a species of spacecraft whose primary purpose will be the destruction of other spacecraft. Like fighter aircraft, these will consist of high thrust-weight ratios, advanced tracking systems for intercept, and as their cost soars and the need to use them for multiple missions rises, weapons capable of destroying enemy platforms. As systems mature, the confrontation of space fighter with space fighter will fill the romantic lore with space stories—while space systems will become senile. But that is well into the future.

For now, the essential issue is battle management—how to command and control space-warfare systems. The first question is the extent to which space combat systems can be automated and the extent to which men must remain in the loop. The second question is where the men who remain in the loop ought to be stationed—in space or on earth.

The key to effective battle management is intelligence—not only about enemy deployments but, even more, about the position and condition of one's own forces. Command and control are impossible without this intelligence. During battle, intelligence has, historically, been hard to come by. As the size of the battlefield increased, the ability of commanders to see the battlefield declined. The natural confusion of the battlefield was then compounded by the fact that intelligence was being delivered to the commander through intermediaries—first by messengers, later by radio messages. Sometimes the messengers were killed or delayed, at other times radio messages failed to get through or were garbled or jammed. Always, the information contained in the message differed substantially from the reality of the battlefield. Sometimes, the reporter did not see the battlefield or did not understand what he was seeing. Sometimes he could not record his observations fully and accurately. Sometimes the commander could not interpret or did not properly understand the message. This was Clausewitz's fog of war.

Nowhere was the fog of war greater than in the air battle. During World War II, the key problem of the air campaign was binding aircraft together into a coherent fighting force to carry out individual missions and building a series of missions into an effective air campaign. Nothing was more difficult than to manage hundreds of bombers and fighters spread out over a thousand cubic miles while under attack from antiaircraft fire and fighter aircraft. The speed of the battle, combined with its violence, meant that circumstances changed second by second. The quantity, nature, and deployment of air defense may have been unanticipated, the number of mechanical failures greater than expected, the number of casualties surprising. Weather conditions may have changed hourly. At each moment, battle management required that air commanders redeploy their forces, shift to secondary targets, detach their fighters, and so on.

In many cases, the mission broke down because unit cohesion collapsed—intelligence flowing to the commander was inadequate, or command decisions were improper or were not transmitted to subordinate forces. There were several solutions to this problem. The first was to impose rigid and inflexible operational directives on missions. Accepting that command and control during the battle were impossible, the second solution was to determine the mission's target, formations, routing, and so on prior to the mission, force everyone to adhere to this preset routine, and allow the chips to fall where they may.

A basic characteristic of the air campaign is its fluidity. Thus, this type of mission planning flies directly in the face of the air war's character. At the same time, the air campaign is parsed into segments, since each mission begins at an airfield and concludes by returning to the airfield. Each mission might be rigidly planned, but the air campaign could retain fluidity by using its inherent pauses to redefine missions in the face of changing circumstances.

This approach assumes that no single mission will determine the course of the war. In turn, this assumes that the quantity of aircraft launched on any single mission will be a small percentage of aircraft available. In this way, even the complete annihilation of a mission will not mean general catastrophe. But limiting the size of missions implies limiting the effectiveness of missions. As the size of the mission increased, the importance of flexible, effective real-time command also increased. There is a paradox here. Where missions are small, it is much easier for a commander to penetrate the fog of war and command his forces. At the same time, such control is usually less necessary. On the other hand, when the forces are large, the need for flexible command is much greater. Yet the ability to command a large force is much smaller.

The tension between the ground planners' desire to control the evolving campaign via preplanned mission orders, and the mission commander's need to manage his forces effectively—and preserve them if possible—is the hallmark of large-scale warfare. And the degree to which planning intrudes on operations is in large part determined by the C^3I capabilities available to the strategic planner. For example, during World War II an army commander such as Patton could manage his forces at the division and regimental levels, using radio communications, aerial reconnaissance, and his own motorized transportation. Confronted by the danger of tactical decisions diverging from the strategic plan, Patton was able to intrude on the lower command levels to impose coherence, accepting that such intrusion would lead to at least some tactical failures. The vastness of ground warfare meant that tactical failures could statistically be absorbed while strategic/operational/tactical coherence would more than make up for such defeats.

In naval warfare, the units of force are much fewer. Where ground campaigns are made up of tens of thousands of soldiers and thousands of tanks and artillery pieces, entire battle fleets can consist of only a few score warships. Tactical error could destroy a substantial portion of the total force available. Under these circumstances, forcing coherence between strategic planning and operational/tactical execution had to take second place to maximum operational and tactical control. Thus, a naval commander such as Nimitz was hard-pressed to collect sufficient tactical intelligence to do much more than provide general strategic guidance at the Battle of Midway. Indeed, when Admiral Yamamoto decided to command Japanese forces at Midway, he accompanied the fleet—and even then could not control his forces at the critical juncture of the battle.

Aerial warfare has, during the postwar era, seen a decline in the number of aircraft engaged in combat and the increasing importance of each aircraft. As this has occurred, the ability to submerge the tactical within the strategic has declined. At the same time, the quality of C^3I for aerial warfare has soared. During Desert Storm, the combination of airborne sensor and data-management systems, such as AWACS, with computerized mission-planning systems permitted theater air commanders to combine campaign strategy with mission execution. The quantity and quality of data made it possible to respond rapidly to changing circumstances. Of course, it is not clear how successful this integrated approach would have been had the Iraqi air defense system not collapsed in the early days of the war. Had that not happened, the old tension between strategy and operations might well have reasserted itself, with the need to preserve combat assets shifting control from planners to operational commanders.

One might imagine that war in space would be structurally similar to war in the air. In fact, this is not the case. Space warfare is far more similar to naval warfare than it is to aerial warfare or ground warfare.

- In aerial warfare, aircraft are tied to ground bases, returning to base regularly and spending most of the time on the ground. Naval vessels operate away from shore facilities for extended periods, from weeks to months, not returning to port for the entire period of deployment. In this extended detachment from bases, spacecraft are much more like naval craft.
- Naval vessels are extremely expensive and therefore relatively few in number. Aircraft are much cheaper—although their cost is increasing—and therefore more numerous. For the foreseeable future, the number of spacecraft is more likely to resemble the number of naval vessels than aircraft.

- Land warfare takes place within an obvious geography and topography. Aircraft, on the other hand, beyond being constrained by gravity and aerodynamic principles, are free of the singular restrictions of topography. Indeed, this freedom is what makes airpower unique among the services. And although the sea and space may appear to be vast and featureless, appearances are deceiving. The sea has characteristics that are both intrinsic and extrinsic, imposed by physical and economic geography. These characteristics ruthlessly shape the pattern of naval operations and warfare. In the same way, space's apparent featurelessness is a thin mask over powerful forces that shape the pattern of space operations and warfare.
- On land, topography and vegetation reduces the contrast between surroundings and weapons. It is much easier to hide a tank than an aircraft carrier, or any warship, since ships stand in stark contrast to their surroundings. The real protection for warships is the vastness of the sea and their own defensive means. Spacecraft also stand in sharp contrast to their surroundings. They are also protected by the vastness of space—and their own defensive capabilities. Increasingly, aircraft are losing their ability to reduce their contrast—hence, stealth aircraft.
- During land operations, it is possible for strategic planners to communicate with operational commanders directly—moving from the front to headquarters as necessary. In naval operations, communications between headquarters and the fleet require radio communications, in code and over long distances. The same is true for space operations. Aircraft constantly return to base—and frequently to headquarters.

This is by no means to argue that battle management in space will be identical with battle management at sea. There are far more differences than similarities between space and the sea warfare. But it does point us in two directions. First, it is not clear that space operations ought to be a branch of air operations. The mere fact that both take place off the ground does not, by itself, make them equivalent. The mesosphere divides the two domains conceptually as well as physically. Aircraft are much more intimately connected to the ground than are spacecraft. As important, their behavior is much more constrained than that of aircraft. The type of thinking necessary to control an air force is different from the type necessary to control a space force.

Second, the naval model of command and control, in which general strategic guidance—as well as operational doctrine—is provided by headquarters, while operational and tactical control is in the hands of fleet commanders, would seem very much to apply to space warfare. The current doctrine places operational control in the hands of ground controllers. Given the relative invulnerability of satellites and the long lead times necessary for maneuver, this is a reasonable procedure. But as space warfare enters a maneuver phase, in which satellites must be shifted rapidly to avoid antisatellites, more direct tactical control will become essential. This will become even more true when anti-antisatellite systems and other active and passive countermeasures come on line.

The weakness of ground-based operational and tactical control is the long line of communications linking the ground operator with the satellite system. A radio transmission takes 0.12 seconds to travel the 22,300 miles from the ground to geostationary orbit—assuming the satellite to be directly overhead. Consider a case where a surveillance satellite in geostationary orbit detected an antisatellite weapon closing on a key communications satellite. At the very best, it would take at least one-quarter of a second for the information to reach a ground controller and for the controller to send a command to maneuver. At the speeds of modern computers and satellites, this is not a trivial amount of time.

In fact, the process would take substantially longer. It would take the sensor satellite seconds to interpret the data it had accumulated on the movement of the antisatellite, switch on its transmission system, and pass the data through several intermediate links before arriving at the ground controller's monitor. Since it is unlikely that the threatened satellite would be located directly overhead, the response would have to pass through several long-range communications links. In all, leaving aside the irreducible time for interpretation and decision-making, the mere process of communications would undoubtedly take several very long seconds, seconds in which the satellite could be lost. It is as if a fighter pilot would receive intelligence a second or two after events happened, and it would take a second or two to communicate his commands to his aircraft. Remote piloting would work in an unthreatening environment, but it would not be advisable in a dogfight.

It is unlikely that, in the near future, an autonomously intelligent spacecraft can be created that would have both the ability to interpret data and the judgment to act on it. While individual projectiles might become brilliant, the battle management system must have a systemic operational intelligence. This means, in other words, that the human being must be kept in the operational decision-making loop, managing satellite constellations, and engaging in defensive and offensive actions.

The only solution is to unite sensor and battle management on a single platform. Aside from eliminating some of the delays caused by data transmission over long distances, there is another vital advantage. Beaming data to earth from space and then beaming commands into space creates long lines of communication that are vulnerable to distortion from a variety of natural causes; disruption by intentional jamming; or even hijacking by sophisticated electronic-warfare systems that might be able to substitute false data to mislead ground controllers or satellites.

Radio beams tend to spread and diffuse over distances, reducing security and increasing the possibility of interference. Moreover, radio beams passing through the atmosphere inevitably encounter atmospheric distortion due to electrical discharges and other natural phenomena. Other communications mediums, such as lasers, solve the problems of spreading signals coverage, but face other problems—such as the inability to penetrate cloud cover. This is not an insurmountable problem, perhaps, as experiments at the Starfire Optical Range are beginning to show, but any solution will be extremely expensive and vulnerable to countermeasures. Finally, ground stations are vulnerable to attack by cruise missiles and other munitions.

By transferring battle management to space, the advantage of absolutely clear line-of-sight communications is achieved, whether direct or through intermediate links. Extremely narrow beams would minimize both security and signals-integrity problems—and would eliminate atmospheric interference as well. Indeed, by using lasers for space communications, electromagnetic interference from solar storms can be eliminated. Most important, instantaneous, human management of sensors would permit timely command decisions.

Conclusion: From Watching to Shooting

We have discussed the manner in which intercontinental, hypersonic projectiles guided from space will be able to intervene in tactical engagements on the ground. To this point, we have assumed that the initial deployment of such projectiles will be on the ground. In the long run, however, a far more efficient deployment of precision-guided munitions will be in space. The same sensor technologies that would guide a cruise missile intercontinentally would also guide projectiles fired at the ground from space. The only difference would be that space-based projectiles would be much less expensive and much more responsive.

A hypersonic cruise missile is an expensive proposition. Fighting gravity all the way, it requires powerful engines to maneuver to the target. Space-based projectiles use gravity as their engine, rather than fight against it. Moreover, they spend less time in the atmosphere than hypersonic cruise missiles. A belt of satellites, in low earth orbit, would provide continuous coverage over a given battlefield. Clusters of unmanned weapons platforms could be held in reserve in geostationary orbit. In the event of a conflict, they could be shifted into a series of elliptical orbits whose perigee is at the latitude of the combat zone. A constellation of such platforms would provide continuous coverage to the battlefield. Requests for fire support, not unlike contemporary requests for artillery missions, would be transmitted by ground commanders to manned space stations in control of both weapons platforms and sensor platforms able to provide initial guidance.

These manned platforms would then order launches—allowing the space-based precision-guided munitions to guide themselves to the precise point requested by ground control—or to points selected by space command where ground combat was not under way. Command platforms would, in addition to running the sensor and weapons systems, operate defense systems to protect space-based systems from attack. Thus, the manned platforms would become headquarters for fleets of weapons, sensors, antisatellites, fighter craft, decoys, communications satellites, and so on. In addition, since repairing satellites is cheaper than orbiting new ones, command platforms would spawn smaller manned repair vessels that would carry out both routine maintenance and emergency repairs of battle damage.

These command platforms—these battlestars, to coin a phrase—will become the center of gravity of future warfare. In the end, the economy of force with which space-based weapons can control events on the surface of the globe will make control of space a military imperative. The ability to see the enemy, and to strike at him with precise, inexpensive, and deadly force, will make control of space as fundamental as control of the air was during the Persian Gulf War.

Control of space will mean the ability to command and control the complex systems that allow space-based fleets to see, to shoot, and to communicate. Command and control will rest with the men and women in space, whose intelligence is necessary to manage these complex fleets spread over trillions of cubic miles. Defending manned platforms will allow continued use of space. Thus, these invaluable assets will themselves become defensive fortresses—fast, agile, heavily armed, each command center linked to the rest by highly secure links such as laser-based communications.

Time will be measured in milliseconds, and decisions will be made in seconds. Thus, just as with the navy, fleet commanders will operate under general guidance from CINCSPACE, but will manage their fleets operationally and tactically. Both the quantity of information and the speed of decision-making will make command decentralization essential, while at the same time, space sectors and space fleets consisting of diverse weapons and other systems will have to be welded together into an exquisitely organized system of warfare.

We are only at the beginning of the process that will conclude with manned command centers commanding fleets of spacecraft. We already have fleets of spacecraft, but they are controlled from earth. But as these spacecraft become more and more vulnerable and as the need for timely decision-making and secure communications increases, personnel will move into space to manage war-making ability. As this occurs, the ability to use weapons to attack and defend in space will increase dramatically. As the use of weapons in space becomes a customary matter, it will follow that space-earth weapons will be introduced.

At that moment, the two traditional forms of surface warfare, ground and naval combat, will be changed beyond recognition—changed in the way that airpower promised to change warfare but never quite managed to do. The ability to see both globally and with extreme precision and to deliver conventional explosives to any point on the earth at any time means that the course of even small engagements can be determined from space. Airpower was limited in its range, in its precision, in the timeliness of its response, and above all, in the weakness of its intelligence. Space power has none of these limits. Its range is global, its precision as good as the best precision-guided weapon, it can deliver munitions in minutes, and above all, its intelligence is spectacularly good. Indeed, one can even imagine a time in the future when infantry forces will be held in reserve on space platforms and delivered to the battlefield in suitable conveyances, turning the tide of battle in a matter of minutes.

It is the rapidity and precision of space-based systems that must transform warfare. It is no coincidence that at the very moment that space is emerging as the center of gravity of modern warfare, both land and sea combat are facing urgent crises. The ability of conventional weapons platforms—tanks and aircraft carriers—to survive in a world of precision-guided munitions is dubious. The same revolution that made space an emerging domain of military power threatens to undermine the foundation of modern warfare—the hydrocarbon-driven, gunned weapon platform. The issue is whether a ship that can travel at twenty knots or a tank that can manage fifty miles per hour can survive in a world of brilliant munitions traveling at ten or twenty times the speed of sound—especially when guided by unblinking eyes, watching and waiting in space, far beyond the reach of the finest tank or the most advanced carrier. The emerging helplessness of the great weapons of the late European empire must be studied with care, as it introduces us to the end of the European epoch, and the beginning of the American.

15

The Return of the Poor, Bloody Infantry

The individual soldier is the hardest thing to find on the battlefield; he is the smallest unit of warfare, and his intelligence makes him naturally stealthy. But, in general, he is also relatively harmless. Ever since the invention of artillery and the tank, the amount of firepower the individual infantryman could wield was limited. Even the machine gun, powerful as it was, could not fire an explosive shell and therefore was inherently inferior to larger explosive rounds.

But imagine that this limit was removed and the stealthiest element of warfare could bring to bear the deadliness of the most advanced weapon system. Imagine if enormous firepower were concentrated in the hands of single individuals, harder to kill than tanks. In other words, imagine if the revolution in weapons, sensors, and battle management were applied to the infantryman—a human system designed to engage the enemy at close quarters, seize terrain, and, in general, operate in close conjunction with weapons systems scattered globally. Such a transformation would be a return to an ancient understanding of war—the ancient logic of the infantryman, the logic of weapon against weapon and life against life.

In his current form, the infantryman has a number of weaknesses. First, if located, he can easily be killed by a wide array of weapons. Second, compared to everything else on the battlefield, his weapons are both weak and inaccurate. Infantry weapons lack enough range to project power adequately and they barely reach the limits of the individual's line of sight. During combat, individual soldiers are easily isolated from one another and their commanders. Commanders communicate with each other by radio, while the infantryman's means of communication is still a loud shout. Perhaps most important, the infantryman's knowledge of his surroundings is limited by what his eyes can see and his ears can hear. He usually has a pretty chaotic sense of what is going on around him; in fact, no one experiences the fog of war more intensely or personally than he does.

The traditional solution to this problem has been training—forging individuals into units, drilling them as realistically as possible on their responses to different situations, creating noncommissioned officers to hold the unit together while officers devise strategies. Sometimes this worked. Usually, victory went to the side whose cohesiveness and coherence collapsed least. Until the technical revolution of the 1980s started to percolate downward, until commanders started considering the implications of advanced technology on the grunt, the infantryman's fate appeared to change not at all.

But the same technology that has made the tank obsolete opens the door to a radically different future. The multispectral sensors, high-speed computers, and brilliant munitions (as well as the advanced materials used in the vain attempt to prolong the life of the tank) raise the possibility of a superior soldier or, to use a phrase from a study by former CINC South Paul F. Gorman, a *Supertroop*.[1]

The idea for a superinfantryman did not originate with military men or engineers. Just as H. G. Wells was the first to imagine the tank, a science-fiction writer, Robert A. Heinlein, writing in 1959 about space warfare in the future, described a "Cap Trooper," dropped by capsule from an orbiting spacecraft to the surface of the planet.[2] Heinlein envisioned the Cap Trooper wearing an armored powersuit, which multiplied his strength and speed tremendously. The helmet would be filled with sensing, communications, and data-management apparatus. He would be armed with a wide range of precision munitions, from antipersonnel devices to atomic weapons. A small number of such troopers would be able to devastate a city if needed.

The idea of a superinfantryman is no longer mere speculation. The U.S. Army has initiated a program known as the Soldier as a System (SAAS), in conjunction with the Marines and the U.S. Special Operations Command. The program, part of a larger thrust called Warrior's Edge, is intended to have two parts. The first, Block I, or The Enhanced Integrated Soldier's System (TEISS), will be deployed in 1999 and will be followed by a second phase, Block II, which is scheduled for deployment in 2010. These projects will involve a wide array of new systems, including advanced weapons for individual soldiers, computer networks at the platoon and company level, helmet-mounted sensors and displays, exoskeletons, and even chemical compounds to improve the ability of soldiers to learn—in short, Heinlein's Cap Trooper.[3]

This new approach to the problem of infantry warfare is not confined to laboratories. Thinking about technology and the infantry has penetrated to the operational level. A report issued by the United States Army Infantry School at Fort Benning, entitled *Infantry 2000*, states:

> The future infantryman requires a system that integrates full body ballistic protection along with NBC [nuclear, biological, chemical], flame, laser and microwave protection. Enhanced productivity will be achieved if we can relieve climactic stress on the soldier. Lethality will be increased with an integrated full solution individual fire control system. It will use a helmet-mounted image display (HELMID) to provide point and shoot accurate fires which will be equally effective day or night or through obscurants and camouflage.[4]

The SAAS program, and the needs described by the Army Infantry School, present a consistent and coherent vision of a revolution in infantry warfare. Until now, the infantryman has been fairly well limited to combat capabilities provided by biology. He could move, see, hear, and so forth only to the extent that his body permitted him to do so. Now the infantryman will be radically transformed, even enhanced, which will initiate a new era of ground combat.

The Infantryman and the Sensor/Data Revolution

Extending the range of vision in space and spectrum is the first task in creating an advanced infantryman. Such an extension has already taken place with the introduction of night goggles as near standard issue to U.S. combat troops. These goggles, which use available light, such as starlight, and enhance it thousands of times, make night vision and twenty-four-hour combat possible. At present, such sights, called I^2 (image intensification), have a number of limitations, particularly in trade-offs between acuity and field of vision as well as general clumsiness. But by gathering light through a lens and converting it into electrons, and passing the electrons through a phosphor plate, thereby multiplying them thirty thousand times, it has become possible for an infantryman to see on a moonless, cloud-covered night.[5] The addition of other types of sensors will further enhance the infantryman's ability to see things hidden to the naked eye.

The revolution in sensor technology that made the nonballistic projectile possible offers even more dramatic possibilities for the infantryman. A report prepared for the U.S. Army by the National Research Council, *Star 21: Strategic Technologies for the Army of the Twenty-First Century*, published in 1992, includes the following set of expectations about the future of sensors:

> Passive optical and infrared systems provide information on direction (bearing) and on spectral distribution and intensity, range, range extent, velocity and direction. Millimeter-wave synthetic aperture radars provide high-resolution images that are responsive to the material properties of targets. These systems can be configured so that the active and passive components share the same optics and thus can provide pixel-registered images in a multidimensional space, which allows multidimensional imagery. Acoustic sensors can provide information regarding frequency and direction of detected signals.[6]

Taken together, the wide array of sensors that were developed to locate targets for precision-guided munitions and cruise missiles allows for a vast and comprehensive sense of reality—providing an enhanced and extended sense of risks and targets. In addition, information from other sensor platforms, such as satellites, unmanned aerial vehicles (UAVs), manned aircraft, ground sensors, and so on, could be transmitted to the infantryman, extending his vision even farther.

The gathering of such data is already a technical reality, with relatively minor problems—including miniaturization, shared use of apertures, and designing data transmission systems large enough to handle vast amounts of complex, graphical traffic—still to be solved. The problem derives from the very success of these sensors—they provide so much data that it would be impossible for the infantryman to read it all in alphanumeric form, let alone to absorb it and act on it during combat. The situation is similar to that faced by fighter pilots, who must simultaneously fly their aircraft, locate threats, fire their weapons, and navigate.

Fighter pilots have solved this problem—not altogether satisfactorily—with the heads up display (HUD), in which all necessary data is displayed on the canopy of the aircraft or on the visor of the helmet.[7] The data displayed is carefully structured to be absorbable by the pilot, in an environment where data overload has become a fatal problem. The Hughes Corp. is already developing a display unit for ground forces' helmets weighing two ounces, which will display graphics or text on the front of a helmet.[8]

While HUD presents essential data within the pilot's field of vision—as if he were viewing a control panel without having to look down—it does not provide him with a fused sense of reality. He does not have something that he can grasp as effortlessly as if he were looking at something outside his window. Instead, the pilot is left to integrate the data. Warfare is spatial—with interactions of location, shape, motion. Grasping spatial relationships rapidly, intuitively, as a gestalt, without pausing for reflection or calculation, is a matter of life and death.

Imagine two swordsmen, dueling in the dark. One has sensors that display a simulated picture of his opponent, properly proportioned, and moving synchronously with the real figure. Now imagine the other swordsman having, instead of a picture, a series of gauges and readouts. To figure out where his opponent was and what he was doing, this other swordsman would have to read the data, integrate it, understand it, then execute his swordplay according to this flow of data. Who would win the fight?

Let us imagine a second scenario. Assume that the visual display was a bit off, while the data display was precise. Would it take longer for the swordsman with the picture to adjust for error than it would take the swordsman with precise alphanumerical data to read it, absorb it, and act on it? It is clearly easier to comprehend a picture—and compensate for errors—than to understand data.

The problem is no longer the gathering of data—indeed, there is a surplus of usable data. Nor is it a problem of rendering data as information—that too can efficiently be done. Rather, the key problem has been the management and display of information in a manner compatible with the normal sense and thought processes of soldiers.

The system must be designed so that each bit of information from a variety of sources is fused into a coherent image that can be taken in by the infantryman as if it were a single reality in the visible spectrum. Assume, for example, that the millimeter-wave radar spots a dug-in tank straight ahead. The infrared sensor detects a squad of enemy infantry at three o'clock. The acoustical sensor detects an aircraft engine overhead. An overhead unmanned aerial vehicle spots some armored personnel carriers off to the left. How could all of this data be usefully displayed?

The term *virtual reality*, which has come into vogue, revolves around the idea of managing data generated by a computer program and displaying it in such a way that it simulates ordinary reality as a human being experiences it. Usually, virtual reality involves encapsulating someone's head in a helmet and using a computer to generate images. But other sensory inputs can also be controlled. For example, sensors and servo-mechanism on the limbs and fingers can exert pressure and tactile sensation so that, after "seeing" an object, a person can "feel" and "pick" it up, actually experiencing its weight and texture. Much has been made of the entertainment value of such a construct, but it has considerable usefulness as a training device, in military and other contexts.

The U.S. Army has already committed itself to virtual reality for training purposes as part of its Force XXI concept. During the 1980s, a program called SimNet was initiated to train tank drivers and gunners in a virtual-reality environment. SimNet has evolved into a much broader initiative—the Distributed Interactive Simulation Environment—and a specific program intended to train infantrymen, the Close Combat Tactical Trainer (CCTT).[9] Infantrymen will be placed in a room and outfitted with helmets that will cover their eyes. They will "see" a combat situation, from terrain to enemies, and they will carry weapons that will have the feel of the real thing. A camera will track the movement of their bodies and adjust the picture accordingly. They will feel as if they were in combat—except that they will be perfectly safe, and the simulation can be run over and over again.

Rensselaer Polytechnic Institute, working with Avatar Partners, has been awarded a contract by the U.S. Army to develop a full-scale artificial-intelligence environment, the Dismounted Infantry Virtual Environment (DIVE), which will include:

- An instrumented room with multiple video cameras for video-based tracking and orientation estimation, which will track the key body joints of a soldier without tethers, bodysuits, or other restrictive equipment.

- An ultralightweight, wireless, head-mounted display including spatial sonics.
- A high-speed, real-time image-generation system, capable of rendering body models of immersed users and combining those models with digitally created environments.
- Virtual weapons and software that will provide the user with the ability to "fire" at simulated targets, with simulated results that account for standard effects such as ordnance type, gun elevation, and wind.
- Networking capability to interconnect individual DIVE modules for squad/platoon-level exercises and to connect them to the Distributed Simulation Internet via standard protocol data units.
- Intelligent agents that respond to human voice commands to simulate an entire squad or platoon under the leadership of a DIVE-immersed human commander.[10]

Virtual reality is a system of data fusion with a presentation that is faithful to the normal experience of reality. That is what makes it such an exciting training tool. But virtual reality can have a more direct use. Instead of being generated by a program, the data could just as easily be generated by actual sensors scanning a literal and not a virtual reality. The interface between the infantryman and the system does not care where the input originated. It could have been a simulation of an infrared sensor—or it could have come from a real infrared sensor.[11] Thus, the technologies being developed for simulations and training could, with different sensor sources, solve the problem of data fusion, not only in a training room but on the battlefield as well. An infantryman could be fitted with a completely opaque helmet—identical to those used in training—inside of which he would see as real images and icons the data that was being fed into the system by the sensors.

For example, he might see arrayed on his screen, in full relief resembling his own optical experience, an airplane thirty miles away, visible through enhanced optical television; enemy troops two hundred yards away and camouflaged, visible in imaging infrared; a camouflaged tank two miles away, visible only to radar; and a fortification visible in ultraviolet. Side-looking synthetic aperture radar located in an overhead satellite would note minefields, while nuclear, biological, and chemical sensors in a UAV would measure air quality and flash a warning if necessary.

The data available to the infantryman will not only be gathered by his own sensors but by those of the rest of the unit as well. Data will also flow from other data-gathering platforms—satellites, high-altitude UAVs, low-altitude remotely piloted vehicles, and ground-based sensors—all linked together in complex laser and electromagnetic nets. A flick of a switch will display the trooper's rear, another will allow the soldier to zoom in on a particular feature. The unit's commanding officer will be able to see the position of each of his men, plus call up a readout of each man's physical condition, available ammunition, and systems integrity.

This nonvirtual reality will, first, extend the infantryman's physical senses into distant spaces. The experience will be undistorted—he will understand what is happening around him as quickly as if he were using his own eyes and ears. Second, the old problem of command and control in combat, as well as some parts of the problem of unit cohesion, will be solved or, at least, eased. The commander's sense of where his men are and what they are doing will be greater than at any time since warfare became a large enough enterprise that it extended beyond the reach of a commander's eyes and voice.

For these developments to take place, the revolution in data management will have to be supplemented by a revolution in communications. The key to the operation of the system will be a high-speed computer, linked to a high-capacity data communication system that will be rugged enough and small enough to be carried in combat.

The U.S. Army's Communications Electronics Command (CECOM) is currently developing such a device, called the soldier's computer, which is planned for validation by 1996, and for introduction under The Enhanced Integrated Soldier's System (TEISS) program by 1999. It is intended to be a multiprocessor computer, with specific chips for such functions as graphics, communications, position location, and voice recognition. It would also have extensive storage capacity based on hard drives, CD-ROMs, and other more advanced memory media such as EEPROM.[12] Indeed, so high are the expectations for the lightweight computer that CECOM hopes that its central processing unit (cpu) will have the same capacity as today's supercomputers, such as the Cray.

The memory will contain extensive mapping data, which, in conjunction with the data flowing in from sensors, will tell the soldier where friendly forces are, where enemy forces are, and his own precise location. In addition, a large graphic capability will match images gathered by sensors with images built into the memory, for target recognition and additional identify-friend-or-foe (IFF) capabilities.

The computer would also manage communications. In part, this communication would be standard voice communication between soldiers based on secure transmission systems such as SINGCARS. More important, it would also include the transmission of graphical data, which would require a cpu for formatting and interpretation.

In addition to being tied to a worldwide communications network, all computers in a unit would be linked together in a single data network, transmitting data via high-speed fiber-optic links (which have the advantage of being jam-proof) or line-of-sight photo-optic or laser links—with UAVs and satellites being used to relay even short-range communications for the sake of signals integrity. In large part, this system was in place during Desert Storm; however, it lacked sufficient capacity to carry the amount of traffic. As we discussed earlier, data caused one of the most severe logistical problems in the war. It will be necessary to develop satellites and UAVs with more data transmission capacity, or there must be a breakthrough in the way data, particularly graphical data, is formatted.

From the squad to the CINC, a commander will be able to view the battlefield from any level of resolution desirable, with data aggregated in any way useful. He would immediately have information available on casualties, ammunition expenditure, and so on, allowing him to make battle plans with enhanced precision. A CINC, for example, might view the developing situation from the standpoint of a particular brigade or focus in on a single battalion and, if desirable, analyze the deployment of a platoon or even a squad. He might generate a map showing the location of all fuel in the theater, from tankers to trucks. A platoon leader could monitor casualties and the precise deployment of his forces, as well as enemy troops, obtain information on the last time a soldier slept in order to gauge fatigue levels, and so forth. The soldier's computer could be the rock on which the next century's army is built.

The Art of Killing in the 21st Century

The purpose of all this, of course, is to kill the enemy. All of the data in the world, no matter how brilliantly managed and displayed, will be of no use if the individual infantryman can't act on it by destroying enemy soldiers. All these various sensors and data management systems must, in the end, converge on the individual soldier's weapon, his means of destruction. Indeed, there has been little or no progress in the weapons of individual soldiers since World War I. They have gotten lighter, less likely to jam, able to fire more rounds, but the machine gun, submachine gun, rifle, hand grenade, and light mortar are all old weapons with fresh veneer. The AK-47, M16, Galil assault rifle, and the rest have not changed their basic design in nearly thirty years.

The U.S. Army recently concluded that the conventional rifled personal weapon has reached the limits of its development. During the 1980s, the Department of Defense undertook the Joint Service Small Arms Program, looking for a successor to the M16, which had been introduced during the Vietnam War. After an eight-year, $50 million search, Program Manager Vernon Shisler announced, "It's now obvious that you can't get much better performance from bullet-type rifles. You can lighten the load using caseless or plastic-cased rounds and gain some improvements with optic sights, but you can't significantly increase performance."[13]

It is difficult to imagine the conventional rifle surviving in the radically changed environment we have been describing. It is a line-of-sight weapon in a world of indirect fire. It fires a dumb, slow projectile in a world of brilliant, hypervelocity projectiles. It fires a nonexplosive projectile in a world of high explosives. In the end, the rifle-bearing infantryman is governed by the same principles that governed the spear hurler and the bowman—first see the target, then try to get your hands to direct your projectile toward it. The failure of the Joint Service Small Arms Program is merely official confirmation that the rifle is at the end of the line.

No matter what improvements are made to the rifle, it is not going to work any better. But the sensor revolution opens a new avenue for improving human control. The *Star 21* report predicts:

> Special helmet-mounted sensors could track the soldier's eye movement to aim personal sensors and weapons. For instance, a soldier might look at a building at a distance. A laser range finder and the navigation system could quickly determine the building's exact location. The soldier could provide audio information about the building through a helmet-mounted microphone. All the real-time information could be stored in the soldier's personal computer or transmitted through the C³I/Rista network.
>
> The helmet and visor conceivably could be used to aim the soldier's personal weapons. Current weapons depend on tight hand-eye coordination for aiming. The problem is that the eye is accurate, but the hand is not. Eye-only aiming is certainly possible with emerging technologies.[14]

The eyes survey a multisensor, multiplatform reality, select a target, and focus on it. A laser sensor would note precisely what the eye was looking at and determine, from a database created by sensors, the precise position of the target. The infantryman could then select a projectile to be fired at the target, blink twice, twitch a finger, or perform whatever action was programmed into the system, and the projectile would be launched at the target. The question that remains is, what sort of projectile would the infantryman be carrying, and how would it be launched?

The Block I plan of The Enhanced Integrated Soldier's System (TEISS) includes an element called the Small Arms Master Plan, which envisions the reduction of the current mix of weapons to three basic types: the sidearm, the individual combat weapon, and the crew-served weapon. The individual combat weapon will be fundamentally different from the rifle—much more powerful, with an explosive charge.[15] Both it and the crew-fired weapon are intended to fire more than one type of munition—including grenades and explosive bullets. While certainly increasing the flexibility and lethality of the infantryman somewhat, the changes envisioned under the first phase of the system do not represent a quantum leap in the firepower of the infantryman, but merely an incremental improvement.

The future of infantry weapons can already be seen in the man-portable antitank weapons currently in use, such as the Javelin, as well as in the new guided mortars. The Javelin can be fired by a single infantryman, who focuses on the target, locks the warhead onto the target point, and launches. He can fire and forget, as the Javelin will guide itself to the point the infantryman focuses on. Other weapons do not even need an initial lock-on. Once fired, they can locate the target themselves, or they can be guided to the target by the gunner or by another sensor platform, such as a UAV or satellite—there are multiple guidance choices. In each of these cases, both the inefficiency of hand-eye coordination and the tyranny of ballistics have been abolished. The problem remaining is wedding them to the individual infantryman, something that does not really require vast innovation or imagination.

The continued miniaturization of warheads and rockets allows more and more of them to be made man-portable—even without strength enhancement. Imagine a series of tubes (one to four) mounted on the back of the infantryman, made of a light, durable material, such as fiberglass insulated with aerogel. The infantryman would observe his surroundings with his and other sensors. The data would be collated and fused by his computer, which would display it graphically on his screen.

His eye would then focus on the target, while his hands selected the type of projectile and warhead he was going to launch—which would depend on the type of target. A laser scanner inside the helmet would identify the target being focused on, the computer would use its gunner's primary sight (GPS) system to locate the target, provide a vector, and order the missile to launch.

The missile would have a two-stage engine, as before. The first stage, a powerful CO_2 motor, would propel the missile upward and away from the infantryman, without back blast. An internal gyroscope would order side thrusters to stabilize the missile at the appropriate angle for the second, explosive-rocket engine to ignite, delivering the missile to its target. The projectile and warhead could be a small, explosive bullet, a grenade, a hypervelocity antitank round, a high explosive, a shrapnel-laden mortar round—whatever was appropriate. A smart sensor would guide it to maneuvering targets. Once the projectile was fired, the infantryman would be free to get on to his next task—getting away from the launch point.

Whichever sort of round was fired, the infantryman would cease being the weakling of the battlefield. He would be able to carry with him the firepower of armored vehicles and have greatly increased range and accuracy. As a result of these developments, infantry warfare will cease to be the statistical game that it has been since the invention of gunpowder. It will no longer be a matter of vast numbers of soldiers firing enormous quantities of highly inaccurate projectiles in the hope that, by saturating a target area, something would be hit. **The massed infantry armies of the past, necessary to produce the swarms of projectiles required to hit even a single target, will have become as obsolete as the tank.**

Obviously, no matter how small projectiles get, the ability to carry a sufficient number will tax the physical capability of the infantryman. That infantryman will need to have substantial assistance, both in the form of robots to carry material and, more important, augmentation of his body's strength. In a way, this augmentation is one of the keys of modern warfare.

One of the virtues of the tank was its ability to mount a heavy gun, carry ammunition, and still move about the battlefield. The tank's weakness, its visibility on the battlefield, could be solved by the infantryman. But, in the end, the infantryman cannot begin to replace the tank unless he can field an equivalent amount of firepower—and that weighs a lot, no matter how much it is miniaturized.

At the same time he must still remain agile. This is an old problem, and the solution is an exoskeleton—a frame that fits to the outside of the body, senses the body's motions and exertions, and multiplies their power.[16]

During the 1980s the Los Alamos National Laboratories was working on a project, code-named PITMAN, to produce an infantry battle dress that would use robotics to amplify human strength.[17] The suit would be built around a computer chip that would memorize the motions of a particular infantryman. This could be done either by placing tiny sensors on his body, the conventional path, or as the chief engineer on the PITMAN project suggested, by attaching electrodes to his skull that could sense magnetic fields generated by the brain prior to and during motion—magneto encephalography—the more exotic approach.[18]

Drawing on a power source,[19] the exoskeleton would emulate and enhance human muscle motions, permitting not only lifting but rapid movement as well, providing strength and mobility at the same time.[20] A man able to lift a hundred pounds, for example, would find that he might lift five hundred or a thousand pounds. The suit, weighing up to a ton, would bear its own weight and the weight of all equipment, including that of armor and projectiles and launchers.[21]

The suit's frame would be made of a strong, lightweight material, such as a graphite epoxy. It would be surrounded by a rigid, advanced material that would protect the soldier not only from enemy projectiles (up to perhaps 20-mm rounds) but from chemical and biological threats. Ideas for this material include combining Kevlar (the current protection against ballistic threats) with a substance such as silk, for a lighter-weight protection. More creatively, projects are under way to bioengineer new materials to control the permeability of clothing so that they would normally be air permeable (and comfortable) but would become impermeable when in a dangerous environment.[22] Under any circumstances, a sealed suit with an air circulation system along with bottled air for use when the outer atmosphere might be contaminated would provide a comfortable and safe environment for the infantryman in his exoskeleton.

The suit itself, essential to tie together the entire system, cannot really proceed until the power-source problem is solved—which, given the quantity of power required, is not an insurmountable problem, as it may be with electrothermal (ET) and electromagnetic (EM) guns.[23] Thus, the infantryman would achieve the firepower that previously required a platform driven by a petroleum engine, along with an accuracy and range beyond anything that direct fire could achieve.

In changing the range of weapons the structure of command is dramatically changed. The chain of command was created to facilitate management of forces too large to be controlled by any single commander. Direct interventions by higher commanders into lower echelons had historically been undesirable because the senior commander could not possibly have sufficient information about the situation on the ground to make reasonable judgments. In the future, a company commander's data display will instantaneously be available at any higher command level. Thus a senior commander can view the situation from the standpoint of the junior commander. Micromanagement, previously a dirty word in the military lexicon, might carry different connotations. The commander could, in addition to managing the entire battle, control the movement of a critical spearhead formation. Whether he would wish to would depend on his personal management style.

But just as data flows become decentralized with the new technology, the possibility of a hypercentralization of command also becomes a possibility. Certainly, the rigid command structures of the mass armies of the last five hundred years will become more fluid, more ad hoc, depending on circumstances and even personalities. It should also be added that the function of the general staff, which had been developed in part to accumulate and manage data for the commander, would have to shift. An interesting evolution to observe will be the extent to which staff function at all levels will be changed by the new technology. With new means available for command and control, the responsibilities of commanders will increase—along with the pressures. We must not expect, of course, that the fog of war that Clausewitz described will be abolished, but it will certainly be driven back.

Beyond Total Wars

Three things distinguish the emerging infantryman—or more appropriately, "individual armor units"—from his predecessors:

- Relative invulnerability—The only defense of the traditional infantryman, who fought without any shielding, was to dodge the bullets. Wearing a protective "suit," the future infantryman will be invulnerable to traditional threats—nonexplosive ballistic projectiles, NBC threats, shrapnel. He will be vulnerable to armor-piercing rounds and direct hits with high explosives.
- Multispectral sensing—The traditional infantryman saw with his eyes. The future infantryman will see in spectra and at distances far beyond human vision. Moreover, sensors on other plat-

forms—UAVs, satellites, reconnaissance aircraft—will extend his visual capacity even farther.

- Non-line-of-sight weapons—Powered exoskeletons and robot ammunition caddies will permit the infantryman to carry large weapons loads.[24] While it is not clear that there are any limitations on the weapons range, assume conservatively that the range of the individual weapon would equal that of a multiple-launch rocket system—about twenty miles[25]—and that of the crew-served weapon would equal that of a Lance missile—about fifty miles.[26] This would permit a small number of soldiers to lay down enormous firepower over a large area.

Consider—again arbitrarily—the makeup of a standard eleven-man squad in the twenty-first century:

1. **Squad leader**—He will carry a personal weapon and massive computing and communications gear that will enable him to communicate to any command echelon—from infantryman in the field to company commander.
2. **Programmer/telecommunication specialists**—Their primary job would be to calibrate weapons and personal gear for satellite grids and to reprogram projectiles for new targets and tasks. In combat they would serve as the target-acquisition team, using multispectral sensing devices to search for enemy air and land threats and targets, transferring data to appropriate weapons systems.
3. **Heavy weapons team**—Supplied with heavy-duty exoskeletons to aid in lifting and follow-on robots to aid in launching, they could simultaneously launch twenty heavy projectiles into combat, using the multimission projectile system.
4. **Personal weapons specialists (plain-vanilla infantryman)**—Armed with ordinary weapons launchers, they advance ahead of the groups to provide perimeter security for the specialist teams and do the dirty work.[27]

Assuming sensor support from unmanned aerial vehicles, a single squad could secure an area twenty miles per side (four hundred square miles) and project explosive power over a radius of fifty miles (nearly eight thousand square miles)—although probably without sufficient ammunition in a **target rich environment without substantial prior planning.** Lest the mind boggle at these numbers, bear in mind that the gap between them and the amount of territory able to be secured by a modern-era squad is no more extraordinary than the range, mobility, and firepower of a Vietnam-era Air Cavalry company compared to its World War II equivalent or the mobility and firepower of an American armored battalion in World War II compared with a Civil War regiment. Such quantum leaps in capability have become commonplace in warfare since the industrial revolution.

Depending on the likely opposition and the quantity of terrain involved, larger numbers of troops may be needed—but it is impossible to imagine the need for five hundred thousand troops as in Desert Storm. First, the radical increase in mobility, firepower, and, above all, accuracy makes the firing line obsolete. Second, the decrease in man-power dramatically reduces the need for logistical support. Where thousands of artillery shells are to be fired in an hour, and tens of thousands of gallons of gasoline are going to be consumed in an afternoon, and thousands of meals need to be prepared and delivered, then thousands upon thousands of truck drivers, cooks, munitions specialists, are needed. However, when dozens of munitions fired by dozens of men moving in battery-powered suits are needed, a few man managing a few robots will be sufficient.

What we are seeing is the end of the GI. The GI, the stamped government-issue interchangeable warrior, becomes obsolete when masses of men are no longer required to fight wars. Ever since the invention of the musket, the purpose of training was to force men into a mold—to drill them, depersonalize them, until they became a unit, until they fired in unison to overcome the inaccuracy of their firearms. The archetypal old sergeant used to tell the recruits that if the Army had wanted them to think, it would have issued them brains, and indeed, too much imagination was the ruin of many a soldier, contemplating his probable fate on the firing line.

The model for the soldier of the future is not the GI of our large-scale wars, but the Special Operations trooper—the Green Beret, Special Air Source, Spetznaz, or, indeed, knight of old. The future soldier will be highly trained and skilled, but not in the rigid way of mass armies. He will have to master technologies that are esoteric in the extreme—communications theory, sensor technology, and so on. As with the Special Forces, the small size of the unit will require each man to become an expert.

Small-unit operations in the past were associated with low levels of destructive force. Small units in the future will be capable of tremendous destructive force. Soldiers will have to have a deep sense of unit loyalty and, simultaneously, a strong sense of personal independence. In a physical sense, the individual's level of isolation will dramatically increase. Visual contact with other troops may be impossible. The data links will keep the unit together—but when those fail, the mission will have to continue.

For the first time in five hundred years, we are about to see a dramatic decrease in the size of land forces, without a decrease in military power. Sociologically, this will mean that members of the military will once again constitute a social elite as they did in the Middle Ages—where the means of war were expensive, the skills esoteric, and the powers of those who mastered the skills great. Mass armies are, ultimately, democratic armies. Small armies, consisting of skilled and courageous men wielding enormous power, represent a challenge to democratic ideals. Meritocracy may well turn into aristocracy.

Modern war became total war because of the inaccuracy of weapons. Mass-producing weapons required near total mobilization of factories and soldiers. The distinction between civilian and soldier was obliterated. Everyone fought or worked, and everything was at risk. War became a social catastrophe more than a political one. Nuclear weapons, which placed absolutely everything in danger, were the logical conclusion of this process.

With precision-guided munitions, the number of men involved in arms factories and armies will decline precipitously—one projectile can be fired for every thousand previously needed. More important, the level of devastation will decline as well. The relatively light damage to Iraq in the six-week bombing campaign, compared, for example, to the damage to Hanoi in the Christmas bombing, is a foretaste of a more moderate sort of war. More precisely, in seeing the end of total war, we see an end to an era where war puts society's very being at stake. Regimes may rise and fall, but as in the premodern era, the life of ordinary men will go on.

Through most of human history, the city-state was the natural political institution. The nation-state emerged only after guns blasted down the city's walls and cities could not produce the cannon or the men for armies. This is what befell the last generation's great powers—Britain, France, Germany—so they coalesced into continental alliances or avoided politics altogether. The end point of the first global system, the post–World War II era, was the continental state—the United States, the Soviet Union, and China. But the weaponry does not require continents. It requires expertise.

With the new technologies of war, smaller nations and cities suddenly become important. Countries like Israel or Singapore, with a few hundred thoughtful scientists and skilled engineers, can produce the instruments of war—sensors, computers, precision-guided munitions, and so on—in the coming centuries. If nothing else, their ability to sell weapons to less gifted but larger nations gives them tremendous political power.

Land warfare is therefore making a quantum shift, not only in technology but also in the consequences of technology. The logic of the first global empire—the logic of mass armies, nation-states, total war—makes little sense in a world of precision-guided weapons. Certainly, the transitions will take generations to work themselves out—to senility, as inevitably happens. But just as Cervantes could see the absurdity of the knight at the dawn of the first global epoch, so we can see the end of the GI and the birth of the Supertroop—at the beginning of the second epoch.

Conclusion

THE PERMANENT DILEMMA— CONTROL AND USE OF THE SEA IN THE AMERICAN EPOCH

We are faced with a problem referred to in the defense community as a problem of net assessment: on the whole, how much and in what areas has the revolution in warfare increased or decreased American power? How must the United States restructure its forces to take advantage of this revolution? What threats are posed to American power by this revolution? There are, as in all things, both simple and complex answers to these questions. Each tells the truth in its own way.

The simple answer is that the revolution in warfare both emanates from the United States and benefits the United States. It is, simultaneously, inevitable and desirable. The more complex answer is that, however beneficial the revolution is to the United States, it also poses a fundamental threat to American interests—threats that will have to be addressed and dealt with in the immediate future.

On the whole, the revolution in warfare has strengthened the United States on land and in the air. On land, the United States had been operating at a great numerical disadvantage, particularly in Eurasia. In the conflicts of Korea, Vietnam, and the Persian Gulf, for example, the United States urgently needed a way for its smaller forces to match and overmatch much larger forces. More accurate weapons helped to compensate in that they decreased the number of men necessary to carry out a mission. The same was true in the air. During any number of operations, the United States had to fly combat missions against forces that were greater in number than those it could bring to bear at a given point and a given time. The development of long-range, hypersonic, brilliant munitions has made it possible to strike at Eurasian territory from the air without having to send American aircraft into the theater of war.

There is another dimension to this. Since World War II, on the ground and in the air, the United States has been in the offensive position, inserting its power into or near enemy territory at intercontinental range. While its motive may well have been to protect its allies, the United States has operated much nearer to enemy territory than to its own—usually at the end of a long supply line. Long-range operations have meant that numerically inferior forces faced superior forces. Under those circumstances, the United States needed to end wars as quickly and painlessly as possible. In spite of its strategy, the correlation of forces dictated that the United States had to adopt a defensive strategy. Rather than controlling the tempo and scope of the war and carrying the fight to the enemy to force a rapid decision, in Korea and Vietnam the United States ended up engaging in attritional warfare on the enemy's timetable. This opened the United States to disaster.

In addition to being strategically aggressive—fighting on the other guy's turf—the United States had to be operationally aggressive as well. It had to initiate conflict at the time and place of its choosing and maintain an operational tempo that could force an early decision. Its aggregate fire control had to be maximized and made thoroughly efficient. This was the achievement of the Gulf War.

During that conflict, projectiles were delivered from intercontinental distances. This meant that there were fewer of them than there were enemy soldiers and that each one had to count—in other words, the probability of a hit and a kill had to be multiplied many times over. Precision-guided munitions, in all of their forms, made this possible. This was the core difference between Vietnam and the Persian Gulf. Doctrine and technology had evolved so that the duration, the tempo, and the cost of war were controlled by the United States rather than by the enemy.

The most pressing mission American forces have today is the ability to maintain the balance of power in Eurasia by periodically intervening with sufficient force to shape the outcome of regional conflicts. However, the precondition for intervention has always been control of the sea—the ability of the United States to move forces to trouble spots in a secure and timely fashion. Ever since World War II, we have taken control of the sea for granted. In none of our wars or interventions since 1945 have we had to contend with a serious naval challenge. Yet, paradoxically, the same revolution that has increased our power on land and air threatens to decrease our power at sea.

PGM and the Sea: The Crisis in Grand Strategy

On land and in the air, the United States is the strategic aggressor. At sea, on the other hand, the United States is a defensive power. Since World War II, the Navy's control of the sea has been absolute. Its power can only decline. Indeed, Soviet strategy during the Cold War was deeply concerned with finding the means for challenging American control of the seas, particularly in the North Atlantic.

As a result of its preeminent position on the world's oceans, the U.S. Navy was, by definition, the most conservative of the American armed services. The Army and Air Force were always facing powerful enemies and were therefore constantly looking for newer weapons and doctrines. They were always hearing footsteps. For the Navy, however, it was a case of "mission accomplished" for the last fifty years. Improvements in weapons systems were viewed as merely an incremental matter, a matter of the Navy's doing what it already did a bit better than before. The U.S. Navy excelled at developing new classes of fighter and attack aircraft, larger and more efficient carriers, and at improvements in antisubmarine warfare. These also suited its task admirably. It did not have to rethink the foundations of naval warfare since the introduction of the aircraft carrier, while the Army and the Air Force not only had to rethink doctrine but had to rethink doctrine because of very real defeats and failures.

Unfortunately for the Navy, the revolution initiated by American air and land forces is inevitably sweeping out to sea. The same weapons systems that permitted precision air strikes against Iraqi forces and permitted outstanding antitank weaponry on the ground have been transferred to the world's oceans. But instead of strengthening the Navy's strategic position—which, given its already overwhelming power, could hardly be strengthened—this evolution threatened the Navy.

Powerful munitions, striking at enemy planes, tanks, and strategic sites with stunning accuracy and from beyond the range of enemy retaliation, transformed the correlation of forces in the air and on the ground. If this new weapons culture were introduced at sea, where U.S. weapons platforms swarmed and outnumbered those of potential enemies, the effect would be reversed. Here, U.S. naval weapons platforms, surface ships, aircraft, submarines, would find themselves very much at risk. Indeed, the Pax Americana, in place at sea since 1945, might well be threatened or undermined as American weapons systems became vulnerable.

The force multiplier that benefited the United States in Eurasia seemed to turn against it at sea. In the air and on land, the problem was to push ahead with more and better intelligent weapons—at least for those parts of the services not committed to the tank or the manned bomber. For the U.S. Navy, the problem was developing defenses against these weapons, with relatively little emphasis being placed on forcing the evolution of antiship and antiair precision-guided munitions. This was both reasonable and ominous. The Navy already had in its weapons inventories the means of sweeping the oceans clear of enemy platforms—air, sea, and submersed. Radical breakthroughs in offensive weapons were not seen as essential.

The Navy's obsession was with protecting itself against enemy attack. This was the rationale behind the Aegis ship-defense system. This system, consisting of sensors, computers, and weapons systems, was designed to destroy attacking projectiles. It was fiendishly expensive and was set an impossible task—the destruction of every incoming projectile. As the USS *Stark* demonstrated when an Iraqi Mirage attacked it without warning, a single error would result in catastrophe.

The threat posed by faster and more intelligent projectiles, whose launch sites can neither be located nor destroyed, imposes a tremendous defensive burden on the American fleet, as greater and greater resources are diverted from the primary offensive mission to the task of self-defense. Indeed, the cost of fleet and ship defense is soaring as a proportion of total cost. Not only is American maritime hegemony threatened by secondary powers, such as Iran, Iraq, and North Korea, but there are increasing signs that the weapons systems that are the foundations of American maritime power—the carrier battle groups—are heading toward senility, the point where the cost of merely keeping them alive is undermining the general capacity of the United States to carry out its strategic mission. We have not yet arrived at this point, but as we shall see, the revolution in smart weapons is rapidly moving the U.S. fleet in this direction.

If this is so, it more than wipes out the good news we have had on land and in the air. The United States, like Britain before us, is at root a naval power. Control of the seas is the foundation of American security, defensive and offensive. If the revolution in precision weapons allows the United States to engage and defeat much larger air and land forces, this ability is meaningless unless the United States can also project its forces across the oceans and supply and sustain them there. If the revolution in precision makes this impossible, then the United States, which initiated the revolution, will turn out to be the net strategic loser. Therefore, we must measure carefully the effect of the new technologies on U.S. naval power, consider possible responses, and decide on the optimal allocation of resources.

The United States was the master of maritime airpower because of its aircraft carriers—a weapon the Soviets tried, but failed, to emulate. American carrier battle groups and amphibious expeditionary forces sailed the globe at will, controlling the sea-lanes and landing expeditionary forces where they wished. Indeed, the operational role of the carriers had less to do with sea-lane control than with transporting airpower to Eurasian shores, simply because carriers were so effective in their policeman role that no one seriously challenged them for half a century.

The threat facing the aircraft carrier today is the long-range, hypersonic missile. These missiles pose two threats to carriers. The first is that they will penetrate the carriers' complex and expensive defenses and sink them; the second threat is that they will do the carriers' job more effectively and more cheaply. This is the classic case of senility. On one side a new weapon system brushes aside the intricate defenses of the older system. On the other side, the same technology that destroys the older system carries out its tasks more efficiently.

One mission of the aircraft carrier was to secure sea-lanes by destroying enemy warships and threatening enemy merchantmen. Its other mission was to bombard enemy territory within the range of its aircraft. Where guns could fire for a distance of a few dozen miles at most, aircraft could range outward for hundreds of miles, patrolling the central oceans as well as coastal waters. Carriers allowed aircraft to operate where there were no airfields. During the great carrier battles of the central Pacific in World War II, American and Japanese naval forces were frequently operating far beyond the range of land-based aircraft. Even if some aircraft had the range to fly from land and attack the carriers, their speeds were so slow that by the time they reached the target the battle would have been long over. The carrier existed as an intermediate platform for aircraft—bringing them within range of enemy forces without putting itself too close. Carriers were necessary, therefore, to supplement the increased but still limited range of aircraft, as well as to decrease the amount of time needed to fly to a target. Range and speed have become the great variables of modern warfare, particularly naval warfare. The carrier brought aircraft within range and into a time frame where the speed of the aircraft did not limit its ability to affect the tactical evolution of the battlefield.

In this, there is a direct parallel between the aircraft carrier at sea and the tank on land. Both were designed to bring a weapon of limited range into contact with the enemy. The tank brought the gun to within range of the enemy. To do this the tank itself had to be transported long distances, protected from counterfire, supplied, and so on. As we saw in part 2, the net result was an extraordinarily expensive weapon whose mission was to fire a shell containing a few dozen pounds of explosives a few thousand feet. The cost of transportation and supply eventually dwarfed the mission.

The aircraft carrier finds itself in an analogous position. Its mission is to deliver manned aircraft to within a few hundred miles of a target, so that each aircraft can carry and drop seven or eight tons of ordnance on an enemy. For thirty-six or forty aircraft to be brought into proximity with a target, a vast flotilla must be launched, supplied, and, above all, protected. In the end, three squadrons, all of which cannot fly at the same time except under special circumstances, are all that an entire carrier battle group could deliver to its target. During Desert Storm, when the efforts of the Navy came to be measured against those of a land-based air force, they didn't appear to be quite up to snuff.

The Navy nevertheless had two superb arguments in favor of the carrier's continued effectiveness:

- If the United States wants to control the sea-lanes by dropping high explosives on enemy ships, carriers remain the only effective way to transport the aircraft needed to carry the explosives within range of strategic shipping lanes and choke points.
- If the United States wants to drop explosives on the littorals of the world's land masses without being dependent on local air bases, only the aircraft carrier can carry out the mission.

The need for carriers to sustain the mission is based on the assumption that the only effective way to drop ordnance on ships or on nations bordering the oceans is to transport aircraft to within range of those targets. If that assumption is true, then the decline in the cost-effectiveness of the carrier, while regrettable, does not reduce American reliance on them. It is what might be called a threshold mission. Sea-lane control and force projection have to be carried out regardless of cost. No calibration is necessary or possible.

The post–Cold War defense debate has not focused on the continued usefulness of the carrier battle group. Rather, it has revolved around the number of carrier battle groups needed to maintain the strategic interests of the United States. For example, the General Accounting Office has argued that the basic political mission of the United States could still be carried out with only ten carrier battle groups rather than the twelve it currently has.[1] By implication, as any of these elements shift, the operational elements of the core strategy will also shift.

Now, there are those who would argue that after the Cold War, neither sea-lane control nor force projection is in the national interest of the United States. We clearly don't share this view. Nor do we argue that the United States cannot afford a strong military force. The United States has vast riches and could, if needed, afford far more than it has without substantial debilitation of its national economy or society. Rather, our argument focuses on an operational question—whether the aircraft carrier remains an effective means for the United States to carry out its national mission.

Certainly, it cannot be argued that antiship missiles have already made the aircraft carrier obsolete. Nor can it be argued that cruise missiles are now available to take over the carrier mission. The carrier can continue to operate, if only because antiship technology has not yet diffused to the point that it is threatened. However, given the calculus of risk and reward that we discussed earlier, it is no longer clear that defense planners can permit carriers to be exposed in waters where enemy antiship missiles are likely to be in place. Certainly, the insertion of carriers requires extraordinarily good intelligence on the location of launchers and airfields, and the capacity to destroy both—which poses a chicken-and-egg question. The precondition for insertion of carriers is the destruction of threats that have traditionally required carrier-based airpower to be available in the first place.

The ambiguity of carrier survival does not extend to cruise missile technology. At this point, cruise missiles have neither the range nor the speed to take over the bombardment role that carrier-based aircraft have. However, even here we see the beginning of an evolution. Consider the problem posed by antiship missiles launched from land and from aircraft, and consider it in the context of operations in the Persian Gulf. Given the likelihood that Iran currently buys surface-to-surface antiship missiles from China, it would be imprudent to move carriers through the Strait of Hormuz in any circumstance where hostile Iranian action was deemed likely. Now, even assuming perfect intelligence on location and type of missile, an air strike by carrier aircraft is dangerous. Since the cancellation of the A-12 project, the Navy has no stealth aircraft. The EA-6 is able to fire antiradiation missiles, but neither the F-18 nor the A-6 is stealthy. Use of either carries a substantial probability of casualties. Moreover, although the Navy has recently improved the capabilities of its precision-guided weapons, they were notoriously weak during Desert Storm on both laser designation of targets and electro-optical guidance.

The optimal solution for attacking targets protected by surface-to-air missiles, as was shown in Bosnia, is the TLAM—the fabled Tomahawk land-attack missile. The Tomahawk's range allows the carrier battle group to stay so far offshore that an intelligence failure concerning the type or capabilities of antiship missiles would not result in a potentially fatal attack on the carrier battle group. It would permit an assault on the control centers of the air defense system, no matter how far inland they might be, as well as on specific launchers. Having suppressed the air defenses, the fleet would then have the option of moving in and launching aircraft or using additional Tomahawks to take out the antiship missiles. At that point, the carrier battle group could pass through the Strait to carry out whatever conventional strike missions were needed.

In both Desert Storm and in Bosnia, a class of missions emerged for which cruise missiles were less suited than carrier-based aircraft. Of course, the next question that must be addressed is what missions could better be carried out by cruise missiles than by carriers. The General Accounting Office, concerned with reducing the cost of America's politico-military strategy, has already noted the possibility of substituting Tomahawks for carriers. Given the proclivities of the GAO call for cuts, and dangerous ones at that, we must harbor a suspicion that their views are premature. Nevertheless, they've expressed a view that is unlikely to be expressed in the naval air-warfare community—that cruise missiles can be as good as, if not better than, aircraft carriers.

The report, issued on April 25, 1995, made the virtues of cruise missiles clear:

> As demonstrated during Operation Desert Storm and the two
> Iraqi raids, cruise missiles have advantages over tactical air-
> craft systems and provide military commanders additional

options for precision strike operations. Cruise missiles can strike many types of targets and can be used in many conditions, such as at night, in a variety of weather conditions, or in heavy air defenses. Cruise missiles can also be used without additional resources—electronic warfare aircraft, fighter escort, and refueling aircraft—required for manned aircraft strikes. Additionally, as the raid on Iraqi intelligence headquarters demonstrated, such strikes do not require the presence of an aircraft carrier battle group. Employing cruise missiles can also avoid possible political constraints, such as obtaining host nation permission to use U.S. aircraft from forward deployed bases or fly through a third nation's airspace. Most importantly, cruise missiles provide the ability to strike targets without risking the loss of aircraft and the death or capture of U.S. air crew members. However, tactical aircraft systems have some advantages over cruise missiles, including their ability to attack mobile or relocatable targets and penetrate more hardened targets, and will therefore retain a key role in offensive air operations. Also, aircraft-delivered munitions are better suited for conducting large-scale or extended campaigns because of their relatively lower costs.

Since the Tomahawk and CALCM [conventional air-launched cruise missile] have broadened the options available to commanders and can be used against many categories of targets struck by manned aircraft, the characteristics (such as range and degree of stealth) of most aircraft and the number of aircraft required for future precision strike weapon systems should be affected. In addition, since Navy warships carrying cruise missiles have shown that they can conduct forward presence missions and crisis response without the presence of carrier-based air forces, they are a viable option for performing those missions. As a result, fewer aircraft carriers may be required, which could result in budgetary savings.[2]

Tomahawk and its equivalents have limitations, but they are not *inherent* limitations. The limits on an aircraft carrier are, indeed, inherent—they are the nature of the beast. To launch and land aircraft, a large ship is required with a flat, sharp deck. The aircraft that are used are manned and suffer from the limitations of manned aircraft. Aircraft carriers are costly, and therefore no nation can afford too many of them. Finally, aircraft carriers are vulnerable to missiles and therefore require extremely expensive and uncertain defensive systems.

The problems noted with cruise missiles are the design characteristics: they are inherently soluble. The report specified a series of operations in which manned aircraft are currently more effective, including "their ability to attack mobile or relocatable targets and penetrate more hardened targets." Both of these can be dealt with through cruise missile design:

- Currently, cruise missiles are pretargeted both by route and by final destination. Targets that can be relocated during time of flight can confuse cruise missiles. However, this weakness is not inherent. Sensors, such as reconnaissance satellites or unmanned aerial vehicles observing the target during flight, can note the relocation of targets and feed targeting information during flight to the cruise missile, allowing onboard computers to redirect the missile.

- Substantial work obviously has to be done on both warheads and the terminal phase of cruise missile flight. A more efficient explosive is clearly needed to maximize the effect of each cruise missile mission. In addition, to attack and penetrate hardened targets, a detachable warhead is needed, with rocket-assisted propulsion designed to increase terminal velocity. Current GBU-27 and GBU-28 hardened penetrating bombs can do this.

- Not mentioned but frequently cited as a mission more suited to aircraft than cruise missiles are area saturation bombings, such as those against an enemy infantry force. Cruise missiles have thus far been used as a single explosive system, rather than as a canister or dispenser. There is no inherent reason why a cruise missile could not dispense area munitions with at least the precision of an attack aircraft. Indeed, a cruise missile, having dispensed its munitions, could then return to its base, eliminating the problem of the costliness of a single use.

The GAO captured the essence of the debate. By 1995, a cruise missile could carry out numerous missions at least as effectively as a carrier, without placing a carrier at risk. Indeed, in November and December 1995, due to rotation problems, the U.S. Navy had sufficient confidence in cruise missiles to leave the Gulf region without a single aircraft carrier—the first time since 1990. Left in their place were two

Tomahawk-capable cruisers, and a squadron of F-16s grudgingly permitted temporary haven in Bahrain—far from the potential action.

The GAO's focus is primarily financial—the cost of the carrier battle group. This is not an inappropriate focus, but it is not sufficient. The high cost of the carrier is a result of its vulnerability to antiship missiles, the cost of aircraft able to operate in a hostile environment and be launched by a carrier, and the fact that a single carrier battle group lacks sufficient punch to be decisive in a Eurasian crisis. The emergence of a cost-effective cruise missile is merely the other side of the coin. The same technology that makes the aircraft carrier so expensive can also sustain the missions that aircraft carriers currently execute. What we are seeing is not so much a financial crisis as the rotation of weapons cultures—the end of the carrier age and the genuine beginning of the missile age. The question is how U.S. defense policy and politico-military strategy ought to respond to it.

Blue Water Power and the End of the Carrier Age

We have projected an evolution of the current primitive cruise missile into an instrument that can strike globally with centimeter precision, powered by scramjet engines, traveling so quickly that it can hit a ship in the center of the Atlantic Ocean from an East Coast base in under five minutes. Multisensor surveillance satellites and high-altitude, long-endurance UAVs could see ships at sea with clarity and precision. Visible light, infrared, radar, ladar, and combinations would make it impossible to hide tons of steel. Whichever direction the ship maneuvered, data fed to the on-rushing missile moving at Mach 20 to 25 would make impact inevitable. Possibly the cruise missile could be shot down, but the ship would be the least likely platform for doing so, sitting virtually stationary, reaching out with its own radar, and having bare seconds to respond after intercept.

A warship using a super-Aegis system could shoot down one or two cruise missiles. But suppose the missile calved, as the multiple independently targetable reentry vehicle (MIRV) did in the dark days of the Cold War. Suppose the missile, seconds before impact, split into two, three, ten warheads. Suppose ten warheads converging on the ship split into a hundred. In the end, any purely defensive system could be overwhelmed. Given the limited number of warships, their cost and strategic importance, it would be inevitable that at some point there would be one more incoming missile than the ships could intercept.

The counter to the cruise missile would have to be a system that could see the missile from the moment it was launched, track it on its path, and select a weapon with which to shoot it down. Just as the hunting of the ship began in space, in the high ground from which the ocean can fully and clearly be seen in all the spectra, so too the hunting of the missile would have to begin in space. Infrared sensors, high in geostationary orbit, would have to spot the missile's first hot burn, then track it. Missiles would have to be sent to intercept the enemy missile. These could be based on ships or on UAVs designed to destroy incoming missiles, such as the Raptor series. But most likely, the missiles would be based in space, traveling in low earth orbit, individually or clustered into pods, triggered into action by a controller—perhaps somewhere on the ground or, to save precious seconds and limit interference, in space—managing both sensors and weapons. The controller would order the missiles' engines to fire and, taking advantage of the gravity well, they would hurtle down toward an intercept.

The question in this scenario, of course, is, what was the warship doing in the middle of the Atlantic? What could it do that could not be done by missiles and space-based systems? It does not take much imagination to see that a warship traveling at twenty or thirty knots is lost in a world of projectiles traveling at twenty thousand miles per hour and that a ship whose radar can see for a few hundred miles—while revealing its position to the enemy—is not a competitor with a sensor that can simultaneously see over one-third of the globe—without revealing its position to anyone.

The paradox of the sea, however, is that while the warship might be forced out of its bluest parts—indeed, it may no longer be needed there—the merchantman must continue to move through those waters. The economy of the world depends on maritime trade, and the economics of maritime trade requires that the merchantman not carry expensive weaponry. This means that maritime powers will have to maintain the sea-lanes, but will not be able to use the traditional methods of sea-lane control—surface warships.

We are moving into a time when it will be much easier to cut the sea-lanes than to maintain them. Ever since the development of a global economy in the fifteenth century, great powers at their apex controlled vast maritime empires—and had large navies dedicated to protecting the merchantmen plying the sea-lanes. Emerging powers, seeking to challenge the predominant power, set their sights on cutting those sea-lanes. From English pirates to German submarines, emerging, challenging powers sought to cut the sinews of imperial power. When they succeeded, it was simply because offensive maritime technology was over-whelmingly more potent than maritime defensive capabilities. With the advent of hypersonic, long-range cruise missiles, we may well be encoun-tering this situation—at precisely the moment when the prosperity of the global system has become utterly dependent on the unimpeded access of all merchantmen to the sea-lanes.

We have had hints of this emerging phenomenon throughout the twentieth century. During both the First and Second World Wars, Germany's challenge to the British Empire focused on the ability of its U-boats to cut the line of supply between the Empire and the British Isles, or between the United States and Britain. In both wars—and par-ticularly the second—the Germans came close. Each time they failed because their enemies had a few too many destroyers or because tech-nology—sonar—was just a little ahead of them, or because they simply did not have enough U-boats. In World War II, the United States, strug-gling for maritime supremacy in the Pacific, showed what the submarine could do in interdicting the world's sea-lanes. During an unrestricted submarine campaign, unparalleled in its savagery, the United States cut the lines of supply between Japan and its empire, bringing its economy to its knees even before the bombing campaign took full effect. Japan never managed to mount a successful counterattack on the U.S. subma-rine campaign.

The submarine was the first systematic attack on the surface war-ship. Its advantage was its stealth. Except for sonar, imprecise for much of World War II, a submarine was invisible. It could wait astride an enemy's supply lines, strike at defenseless merchantmen, and cut an empire in half. The Anglo-Americans fought a desperate battle to hold open the supply lines in the North Atlantic, and at some points from 1941 to 1943, they seemed to be losing the battle. However, using specialized surface warships—destroyers that were armed with depth charges, loaded with sonar, and extremely fast, compared to the eight or nine knots a diesel submarine could make while submerged and operat-ing by battery—the Allies managed to hold open the lines of supply, both on the North Atlantic route and the route to Murmansk.

The threat from the air was also significant. If the Germans had possessed air bases in the Atlantic, in the Azores or Iceland, or if they had had aircraft carriers, they most likely could have shut down the Atlantic. Indeed, in the run to Murmansk, where the Germans were able to use both submarines and aircraft, they did manage to shut down the flow of supplies for a period. In the Pacific, in the islands between Hawaii and the Philippines, Japanese land-based aircraft made resupply in early

1942 virtually impossible. During the island-hopping campaigns of 1943–44, the United States, particularly using carrier airpower, managed to seal off the Gilberts, Marshalls, and Marianas so completely that the Japanese found it impossible to supply or withdraw.

Pressure on the sea-lanes has been growing throughout this century. During the Cold War, the threat to the sea-lanes grew, but it was not fully apparent. From the Pacific campaign onward—Korea, Cuba, Vietnam, the Persian Gulf—limits on U.S. sea-lane interdiction were political and not technical. To be more precise, the United States could close off virtually any body of water that it chose. It did so rarely because it feared military countermeasures on land and because the political or economic prize did not equal the effort. An important question to ask at this point is not whether the U.S. fleet could shut down a given body of water, but something more hypothetical. Had the Soviets been able to construct a fleet identical to the American fleet, would they also have been able to shut down any given body of water? Would the two navies have neutralized each other and neither been able to shut down any body of water— or would one navy have inevitably destroyed the other, returning to the status quo ante?

The power to interfere with ocean commerce has been growing faster than the ability to keep it open. Ever since the introduction of the attack submarine and the attack aircraft, it has been easier to close the sea-lanes than to keep them open. Even against a relatively small bomber force and attack submarines, American planners had serious doubts whether the North Atlantic could be held open in the event of a war with the Soviet Union. At the very least, shipments would be lost or delayed, very possibly representing the margin between defeat and victory. Torpedoes homing in on sound or magnetic sources and missiles homing in on radar images made it possible for the Soviets to entertain hopes of closing off the North Atlantic.

The advent of the long-range, high-speed, intelligent missile divides the problem of sea-lane control as it has never been divided before. On one side, it becomes much easier to control, or close down, sea-lanes. On the other side, keeping the sea-lanes open becomes a tremendous problem, one that can be simply posed: How can goods be moved across an ocean without exposing the vessel to missiles that could be fired by virtually any country in the world? This is, of course, an extreme formulation of the problem, and it will undoubtedly be many years before we reach that point. Nevertheless, it is the point on which the future of naval warfare will turn.

The United States faces four alternatives:

- Accept the closure of sea-lanes. In effect, this would return the world to its condition prior to the fifteenth century. The world would become fragmented into smaller regional entities with overland access limited by normal physical and political impediments, limited coastal movement, and a complete halt in transoceanic commerce. While this may seem like a postmodern nightmare or an impending return of the Dark Ages, it is not, on sober consideration, altogether preposterous, since diffused technology would mean that the nation with the least to lose from disruption and the most to gain from blackmail would have the strongest hand. An international system that depends on the self-restraint of those who have much to gain from disruption requires constant intervention against potential disrupters—on land, and on the least favorable terms to the peacekeeper. This is inevitably a wearying enterprise. The Romans failed at it and there is no reason to believe that we have greater staying power.

- Move goods by means that are not susceptible to interdiction by missiles. Submarines come immediately to mind, save that the enormous cost of submarine shipments would itself undermine the global economy, while the submarine itself is no longer safe from intelligent projectiles. Air transport is obviously too expensive for commodity shipment, while lighter-than-air schemes usually fail because they are even more vulnerable than surface vessels.

- Create ships that cannot be sunk. This is a strange strategy but not as preposterous as it sounds. One of the problems with ships is that they are too small relative to the explosives they face. Hypersonic missiles will have severe size constraints, and though conventional explosives can be improved, they have limits as well. One solution would be the construction of extremely large, if slow, ships in which much of the cargo space was submerged, with a superstructure able to take multiple hits without losing its operational capability.

- Eliminate the threat of cruise missiles by eliminating the strategic sensing systems that must be in place to fire them effectively. In effect, to control the sea-lanes, one must control space. The smallest and most belligerent country may well have purchased hypersonic missiles, but it is unlikely that it has achieved the spaced-based reconnaissance capabilities necessary to use them. Indeed, while a cruise missile can be hidden and launched from

just about anywhere, satellites need launch facilities that could
be located and destroyed should a minor power become genuinely ambitious.

We have now returned to the same strategic conclusion as we
reached in our discussion of the future of air and land power: the centrality of space. The finest long-range, hypersonic projectile is worthless
without intelligence concerning the location of its target. Targeting
information must come from the high ground. One choice is in the
upper limits of the atmosphere—at seventy thousand to eighty thousand
feet using UAVs. But no matter how efficient such vehicles might be, and
no matter how long they could loiter using only solar power, anything
inside the atmosphere is subject to counteraction by less sophisticated
technologies than those required for space. In other words, secondary
and tertiary powers could eliminate UAVs.

Moreover, UAVs do not have global reach. Even if they were able to
travel to a target area, the data would have to be transmitted back to the
launch site. An unmanned device twelve to thirteen miles up would
require multiple links to backlink to the launch site. Each of those links
must be put into position and protected against jamming. Under the
best of circumstances, these vehicles will allow a country to gather intelligence in its immediate region, but will not provide it with anything
resembling a global reach.

Therefore, the key will be space-based reconnaissance platforms
linked to missiles on the ground or, preferably, in space. A power that
has and retains a global reconnaissance capability will be able to choke
off the sea-lanes. It therefore follows that we are entering an epoch in
which sea-lane control will depend less on oceangoing naval vessels than
on space-based vessels. Ships will be convoyed from space, protected
from dangers from space, routed from space, and ultimately destroyed
from space.

Previously, controlling the world's oceans required the ability to
defeat enemy ships at sea, blockade enemy ports and coastlines, and control the world's choke points. The physical ability or inability to move
merchantmen from the Atlantic coast of the United States to Europe,
the ability to move through the Mediterranean, traverse the Panama and
Suez Canals, pass by Singapore, will determine the wealth of nations and
the patterns of international trade and power. It is not the geography of
global trade that is changing; it is simply the means whereby explosives
are delivered to these points. Where, previously, warships were used for
all of those missions, it is no longer necessary to sail a flotilla to the Strait

of Gibraltar or of Malacca. The explosive power of ships' guns and of carrier aircraft will be delivered to those same points by long-range projectiles, targeted and triggered from space.

Obviously that simple change has tremendous strategic significance. The center of gravity of naval warfare will no longer rest with the carrier battle group. It will shift to the command and control systems that can deliver projectiles globally. The nation that controls space will control the world's oceans far more completely than it can control the world's land masses. Hypersonic missiles are always going to be in short supply, compared to the number of possible land targets. As we recall, it is the relatively small number of vessels at sea that makes the power of space-based reconnaissance controlling hypersonic weaponry so impressive. There is a fit between the number of vessels and the probable number of projectiles. Although they would certainly be able to influence the outcome of war on land, missiles could never saturate the landscape. At sea, they would certainly be able to both saturate the seascape and determine the outcome of war.

Thus, where space-based systems will be a critical factor in the evolution of land warfare, they will, almost solely, be able to determine the outcome of conflict at sea. At the very least, they will be able to impose patterns of trade, helping friends and penalizing enemies. That will make control of space a critical economic advantage, which when connected with the obvious military function will make space control a political necessity.

Whoever controls space, therefore, will control the world's oceans. Whoever controls the oceans will control the patterns of global commerce. Whoever controls the patterns of global commerce will be the wealthiest power in the world. Whoever is the wealthiest power in the world will be able to control space.

Power Projection in a Transitional Epoch

It is important that we not end our consideration of naval warfare on the soaring note of space control, but on the more immediate and mundane level of amphibious warfare. In due course, wars will be fought in space. But in both the short and long term, American troops will have to "kick in the door" on hostile beaches and have the situation well in hand, whatever the status in space. Traditionally, sea power *has provided the preconditions* for the projection of power. It has not by itself been the projection of power. Sea-lane control made it possible for the Allies to invade

Normandy and impossible for the Germans to invade England. It was the prerequisite—necessary but insufficient. The infantry still had to go in and take the ground.

The last assault carried out by U.S. troops on a strategically hostile beach was at Inchon in Korea in 1950. In Vietnam, the United States carried out several tactical amphibious assaults. But in all other cases, it either refrained from amphibious operations, as in the Persian Gulf, or attacked against trivial opposition, as in Grenada. The U.S. Marine Corps is undoubtedly the finest large-scale amphibious force in the world—yet it has refrained from an actual assault on contested beaches for nearly half a century. During that half century, the dangers of amphibious assault have soared, while the technology has remained fairly stagnant. Our discussion of the infantry is critical here, with the added element that we must consider how to get them ashore and keep them fed and armed.

We must recognize that amphibious warfare is merely part of a general grand strategic system that the United States employs for political and geographical reasons. The general name for this is power projection, which is divided into five parts:

- The ability to marshal troops and supplies either in the United States or in some large, forward strategic base such as Europe or a fleet.
- The ability to transport these troops and supplies into the theater of operations.
- The ability to force open the target country by direct assault if necessary.
- The ability to reinforce initial forces with additional follow-forces.
- The ability to sustain forces in extended combat operations, with supplies and reinforcements.

The gap between the decision to intervene and the buildup of supplies in a theater to begin offensive (or even defensive) operations can be dangerously long. Yet it is built into U.S. strategy. For example, over a year elapsed from the decision to invade France to the completion of the buildup in May 1944. In Desert Storm, it took six months from the decision to intervene to the ability to mount an assault. In Korea, the length of time it took to build up nearly cost us the war.

During the Cold War, the United States set itself a basic force requirement called the 2½ war doctrine. Under this doctrine, U.S. forces were expected to be able to fight two major wars and one minor operation simultaneously. So, during the 1960s, U.S. forces were tasked with fighting the Vietnam War and were prepared to repel a Soviet invasion of Western Europe; at the same time, they were ready to carry out a secondary deployment, such as reinforcing Israel or invading a Caribbean island.

Following the Cold War, the Department of Defense conducted the Bottom Up Review, which, under Secretary of Defense Les Aspen, arrived at a follow-on doctrine known as the 2 Major Regional Conflict (2MRC) scenario, and it currently drives U.S. strategy and doctrine. The essence of the strategy, which has been under constant attack for being impractical, is that the United States will be able to fight one regional conflict (say, defend Saudi Arabia) while holding on with minimal forces in a second conflict (say, South Korea). When the Saudi situation stabilized, forces would be transferred to South Korea, increasing the tempo of that war.

The problem is not that the United States lacks the forces to carry out this strategy, but that this strategy places tremendous demands on its logistical capabilities. It would have two major insertion and supply operations under way, followed by a massive troop transfer. Critics argued that U.S. sealift capabilities simply were not up to the task. We could not get enough forces into a theater in time for one rapidly developing conflict, let alone to stabilize two widely separated forces.

As the size of U.S. forces draws down, the paradox is that logistical demands increase. Smaller forces mean that predeployment of those forces leaves few reserves for the unexpected. As forces decrease, the need to maintain those forces in a central reserve increases. This means that the ability to rapidly move forces from the continental United States to the theater and sustain them once there becomes the strategic bottleneck. The finest army in the world is worth little if it can't be delivered to the battlefield.

The United States has an ongoing mission of maintaining stability in areas both distant from it and separated by oceans. Whether it's small-unit peacekeeping operations or multidivisional forces to repel invasions, all U.S. military operations require the ability to deliver and supply forces. These operations are frequent, and overlapping. They are also often unexpected in terms of locale and timing. U.S. forces must be in a position to intervene, with little notice, on a global scale. Consider examples such as Kuwait, Liberia, Somalia, Haiti, or Bosnia. These varied in size, intensity, and duration, but all had in common the fact that they were largely unexpected. From a long-term strategic standpoint, the

United States has to assume that it might be called on to intervene virtually anywhere with relatively little notice.

As is evident from the chart below, the time required for even a fast sealift from the continental United States to a Eurasian theater is unacceptable if the goal is to stop enemy incursions as they develop. Bear in mind that this time does not include the transfer of supplies from depots or ports or loading in port and assumes flank speed without a breakdown. It is clear that, using traditional sealift strategies, the United States would not be in position to repel attacks. Instead, it would be forced into a costly and difficult counterattack.

TIME TO THEATER AT MAXIMUM POSSIBLE SPEED

Route	Distance	Mercy Class Hospital Ship 15 knots	Bob Hope Class Fast Sealift Ship 24 knots
Norfolk–Oman	11,604 miles	33 days	20 days
Norfolk–Somalia	9,735	26	17
San Diego–Singapore	7,736	22	12
San Diego–South Korea	5,344	13	8

The obvious solution to this dilemma is the predeployment of U.S. forces throughout the world, on the order of the U.S. presence in Western Europe or Korea. The problem with this precommitment strategy is that it requires a large standing force in areas of expected crisis—which is usually not the place where real crises occur. Following the Cold War, the United States began reducing its force size. Paradoxically, with the end of the Cold War, the scope of potential U.S. commitments actually grew, and U.S. interventions have been more frequent and diffuse than previously.

The United States therefore faces growing strategic responsibilities with diminished manpower capabilities. Lacking the manpower necessary for predeployment, its strategy must be to hold its forces in theater and strategic reserve and deliver them to target areas in time to deal with the crisis. To cope with a potential crisis in Liberia, for example, the United States has amphibious forces already positioned in theater— Marine expeditionary forces are deployed with U.S. fleets—which allows for intervention without a major transfer of forces from the strategic reserve.

As a crisis escalates, requiring forces beyond what a Marine expeditionary unit can deliver, the only alternative is to reach into the strategic reserves. In Bosnia, for example, a combination of forces from theater—Germany and the Sixth Fleet—and forces from strategic reserve—the continental United States—were made available. During Desert Storm, where available theater forces were insufficient for the defensive, let alone an offensive mission, the only solution was the rapid transfer of forces from the continental United States to the affected region.

This problem is not new. Ever since the 1960s, the United States has prepositioned equipment in Europe. In annual exercises called REFORGER, U.S. troops flew to Europe, picked up what they needed, and deployed. During Desert Storm, U.S. troops found supplies waiting in King Khalid City—a vast depot in Saudi Arabia used by U.S. military personnel. Marines had their equipment prepositioned in large ships docked at Diego Garcia in the Indian Ocean. Prepositioning alone made it possible to begin to defend Saudi Arabia within thirty days of the war's breaking out.

Immediately after Iraq invaded Kuwait, Secretary of Defense Richard Cheney and his staff flew to Riyadh. Their mission was to obtain approval for U.S. troops to deploy in Saudi Arabia and utilize the prepositioned equipment. Permission was forthcoming, but only after several days. Had the Saudis reached the same conclusion as the Jordanians, the prepositioned goods would not have been available. Prepositioning on land makes the United States hostage to the wishes of the host government. In unstable areas where the shape of conflict is unpredictable, betting on any one government is risky business.

A much better solution is found in offshore prepositioning—where the United States controls its own capabilities. Current offshore prepositioning is primitive at best and nowhere near the scale necessary to sustain multidivisional deployment. The Marine prepositioning ships are designed to support expeditionary warfare in the subdivisional range. They are suitable for small-scale interventions or the initial phase of larger interventions, but they do not come close to strategic requirements.

During the Gulf War, their defects became apparent. First, they were not efficiently designed for loading and unloading, which created unnecessary delays and problems in maintaining equipment. Second, there were too few of them. Many more were needed to sustain operations while the strategic buildup took place. Finally, and most important, they were ships and thus vulnerable to attack. After Desert Storm, one suspects that any sensible enemy will open its operation with attacks on these vulnerable vessels.

What we learned from the Marine prepositioning ships was the absolute necessity of having supplies held near theater, under the political control of the United States, available for use within hours or days. These supplies must be designed to permit U.S. troops to be moved into position either through landing in friendly airports or by amphibious operations. While the Marine force would carry sufficient supplies for "kicking in the door," the prepositioned supplies would be used to sustain operations for between thirty to sixty days—including land and air operations—and would reduce the long and increasingly vulnerable supply line from the continental United States to Eurasia. This would mean that strategic supply from the United States, however constrained, would be used to construct the strategic reserve rather than be used during immediate operations. With supplies prepositioned, constraints or temporary interdiction could be tolerated better, and the opportunity of passing from the defense to the offense would present itself much more quickly. Defense could occur in days and counteroffensive in weeks, rather than the current unacceptable reality.

The issue is, of course, where the United States could find logistical bases under its control and suitably positioned. As we have pointed out, securing and maintaining bases in critical areas makes the United States hostage to political forces in the host country or forces it to maintain bases in hostile countries. One Guantánamo is enough.

The real solution is for the United States to create forward logistical bases at points of its own choosing on the high seas. In effect, it ought to build a cluster of man-made islands at strategic points around the periphery of Eurasia.

The forward logistical base would have to be able to:

- Store supplies and equipment sufficient to maintain at least one mechanized division in high-intensity combat for thirty days.
- Provide docking facilities for a wide variety of wet and dry supply ships and rapid off-load capabilities for a specialized fleet of amphibious craft.
- House at least a brigade of troops for several months.
- Withstand hits from several high-explosive projectiles and remain operational—not merely survive.
- Land supply aircraft and vertical short takeoff and landing fighters to maintain a minimal combat air patrol.
- Have a minimal air defense system not designed to be leak-proof but designed to reduce hits to sustainable levels.
- House and fire antiship, antisubmarine, and land-attack cruise missiles.

- Surround itself with passive antisubmarine warfare systems, from nets to mines.
- Be movable, although not necessarily mobile. Like communications satellites during Desert Storm, it should have the ability to redeploy from one theater to the other in a matter of weeks. A maximum speed of two knots would be sufficient.

Such a system would have to be huge and tactically immobile. This would be part of its charm. Because it would be vast, it could sustain impacts that would disable or destroy a ship. Not only would it dissipate shock more effectively, but its size would make redundancy of systems meaningful. Size would also permit the construction of compartments that could be sealed off, such that the destruction of one section would not mean the destruction of the rest.

Moreover, its tactical immobility would have little meaning. From the standpoint of a missile coming in at Mach 5 or faster, the difference between a stationary platform and a ship attempting to maneuver at thirty knots would be academic. A ship could not maneuver away from the missiles we have described nor could it hide from their fused, multispectral sensors nor spoof all the brilliant processors. Any vessel or platform at sea will be hit. The trick will be to take the hit and still be able to operate. Marine prepositioning ships will not be able to do so, but huge, stable deep-sea platforms will.

The problem of force projection in an age of precision-guided munitions requires a systems approach—in which the depot organization in the United States is intimately connected with the grunt squeezing off a shot at the enemy. Indeed, everything discussed here requires a systems approach.

These platforms are multipurpose entities. In addition to serving as forward logistical bases, they can also serve as strategic firebases, both for sea-lane control and for fire support for amphibious forces. Properly sited, these platforms can project sufficient firepower to deter approaching ships and planes as well as enemy aggression on land. But with or without these other elements, the ability to supply U.S. forces through the opening phase of a war, without requiring prior supply from the United States or cooperation from regional allies, is quite enough. It will allow the United States to control the tempo of war, stabilizing circumstances regardless of enemy intentions, moving to the offensive well before history would predict, allowing strategic resupply to go to a strategic buildup rather than being frittered away on desperate defensive tactics.

The most vulnerable portion of the bridge between the United States and Eurasia will be the last few hundred miles from the forward logistical base to the shore—and as the range of missiles increases, the vulnerability of supply craft will be continual. Therefore the forward logistical base requires an addition—a new class of tactical amphibious craft able to move with speed and agility from the base to the landing area or port. These will have to be small and numerous. Small size will increase the possibility of evading missiles. Large numbers will ensure that the number of craft outstrips the number of effective missiles in the enemy's hand. Just as missiles will saturate Aegis, so these quick, agile amphibious craft will outlast the missiles. These may well be robotic platforms, designed to cut down the number of human casualties the inevitable missile kills will exact.

The threat to American control of the seas can no longer be solved at sea. In the same way that the advent of aircraft shifted sea-lane control to the air, the advent of space power shifts sea-lane control into space. Put differently, the longer the range of munitions available, the greater their dependence on reconnaissance and intelligence. The less that can be seen from the launch site, the more information will be required from the field.

In the future, projectiles with tremendous speed and range will be available. But they will only be effective if those firing them know what they are firing at. Thus, space-based reconnaissance, which alone can provide the sort of global vision the new weapons demand, becomes the foundation of war in general and of naval warfare in particular. With modern weapons, the ability to hit whatever is seen or sensed means that it is essential that no one be able to see or sense your vessels.

Even as newer weapons technologies diffuse, allowing nations like Iran or North Korea to fire long-range, precision-guided munitions at American surface vessels, the United States will be able to maintain sea-lane control so long as advanced reconnaissance capabilities do not diffuse. It is essential to American national security that the United States retain control of space and deny its use to its enemies.

It is also essential that the United States control less strategic reconnaissance capabilities. This is particularly the case with the poor man's satellite, the unmanned aerial vehicle. While not providing global coverage, a UAV at seventy thousand feet will certainly be able to provide theater-level coverage—the sort of thing that will threaten U.S. supply efforts from the forward logistical base to the shore.

Unfortunately, there is no way to stop the diffusion of UAV technology. It is both too simple and in the hands of too many nations to prevent diffusion. This means that the United States will have to emphasize defeating UAVs. Shooting them down, jamming their transmissions, confusing their sensors—this must be a central thrust of American research and development. Indeed, anti-UAV warfare goes hand in hand with space control systems.

An interservice Space Command is a critical first step to an integrated war-fighting strategy in space. But space is too important to be a subordinate part of the funding process. Its uniqueness and its singular importance argue for a separate Space Force sometime in the next century.

This will, of course, raise the question of the relationship between the space force and the other armed services. Just as the advent of the Air Force deepened the competition for scarce resources rather than ended it, the advent of the Space Force will compound rather than simplify the problem. The Space Force, like the Air Force before it, will claim that it will be able to deal with conflicts without support from the other services. As with the Air Force, this claim will be nonsensical. The Army will hold the ground; the Navy will control the sea; the Air Force the air. But all will depend on the Space Force for the reconnaissance, fire support, and coordination that none will be able to carry out by itself. Two hundred years from now, budgetary battles between the armed services will continue to rage, as important or unimportant as they are today.

In due course, control of space will be challenged by other, emerging powers. At that point, the geography of space will provide us with a new geopolitics. The future of war will then be determined by the surface of the moon, its craters and mountains, the various libration points, the different characteristics of orbital altitudes, as it was determined in the past by the shape of land masses and the terrain or climate.

Postscript

We are entering a dramatically new global epoch in which the United States holds, and will for the foreseeable future continue to hold, center stage. A new reality creates a new power. A new power creates a new type of war. The emerging power of the United States means a great deal not only to American citizens but to the world as a whole. The nature of the American regime and the manner in which it will use its power in the coming centuries will shape human history as other great powers have shaped it in the past.

In due course, American power, like that of all great powers, will decline. As we have argued, rumors of America's demise are premature. This is the beginning of the American epoch, not the end. We can see this when we look at the American way of war. It is not an old, declining military culture but an extraordinarily young and creative one, clumsy but robust. What it will look like in 250 or 500 years is beyond us. But just as by the year 1500 it was apparent that the European experience of power would be its domination of the global seas, it does not take much to see that the American experience of power will rest on the domination of space.

We can no more grasp the meaning of the American epoch today than a European in 1500 could have grasped the ultimate meaning of Europe's rise. What we can do, what we have tried to do here, is examine the structure of one dimension of America's rise.

Just as the gun shaped European power and culture, it appears to us that precision-guided munitions will shape American power and culture. Just as Europe expanded war and its power to the global oceans, the United States is expanding war and its power into space and to the planets. Just as European power redefined relations among nations, so American power is redefining those relations. Just as Europe shaped the world for half a millennium, so too the United States will shape the world for at least that length of time. For better or worse, America has seized hold of the future of war, and with it—for a time—the future of humanity.

Notes

Introduction: The Dawn of the American Epoch

1. U.S. Department of State, *Focus on East Asia and the Pacific,* Department of State Dispatch 4, no. 16 (April 19, 1993).

2. Robert Reich, *The Work of Nations* (New York: Knopf, 1991), 3. A more popular and sweeping argument in this vein can be found in Kenichi Ohmae, *The Borderless World* (New York: Harper Perennial, 1990).

3. George Shultz:

    Borders don't mean what they used to mean. . . . The concept of absolute sovereignty is long gone. As national boundaries blur, sovereign power is dispersed, and new players vie for international influence. . . . Sovereignty, statehood, and the nation may be becoming disentangled in important ways.

    Valéry Giscard d'Estaing:

    The period which we call the Cold War was a period entirely dominated by military issues. We lived in the expectation of war. . . . But economic issues are now taking the lead.

    Shimon Peres:

    Markets are becoming more important than countries. Economics is beginning to be as important as strategy.

    Shultz, Giscard, and Peres in *Los Angeles Times,* December 11, 1990.

4. Norman Angell, *The Great Illusion* (New York: G.P. Putnam & Sons, 1910), 64.

5. *International Historical Statistics: Africa and Asia*
    International Historical Statistics: The Americas and Australasia
    World Bank, *World Tables*
    United Nations, *Statistical Yearbook*
    State, Economy and Society in Western Europe: Data Handbook
    International Monetary Fund, *Direction of Trade Statistics*
    Central Intelligence Agency, *The World Fact Handbook: 1990*
    League of Nations, *Review of World Trade*

6. World Bank, *World Tables.* Figures for 1913 and 1929 were in Gross National Product. Later figures were in Gross Domestic Product.

7. International Monetary Fund, *Direction of Trade,* 1991

8. U.N. Economic and Social Council, Commission on Transnational Corporations
    Statistisches Jahrbuch 1989 für die Budesrepublik Deutschland
    Japan Statistical Yearbook
    Annuaire Statistique de la France
    Annual Abstract of Statistics (British)
    World Bank, 1990
    Feis, *Europe the World's Banker*

9. The major wars are:
1776–83: Revolutionary War
1812–14: War of 1812
1836: Mexican War
1861–65: Civil War
1898: Spanish-American War
1917–18: World War I
1941–45: World War II
1950–53: Korean War
1963–72: Vietnam War

10. Quoted in *The Economist,* December 7, 1991, 51.

1 David's Sling

1. See S. Goldman, commentary, *Samuel* (London: Soncino Press, 1949), 99, for conversion of weights.

2. A. Bailey and S. G. Murray, *Explosives, Propellants & Pyrotechnics* (London: Brassey's, 1989), 3.

3. It has been generally agreed that RDX, first developed in 1920, has been the most cost-effective, particularly when combined with TNT in a 60–40 mix. In short, improvements in explosives are rare: gunpowder, TNT, RDX, have been the basic for over seventy years. See Bailey and Murry, *Explosives,* 6–7.

4. On armor thickness, see Chris Ellis and Peter Chamberlain, *Fighting Vehicles* (London: Hamlyn Publishing Group, 1974), 95.

5. This is a dramatically expanding problem. The U.S. Army's Field Manual 100-5 notes that where Patton's Third Army required 350,000 gallons of gasoline per day, today a single armored division requires over 600,000 gallons. Interestingly, the Red Ball Express that supplied Patton with his petroleum required 300,000 gallons per day to supply Patton's 350,000. P. 60.

6. Martin van Creveld, *The Transformation of War* (New York: Free Press, 1991), 106.

7. James F. Dunnigan, *How to Make War: A Comprehensive Guide to Modern Warfare* (New York: William Morrow, 1988), 464.

8. While new cruise missiles costing less than the Tomahawk are likely, the Tomahawk is cost-effective against even single tanks. An M-1A2 Abrams tank currently costs about $5.4 million per unit (*Defense Daily,* September 18, 1995). A Tomahawk costs from $1 million to $1.3 million. A hit probability of 50 percent still means that it would cost less than half as much to destroy an Abrams than to build one. Given the fact that much lower cost systems, such as the Maverick, are currently in use against tanks, the cost-effectiveness of precision-guided munitions is even more lopsided.

2 Soldiers and Scientists

1. Gordon W. Prange, *December 7, 1941* (New York: Warner, 1988), 98.

2. Richard H. Kohn and Joseph P. Harahan, eds., *Strategic Air Warfare: An Interview with Generals Curtis E. LeMay, Leon W. Johnson, David A. Burchinal, and Jack J. Catton* (Washington, D.C.: Office of Air Force History, 1988). General Short and Admiral Kimmell were, respectively, Army and Navy commanders at Pearl Harbor on December 7, 1941.

3.   Karl von Clausewitz, *On War,* ed. Anatol Rappaport (London: Penguin Classics, 1968), 270.

4.   Ibid, 138.

5.   See Clausewitz's superb discussion of danger on the battlefield and the soldier, in Book I, Chapter 4: "Of Danger in War."

6.   An excellent recent history of the development and role of radar on all sides in World War II is David E. Fisher, *A Race on the Edge of Time: Radar—the Decisive Weapon of World War II* (New York: McGraw-Hill, 1988).

7.   See Rutherford's letter on the death of the young physicist Mosley on the battle-field, in Sir Solly Zuckerman, *Scientists and War: The Impact of Science on Military and Civil Affairs* (New York: Harper & Row, 1967), 13.

8.   On the British utilization of basic scientists during World War II, see J. G. Crowther and R. Whiddington, *Science at War: An Account of Scientific Work Administered by the Dept. of Scientific and Industrial Research of Great Britain* (New York: Philosophic Library, 1948).

9.   Immanuel Kant, *Critique of Pure Reason,* trans. Norman Kemp Smith (New York: St. Martin's Press, 1929), 576.

10. Clausewitz, *On War,* 203.

11. Robert P. Creas and Nicholas P. Samios, "Managing the Unmanageable: Government Management of Basic Research," *Atlantic Monthly* 267, no. 1 (January 1991): 80.

12. Lincoln R. Thiesmeyer and John E. Burchard, *Combat Scientists* (Boston: Little, Brown and Company, 1947), 9.

13. Bernard Brodie, *The Absolute Weapon* (New York: Harcourt, Brace and Company, 1946), 31, argues against the dangers of historical parallels being applied to nuclear weapons.

14. J. F. C. Fuller, "The Atomic Bomb and Warfare of the Future," in *Army Ordnance,* January-February 1946, 34, cited in Lawrence Freedman, *The Evolution of Nuclear Strategy* (New York: St. Martin's Press, 1983), 25.

15. S. K. Allison, "The State of Physics: Or the Perils of Being Important," *Bulletin of the Atomic Scientists* 6, no. 1 (January 1950): 3.

16. Minutes of the meeting of the Interim Committee, May 31, 1945, in Robert C. Williams and Philliom L. Cantelon, *The American Atom: A Documentary History of Nuclear Policies from the Discovery of Fission to the Present: 1939–1984* (Philadelphia: University of Philadelphia Press, 1984), 58–60.

17. In Fred Kaplan, *The Wizards of Armageddon* (New York: Simon & Schuster, 1983), 56.

18. Numerous other institutions were created on the Rand model, most of them beholden to one or the other of the services. The Navy had the Center for Naval Analysis, which grew out of antisubmarine-warfare work conducted by the Navy in World War II and was administered by the Franklin Institute, which also ran a famed scientific museum. The Army had the Research Analysis Corporation and several other institutions. In due course an entire web of such corporations, think tanks, and university laboratories under contract proliferated, until a vast and cumbersome bureaucracy and complex emerged. See Bruce L. R. Smith, *The Rand Corporation: Case Study of a Nonprofit Advisory Corporation* (Cambridge, Mass.: Harvard University Press, 1966), 1–6.

19.   See J. R. Goldstein, *Rand: The History, Operations and Goals of a Nonprofit Corporation* (Santa Monica, Calif.: Rand Corporation, P-2236-1, February 23, 1961), 3.

20.   Ibid., 8.

21.   Bernard Brodie, "Scientific Strategies," in Robert Gilpin and Christopher Wright, *Scientists and National Policy Making* (New York: Columbia University Press, 1964), 240.

22.   Philip M. Morse and George M. Kimball, *Methods of Operations Research* (New York: John Wiley & Sons, 1950), 1.

23.   Zuckerman, *Scientists and War*, 17.

24.   An excellent contemporary overview of the achievements of operations research in World War II can be found in Crowther and Whiddington, *Science at War*, 91–120.

25.   Herman Kahn, *On Thermonuclear War* (Princeton, N.J.: Princeton University Press, 1961), viii.

26.   In Charles E. Kirkpatrick, *An Unknown Future and a Doubtful Present: Writing the Victory Plan of 1941* (Washington, D.C.: Center of Military History, U.S. Army, 1989), 64. Wedemeyer was a major at the time of the preparation of the plan and was later promoted to general.

27.   Ibid., 14.

28.   In William Kauffman, *The McNamara Strategy* (New York: Harper & Row, 1964), 169.

29.   E. S. Quade, ed., *Analysis for Military Decisions* (New York: American Elsevier Publishing, 1970), 7.

30.   Charles J. Hitch, assistant secretary of defense under McNamara, the famed Plans, Programs and Budgets System. He describes the military implications of this process in an address to the Operations Research Society of America. See his "Plans, Programs and Budgets in the Department of Defense," *Journal of the Operations Research Society of America* 11, no. 1 (January-February 1963): 1–17.

31.   Indeed, the Army waged a rather remarkable struggle against the use of nuclear weapons on a wide range of grounds, including moral arguments that were later to be common in the antinuclear movement. That the intent was self-serving—to defend the Army's budget against the Air Force's—does not detract from the remarkable nature of this argument. For an overview of this period and argument, see A. J. Bacevich, *The Pentomic Era: The U.S. Army Between Korea and Vietnam* (Washington, D.C.: National Defense University Press, 1986).

32.   For a discussion of the geopolitical dimension of the Vietnam War, see Larry Berman, *Planning a Tragedy: The Americanization of the War in Vietnam* (New York: W. W. Norton, 1982), 8–17.

33.   Walter W. Rostow, *Stages of Economic Growth* (Cambridge, Mass.: MIT Press, 1960).

34.   In particular, an important counterpoint to Barrington Moore's study, which argued the opposite, that capitalism could no longer generate sufficient development capital to permit a society to develop.

35.   Walt W. Rostow, *The Diffusion of Power* (New York: Macmillan, 1973), 168. See also Douglas S. Blaufarb, *The Counterinsurgency Era* (New York: Free Press, 1977), 57–58.

36.   Blaufarb, *Counterinsurgency Era*, 72.

37.   William A. Nighswonger, *Rural Pacification in Vietnam* (New York: Frederick A. Praeger, 1966), 235.

38. Blaufarb, *Counterinsurgency Era,* 206–7.

39. Richard D. Wollmer, *An Interdiction Model for Sparsely Traveled Networks* (Santa Monica, Calif.: Rand Corporation, RM-5539-PR, April 1968).

3 False Dawn

1. Haywood S. Hansell Jr., *Strategic Air War Against Japan* (Montgomery, Ala.: Air Power Research Institute, Air War College, Maxwell Air Force Base, 1980), 67.

2. John H. Bradley, *The Second World War: Asia and the Pacific: The West Point Military History Series* (Wayne, N.J.: Avery Publishing Group, 1989), 251–52.

3. Bernard Brodie, *The Absolute Weapon* (New York: Harcourt, Brace and Company, 1946), 24–25.

4. Thomas A. Keaney, *Strategic Bombers and Conventional Weapons: Airpower Options* (Washington, D.C.: National Defense University Press, 1984), 7.

5. Hensell, *Strategic Air War,* Appendix B.

6. Carl Berger, *B-29: The Superfortress* (New York: Ballantine Books, 1970), 50.

7. Brodie, *Absolute Weapon,* 27.

8. Keaney, *Strategic Bombers,* 7.

9. Ibid., 27, 68. Also, see Harry Borowski, ed., *A Hollow Threat: Strategic Air Power and Containment Before Korea* (Westport, Conn.: Greenwood Press, 1982).

10. David Mondey, *American Aircraft of World War II* (New York: Hamlyn/Aerospace, 1982), 32.

11. In Lawrence Freedman, *The Evolution of Nuclear Strategy* (New York: St. Martin's Press, 1983), 84.

12. Chuck Hansen, *The U.S. Nuclear Weapons: The Secret History* (New York: Orion Books, 1988), 58 ff.

13. Chris Bishop and David Donald, *World Military Power* (New York: Military Press, 1987), 16.

14. Honoré M. Catudal, *Soviet Nuclear Strategy from Stalin to Gorbachev* (London: Mansell Publishing, 1988), 38.

15. Ray Wagner, ed., *The Soviet Air Force in World War II: The Official History,* trans. LeLand Fetzer (Melbourne: Doubleday, 1974), 18–24.

16. The early success of the *Sputnik* launch and the much larger size of Soviet launchers both derived from the same geopolitical fact: the United States had allies all along the Soviet periphery; the Soviets had none. A host of Soviet policies, from the decision to build larger launchers to support for Fidel Castro, derived from this fact.

17. Robert Oppenheimer, "Atomic Weapons and American Policy," *Foreign Affairs* 31 (July 4, 1953): 529.

18. There were neither suitable targets nor sufficient bombs available. See Michael Schaller, *Douglas MacArthur: The Far Eastern General* (Oxford: Oxford University Press, 1989), 219.

19. In Allen S. Whiting, *China Crosses the Yalu* (Palo Alto, Calif.: Stanford University Press, 1960), 135.

20. Wilfrid L. Kohl, *French Nuclear Diplomacy* (Princeton, N.J.: Princeton University Press, 1971), 95.

21. On Israeli nuclear policy, see Shai Feldman, *Israeli Nuclear Deterrence* (New York: Columbia University Press, 1982).

4. Fundamentals

1. See, for example, Sanjay Singh, "Indian Ocean Navies—Learn from War," *Proceedings of the U.S. Naval Institute,* March 1992, 51–54.

2. Yihong Zhang, "China Heads Toward Blue Waters," *International Defense Review,* November 1, 1993, 879.

3. For example, the Indians are currently negotiating to buy a Russian V/STOL (vertical or short takeoff and landing) carrier, the *Gorshakov* (Vivek Raghuvanshi, "India Tightens Russian Ties as Carrier Talks Continue," *Defense News,* July 17, 1995, 14). The *Gorshakov* was no match for U.S. forces under the Soviets. Without a radically new design for its aircraft, even using Harriers aboard the *Gorshakov* would not permit the Indian Navy to challenge the U.S. Navy in the Indian Ocean.

4. For a discussion of the future of U.S.-Japanese relations, see our *The Coming War With Japan* (New York: St. Martin's Press, 1991).

5. Japan's fleet is clearly the most powerful in Asia even though its tonnage is far less than that of the Chinese fleet. And it is building an oceangoing hovercraft that could be used as a platform for VTOL (vertical takeoff and landing) aircraft. Pierre-Antoine Donnet, "Japanese Fleet Largest in Asia," Agence France Presse, October 17, 1994.

6. A carrier costs $4.6 billion. Accompanying ships and aircraft cost $14 billion. Annual operating cost is estimated at $900 million. *New York Times,* April 27, 1994, sec. A, p. 16.

7. For a discussion of the spread of diesel submarines, see David Foxwell and Helmoed-Romer Heitma, "Submarine Modernisation," *Jane's Defence Systems Modernisation,* July 1, 1995, 15.

Introduction: From Ballistics to Brilliance

1. *Al Ahram Weekly,* cited in *Money Clips,* published by Gulf Cooperation Council, October 21, 1993.

2. Chaim Herzog, *The War of Atonement* (Boston: Little, Brown, 1975), 188–89.

3. Martin Van Creveld, *Technology and War* (New York: Free Press, 1991), 12–14.

4. Hans Delbruck, *Warfare in Antiquity* (Lincoln: University of Nebraska Press, 1975), 53, 67–71.

5. From Gunpowder to Petroleum

1. On the early history of artillery and its role in siege operations, see Ian Hogg and John Batchelor, *Artillery?* (New York: Ballantine Books, 1972), 3–6.

2. Lynn Montross, *War Through the Ages* (New York: Harper & Brothers, 1960), 182.

3. Ibid., 194–95.

4. Shrapnel was named after its inventor, Lt. Henry Shrapnel, who invented it in 1784. It is worth noting the nearly five-century gap between the introduction of artillery and the introduction of an efficient means of turning it into an antipersonnel device. B. P. Hughes, *Firepower: Weapons Effectiveness on the Battlefield, 1630–1850* (New York: Charles Scribner's Sons, 1974), 34.

5. Ibid., 35–36.

6. Ibid., 34.

7. Steven Ross, *From Flintlock to Rifle: Infantry Tactics, 1740–1866* (London: Associated University Presses, 1979), 25.

8. Ibid., 31.

9. Stockholm International Peace Research Institute, *Antipersonnel Weapons* (New York: Crane, Russak & Co., 1978), 10.

10. Chris Bishop and Ian Drury, eds., *Combat Guns* (Secaucus, N.J.: Chartwell Books, 1987), 199–201.

11. On the principle of the offensive embedded in both French and German plans, see J. F. C. Fuller, *Military History of the Western World,* vol. 3 (New York: Da Capo, 1956), 188–89.

12. John English, *On Infantry* (New York: Praeger, 1981), 15. Some have argued that, on the whole, the rate of loss was not much higher in World War I than in previous wars. See, for example, Trevor Dupuy, *Understanding War* (New York: Paragon, 1987), 169. The apparent contradiction between the Battle of Loos and Dupuy's figures is definition. The division, which Dupuy measures, contains increasingly large numbers of support troops who are rarely, if ever, exposed to fire. This is made necessary by the dramatic rise in logistical and technical needs requiring specialists and other troops to manage. The lethality of weapons had dramatically increased. The proportion of troops exposed to that lethality declined dramatically. Thus, the percentage of attrition on the battle line increased enormously while the total attrition rate of the army remained constant.

13. Chris Ellis and Peter Chamberlain, *Fighting Vehicles* (New York: Hamlyn, 1972), 12–17.

14. Richard M. Ogorkiewicz, *Technology of Tanks,* vol. 1 (London: Jane's Information Group, 1991), 5–6.

15. For a discussion of the first use of the tank, see Robin Prior and Trevor Wilson, "The Dawn of the Tank," *RUSI Journal,* autumn 1991. It contains a discussion of the use of the tank in the Battle of the Somme on September 15, 1916. In fact, this, as much as the Battle of Cambrai, might be called the beginning of the age of tanks.

16. Ogorkiewicz, *Technology of Tanks,* vol. 1, p. 8.

17. The British took a similar view, building tanks that were designed to deal with enemy troops or fortifications, but not with enemy tanks. As late as 1931, a British manual on tank-antitank operations paid little attention to the problem of major antitank operations: "The presence of considerable bodies of tanks on both sides in a suitable theater of war is likely to cause considerable modification in operations generally. Such formations do not, however, exist at the moment, and as the subject is purely theoretical, it will not be discussed further in these pages." (This quote is drawn from a British handbook entitled *Mechanical and Armored Formations,* also called the Purple Book. Quoted in J. P. Harris, "British Armor 1918–1940: Doctrine and Development," in J. P. Harris and F. N. Toase, *Armored Warfare* [New York: St. Martin's Press, 1991], 39.)

18.   John Ellis, *Brute Force: Allied Strategy and Tactics in the Second World War* (New York: Viking, 1990), 5.

19.   Richard Simpkins, *Tank Warfare* (New York: Brassey's, 1982), 41.

20.   See Christopher F. Foss, *World War II Tanks and Fighting Vehicles* (London: Salamander Books), 1981.

21.   U.S. lieutenant general Leslie McNair had perhaps the clearest understanding of what was happening. He understood the uses of armor:

> *In my view, the essential element of armored action is a powerful blow delivered by surprise. While the armored units may be broken up and attached to division and army corps, it is readily conceivable, and indeed probable, that the entire force, under a single command, may be drawn against a decisive point.*

But McNair also realized that armor, if diverted into a purely antiarmor role, would not be able to strike against the rest of the enemy army—armor would cancel out armor. McNair recognized that the gun-armor spiral would lead to a dead end. In 1943, McNair said:

> *An increase in armor or gun power can have no purpose other than to engage in tank to tank action—which is unsound. Moreover, such a tank would be disadvantageous in carrying out the primary mission of armor—to defeat those elements of the enemy which are vulnerable to tanks.*

Harris and Toase, *Armored Warfare*, 135, 139.

22.   Ellis, *Brute Force*, tables 61, 62.

23.   At the time of their collapse, the Soviets had about fifty-three thousand tanks in their arsenal. Although some were obsolete models, the bulk were fully usable. See T. R. W. Waters, "The Traditional Soviet View," in Harris and Toase, *Armored Warfare*, 193.

24.   T. W. Terry et al., *Fighting Vehicles* (London: Brassey's, 1991), 125.

25.   The formula for calculating the slope's effect on thickness is $T_\beta = T_N \cos \beta$. Ibid., 134.

26.   Ogorkiewicz, *Technology of Tanks*, vol. 1, p. 67.

27.   P. R. Courtney-Greene, *Ammunition for the Land Battle* (London: Brassey's, 1991), 117.

28.   On the relationship between explosive jets and shell casing, see M. Chick, T. J. Russel, and L. McVay, "Terminal Effects of Shaped Charge Jets," *Proceedings of the 11th International Symposium on Ballistics* (Brussels, May 9–11, 1989), 407–17. Recently, work has been done on titanium liners.

29.   Courtney-Greene, *Ammunition for the Land Battle*, 122–23.

30.   Ibid., 133.

31.   The Panzerfaust did, however, have its successes. It is estimated that from January 1, 1945, to May 1, 1945, the Panzerfaust accounted for 24 percent of all Soviet tanks destroyed. See David Saw, "The Art of Anti-tank Warfare: Defeating the Armor Threat," *Defense News*, November 28, 1994, 23. This figure may be misleading, as the Germans experienced a precipitous decline in other means for combating Soviet tanks, which artifically inflated the Panzerfaust's success.

32.   Ogorkiewicz, *Technology of Tanks*, vol. 1, p. 206.

33.   Ibid., 206–7.

34. A newer missile, the Javelin, already used by the Marine Corps, is being mounted on Bradley AFVs and Humvees. It is expected to have much greater range than the TOW. In addition, it will be a fire-and-forget system, unlike the TOW, which requires the gunner to remain in place until the missile hits the target. For more information about the mounted Javelin, see "TI-Lockheed Martin to Offer Improved Javelin to Replace Tow," *Armed Forces Newswire*, March 29, 1996.

35. Ogorkiewicz, *Technology of Tanks*, vol. 1, pp. 210–12.

36. AT-3 is the U.S. designation of the missile. Sagger is the NATO designation.

37. *Jane's Weapons Systems* (London: Jane's, 1987–88), 144.

38. Studies at Los Alamos and Livermore revealed that the chemical bonds holding ceramics together shatter under high-energy impact. Livermore suggested packing ceramic armor in a Kevlar wrapper so that the shattered material would be stopped from expanding and scattering on impact—a makeshift idea at best. See Malcolme W. Browne, "Plastics and Ceramics Replace Steel as the Sinews of War," *New York Times,* July 18, 1989, sec. C, p. 1.

39. Attempts have been made to combine ceramics with abalone shells and wood to make ceramics harder. See Bill Dietrichs, "A Rock Hard Science—Materials Science," *Seattle Times,* April 13, 1992, Cl. In the meantime, Ceracom, a company located in California, has been given a contract to develop ceramics preshaped to desirable forms at acceptable costs. The Japanese have also been working intensely on increasing the flexibility of ceramics.

40. R. M. Ogorkiewicz, "Advances in Armor Materials," *International Defense Review,* April 1991, 351.

41. John A. Rovinian and V. Aralanian, *Combat Vehicle Technology Report* (Warren, Mich.: U.S. Army Tank-Automotive Command, Defense Technical Information Report AD-A252-258, May 1992), 39.

42. For a full discussion of both active and passive appliqués, see J. H. Brewer, "Appliqué Armor," *International Defense Review* 26 (January 1993): 62–64.

43. Greg Stewart, "Marines Add Reactive Armor to M-60A1s," *Armor,* March-April 1990, 16–17.

44. Saw, "Art of Anti-tank Warfare," 23.

45. The upgrading of the M-1A1 to M-1A2 standards requires R&D costing about $225 million, to upgrade the gun, provide depleted-uranium armor, install NBC protection and advanced fire control. In other words, a tank introduced to service during the early 1980s was within ten years obsolete, and capable of continued operation only through what will inevitably be a multibillion-dollar upgrade. On the upgrade details, see *Aerospace Daily* 163, no. 6 (July 9, 1992): 53A.

46. The charge has been made by the Iraqis on several occasions. Dr. Eric Hoskins, a member of the Harvard Study Team that surveyed postwar Iraq, claimed that the low-level radioactivity released by depleted-uranium rounds might be linked to a postwar increase in childhood cancer and other maladies detected in Iran. If true, this would undoubtedly limit future use of the round. The current cleanup crisis, as American military bases confront past pollution, is too recent to permit ongoing use of this weapon. See Eric Hoskins, "Making the Desert Glow," *New York Times,* January 21, 1993, 25. This has led to some discussion of replacing uranium rounds and armor with more expensive and less efficient tungsten or titanium. On titanium's use in hulls, see George Taylor, "Titanium May Hold Key to Lighter Combat

Vehicles," *Army Research, Development and Acquisition Bulletin,* January-February 1991, 35–36.

47.   On the development of the T-80 see Gilberto Villahermosa, "T-80: The Newest IT Variant Fires a Laser Guided Missile," *Armor,* July-August 1986, 36–39.

48.   "The Soviet Army—Armor and Electronics," *Defense Electronics,* February 1989, 76.

49.   Associated Press, "Army Ends Most Expensive Weapons Program," October 8, 1992. More recent evidence holds that the Russians are actually moving ahead with tank development.

50.   Terry L. Metzgar, "Electrothermal Guns: Next Step on the Road to Hypervelocity," *National Defense,* September 1990, 20.

51.   John Lancaster, "Arms Research Booms as Production Lines Halt," *Washington Post,* December 10, 1991, A1.

52.   Metzgar, "Electrothermal Guns," 20.

53.   Announcement from Los Alamos on a tabletop model of the artillery of the future, *Business Wire,* January 27, 1992. On the mechanics of fluids in electrothermal operations, see W. Oberle and S. Buntz, U.S. Army Ballistic Research Laboratory, "A Theoretical Evaluation of Critical Factors in Electrothermal Gun Performance," *Proceedings of the 11th International Symposium on Ballistics* (Brussels, May 9–11, 1989). They note that the choice of fluids is critical to the performance of the weapon—and they argue for water.

54.   "Royal Ordnance Pursues ETC Gun Technology," *Defense News,* February 8, 1993, 27.

55.   Caleb Baker, "Pulse to Power Future Weapons," *Defense News,* July 16, 1990, 20.

56.   One of the many payoffs of the late Star Wars program—Strategic Defense Initiative (SDI)—which we will discuss in a later chapter, was that it looked deeply at the problem of creating and storing energy. This was necessary for generating the energy beams and hyperkinetic missiles that SDI planners dreamt of. The Defense Nuclear Agency, which had considerable responsibility for Star Wars engineering, has been in the forefront of developing the energy-production and storage systems necessary for the electrothermal guns, through a project code-named MILE RUN. The ultimate goal of MILE RUN was to produce a 150-kg, 2-to-5 megajoule capacitor by 1994—an extraordinary achievement, if it works. Metzgar, "Electrothermal Guns," 21.

57.   Anthony J. Sommer and Thaddeus Goar, *Army Research, Development and Acquisition Bulletin,* March-April 1992, 41.

58.   R. M. Ogorkiewicz, "Future Tank Guns: Part II: Electromagnetic and Electrothermal Guns," *International Defense Review,* January 1, 1991, 62.

59.   Christopher Bellamy, "Pounds 12m Supergun Brings Star Wars to Scotland," *The Independent,* February 11, 1993, 8. The project is being conducted jointly with the United States.

60.   Vago Muradian, "Budget Cuts May Stymie U.S. Army's Electric Tank Gun," *Defense News,* July 26, 1993, 16.

61.   Craig Koerner and Michael O'Connor, "The Heavily Armored Gun Armed Main Battle Tank Is Not Optimized for Mechanized Warfare," *Armor,* May-June 1986, 12.

62.   Lt. Gen. William Pagonis and Maj. Harold E. Raugh, "Good Logistics Is Combat Power: The Logistics Sustainment of Operation Desert Storm," *Military Review,* September 1991, 36–37.

63. Loren Steffy, "LTV May Spearhead Missile Project," *Dallas Times Herald,* April 4, 1991, sec. B, p. 1.

64. David A. Fulghum, "Multi-Role Fighter May Be Equipped with Derivative of Army LOSAT Missile," *Aeronautical Engineering* 138, no. 19 (May 11, 1992): 55. The Air Force cooled on this scheme in 1993 under budget pressure.

6 Sensing Senility

1. For an excellent study of sensor and guidance systems in missiles, including antitank missiles, see James W. Rawles, "A Missile Guidance Primer," *Defense Electronics,* May 1989.

2. Sean Naylor, "The Army Develops Advanced Rounds for Abrams Tank," *Defense News,* September 6, 1992, 42.

3. Information provided by SAAB Missiles.

4. A project code-named THIRSTY SABER proposes designing a cruise missile similar in capability to the Tomahawk, except that having flown a mission, it would be able to return to base—an unmanned bomber with BATs on board. See "DARPA Aims at Rts with Thirsty Saber," *Aerospace Daily,* October 9, 1991, 53. In 1993, THIRSTY SABER emerged as Multi-Sensor Targeting and Recognition System—MUSTRS—and was being tested by Martin Marietta. See "Marietta, ARPA plan Captive-Carry Tests of MUSTRS Prototype," *Aerospace Daily,* March 16, 1993, 12. The cruise missile itself would be armed with synthetic aperture and millimeter wave radar, and imaging infrared using focal-plane arrays, making the cruise missile capable of autonomous target recognition. Both the projectile and the submunition would be intelligent and maneuverable. See George Leopold, "Military Focuses on Sensors, Target Recognition," *Defense News,* February 8, 1993, 13.

5. "BAT Begins First of 20 Flight Tests," *Aviation Week and Space Technology,* July 5, 1993, 26.

6. "Northrop evaluating BAT for MLRS rocket as well as ATACMs," *Aerospace Daily,* October 21, 1993, 130.

7. "BAT Comes out of the Closet," *International Defense Review,* July 1991, 685.

8. Westinghouse Electric Systems has proposed a millimeter wave radar and an infrared sensor that would use the same aperture—a significant breakthrough. "Improved BAT Submunitions Planned for New ATACMS," *Aviation Week and Space Technology,* December 13, 1993, 54.

9. *Aerospace Daily,* October 21, 1993, 130.

10. There are plans to upgrade the JSTARS aircraft to a Boeing 767.

11. Information provided by the Grumman Corporation, main contractor for JSTARS. Also see Peter Grier, "Joint STARS Does Its Stuff," *Air Force Magazine,* June 1991, 40–41.

12. Information provided by the Grumman Corporation.

13. Mark Hewish, "Electronic Payloads for UAVs," *Defense Electronics and Computing Supplement to International Defense Review,* October 1992, 1035.

14. Bradford M. Brown and Robert Glomb, "Unmanned Aerial Vehicles," *Army Research, Development and Acquisition Bulletin,* March-April 1991, 10.

15. John Boatman, "Matching Systems to Missions," *Jane's Defence Weekly,* May 16, 1992, 854.

16. National Research Council, *Star 21: Strategic Technologies for the Army of the Twenty-first Century* (Washington, D.C.: National Academy Press, 1992), 177.

17. Barbara Starr, "Solar RAPTOR to Show Its TALONs," *Jane's Defence Weekly,* September 12, 1992, 19.

18. Philip Finnegan, "Air Force Seeks Designs for Mach 20 Strike, Spy Drone," *Defense News,* January 8, 1990, 1.

19. The emergence of a new form of mathematical modeling—fractals—is an extremely important event. Rather than expressing data as a series of binary numbers— an enormously long string of these numbers is necessary for expressing a high-resolution picture—it transforms the numeric string to a set of equations, much as a curve or line can be expressed as a series of data points, or as an elegant equation. This dramatically reduces transmission and analysis time, but is still a form of data and, as such, is unusable by the field commander.

20. U.S. Army Communications-Electronics Command, *Command Control, Communications and Intelligence Project Book* (Fort Monmouth, N.J.: 1992), 1–2.

21. Mark Tapscott, "New Pictures Emerging in Battlefield Intelligence," *Defense Electronics,* April 1993.

22. Senate Armed Services Committee Hearing on the Conduct of the Gulf War, June 12, 1991. Schwarzkopf: "We just don't have an immediately responsive intelligence capability that will give the theater commander near, real-time information that he personally needs to make a decision."

23. Mark Hewish, "Tank Breaker," *International Defense Review,* September 1982, 1212–15.

24. James S. Goldman, "FMC Corp. Developing a 'Stealth' Fighting Vehicle," *Business Journal* (San Jose, Calif.), May 27, 1991, 1.

25. Quoted in Neil Munro, "U.S. Army Adapts EW to Shield Its Tanks," *Defense News,* August 3, 1992, 26.

26. Vago Muradian, "DoD Probes Smart Tank Armor," *Defense News,* March 1, 1993, 1.

27. "Work on Top Attack Threat Detection System to Be Continued by Delco," *Aerospace Daily,* December 31, 1992, 481.

28. John Rhea, "Smart Skins," *Air Force Magazine,* March 1990, 90–94. Also see Tim Studt, "Smart materials: creating systems that react," *R&D,* April 1992, 54.

29. National Research Council, *Star 21,* 80.

30. "Army seeks sensors, telepresence technology to make next-generation tank 'transparent' to crew members," *Aerospace Daily,* May 14, 1993, 282.

31. John A. Rovinian and V. Aralanian, *Combat Vehicle Technology Report* (Warren, Mich.: U.S. Army Tank-Automotive Command, Defense Technical Information Report AD-A252-258, May 1992), 41.

32. Phillip L. Bolte, "Coordination and Setting Priorities Enhance Combat Vehicle Survivability," *National Defense,* November 1991, 46.

33. Ibid., 42.

34. "Hill Exposes DoD's Counter-Stealth Unit," *Defense & Aerospace Electronics,* June 8, 1992, 1.

35. "Program Acquisition Cost Summaries," *Defense News,* April 13, 1992.

36. Peter Gudgin, *Armor 2000* (London: Arms and Armor, 1990), 122.

37. On electronic warfare and armor defense, see Munro, "U.S. Army Adapts EW," 1.

38. National Research Council, *Star 21,* 46.

7 The Rise and Fall of the Gunboat

1. Frank Howard, *Sailing Ships of War: 1400–1860* (New York: Mayflower Books, 1979), 38.

2. In this vein, it would be interesting to speculate on the interests and geographical possibilities that might compel space explorations. Military advantage and economic advantage would have to be present to justify the traditionally astronomical costs of exploration.

3. A treatment of the relationships between European exploration and the inclusion of the colonial world as a periphery in a European system can be found in Alan K. Smith, *Creating a World Economy: Merchant Capital, Colonialism and World Trade, 1400–1825* (Boulder, Colo.: Westview Press, 1991), 123–50.

4. Ibid., 82.

5. In Daniel J. Boorstin, *The Discoverers* (New York: Random House, 1983), 177.

6. Quoted in Carlo M. Cipolla, *Guns, Sails and Empires: Technological Innovation and the Early Phases of European Expansion* (London: Minerva Press, 1975), 86.

7. Philip Pugh, *The Cost of Seapower* (London: Conway Maritime Press, 1986), 166–69.

8. Robert Gardiner, ed., *Steam, Steel and Shellfire* (London: Conway Maritime Press, 1992), 168–69. The rise in the weight of guns was not, of course, absolute. Smaller guns for smaller ships were constantly built throughout this period.

9. It should be noted that some have argued that the battleship was obsolete even before the aircraft carrier overtook it. See, for example, Robert L. O'Connell, *Sacred Vessels: The Cult of the Battleship and the Rise of the U.S. Navy* (Boulder, Colo.: Westview Press, 1991). For a critique of this point of view, see Jon Tetsuro Sumida, "Technology, Culture and the Modern Battleship," *Naval War College Review,* autumn 1992, 82–90.

10. Pugh, *The Cost of Seapower,* 152–55.

11. See Siegfried Breyer, *Battleships and Battle Cruisers, 1905–1970,* trans. Alfred Kurti (London: MacDonald and Jane's, 1973); and John C. Reilly Jr. and Robert L. Scheina, *American Battleships, 1886–1923: Predreadnought Design and Construction* (Annapolis, Md.: Naval Institute Press, 1980).

12. Herwig H. Holger, *"Luxury" Fleet: The Imperial German Navy, 1888–1918* (Atlantic Highlands, N.J.: Ashfield Press, 1987), tables 17, 18.

13. Thelma Liesner, *One Hundred Years of Economic Statistics* (London: Economist Publications, Facts on File, 1989), table G1, p. 202.

14. Reilly and Scheina, *American Battleships,* 116–33.

15. Breyer, *Battleships and Battle Cruisers,* 189–93.

16. American planes were launched from carriers by steam-driven catapults, hurling

them off. They landed as normal aircraft, then were snagged by wires. The cost of catapults, the cost of training pilots in carrier landings, were so enormous that most nations didn't try. The British abandoned catapult carriers, the Soviets settled for VTOL carriers, and the French used catapults, but with planes that were too old and two few.

17. For a discussion of fast-attack craft, see David Miller, "Corvettes No Longer a Poor Navy's Option," *International Defense Review,* March 1, 1995, 53.

18. John P. Cordle, "Welcome to Our World: There's Much to Learn from the Swift Boat Navies," *Proceedings of the U.S. Naval Institute,* March 1994, 63–65.

19. Stuart F. Brown, "The Secret Ship," *Popular Science,* October 1993, 93.

20. Barbara Starr, *"Sea Shadow* Emerges into the Daylight," *Jane's Defence Weekly,* April 24, 1993, 5.

21. One of the fundamental and still outstanding issues is the full measure of stealth in all of its spectral aspects. See Ken Graham, "Measuring Stealth Effectiveness," *Navy International,* November/December 1993, 370–72.

22. James W. Crowley, "Navy's Former Stealth Ship Now Classified as a San Diego Resident," *San Diego Union-Tribune,* July 7, 1995, B-1.

23. Joris Janssen Lok, "Sweden's Stealthy Advances," *Jane's Defence Weekly,* April 9, 1994, 24.

24. Scott Truver, "Advanced Naval Vehicles in the US Navy," *International Defense Review,* July 1990, 769–72.

25. Edward J. Walsh, "Navy Labs Pursue Surface Warfare Vision," *Sea Power,* August 1991, 42–47. T-AGOS, acoustic surveillance ships, have been built for the U.S. Navy. Their SWATH hull structure minimizes contact with the water and therefore improves the acoustic environment for their particular mission. The issue is whether these hulls are suitable for extended-endurance, long-range sea-control missions, and if the are, whether increased maneuverability will be sufficient to provide them with increased survivability. That remains doubtful.

26. Alan G. Maiorano, "The Right Ship," *Proceedings of the U.S. Naval Institute,* July 1994, 35–39.

27. Leonid Afanasief and John P. Mabry, "The Design of the FF-21 Multi-Mission Frigate," *Naval Engineers Journal,* May 1994, 150–62.

28. John Boatman, "USN to Consider Next Multi-Mission Ship," *Jane's Defence Weekly,* March 12, 1994, 12.

8   The Aircraft Carrier as Midwife

1. Philip Pugh, *The Cost of Seapower* (London: Conway Maritime Press, 1986), 201.

2. Paul Stillwell, *The Battleship New Jersey* (London: Arms and Armor Press, 1986), 294–305. See also Siegfried Breyer, *Battleships and Battle Cruisers, 1905–1970,* trans. Alfred Kurti (London: MacDonald and Jane's, 1973).

3. Norman Friedman, *U.S. Battleships* (Annapolis, Md.: Naval Institute Press, 1985), 250–52.

4. Walter Karig, *Battle Report: Pearl Harbor to Coral Sea,* vol. 1 (New York: Farrar & Rinehart, 1944), 116.

5. Ibid., 119.

6. Norman Friedman, *US Naval Weapons* (London: Conway Maritime Press, 1983), 117.

7. Ibid., 157.

8. See the discussion of phased-array radar in chapter 13.

9. On the evolution of the CIWS, see David Foxwell, Mark Hewish, and Rupert Pengelley, "CIWS: Naval Gunnery in the Anti-Missile Role," *International Defense Review,* January 1991, 47–53.

10. Ibid., 50.

11. Peter Felstead, "Russia Still Finds Cash for Silver Bullet Solutions," *Jane's Intelligence Review,* October 1, 1995, 1.

12. "Sunburn for Sale," *Military Robotics,* October 5, 1995.

13. Giovanni de Briganti, "French Eye Supersonic Anti-Ship Missile," *Defense News,* January 10, 1995, 4.

14. See, for example, Clifford Beal, "Tortoise and Hare: The Supersonic Missile Debate," *International Defense Review,* May 1, 1994, 57.

15. Benjamin F. Schemmer, "Six Navy Carriers Launch Only 17% of Attack Missions in Desert Storm," *Armed Forces Journal International,* January 1992, 12–13.

9   First Thoughts on Airpower

1. Lee Kennett, *A History of Strategic Bombardment* (New York: Charles Scribner's Sons, 1982), 5.

2. Ibid., 1.

3. Our gratitude to Jeffry Erdley, second lieutenant, Armor, for providing us with this calculation.

4. Kennett, *History of Strategic Bombardment,* 1.

5. Giulio Douhet, *Command of the Air* (Maxwell Air Force Base, Ala.: Air University Press, 1986), 12.

6. William Mitchell, *Winged Defense* (New York: G. P. Putnam, 1925), 5–6.

7. Douhet, *Command of the Air,* 55.

8. Alan Clark, *Aces High: The War in the Air Over the Western Front 1914–1918* (London: Fontana/Collins, 1974), 25.

9. Enzo Angelucci, *Military Aircraft: 1914 to the Present* (New York: Crescent Books for Rand McNally, 1990), 37.

10. In Robert Frank Futrell, *Ideas, Concepts, Doctrine: Basic Thinking in the United States Air Force 1907–1960* (Maxwell Air Force Base, Ala.: Air University Press, December 1989), 82.

11. Billy Mitchell said, "No longer will the tedious and expensive process of wearing down the enemy's land forces by continuous attacks be resorted to. The air forces will strike immediately at the enemy's manufacturing and food centers, railways, bridges, canals, and harbors. The saving of lives, manpower, and expenditures will be tremendous for the winning side." (Mitchell, *Winged Defense,* xv–xvii.) See also,

Barry Watts, *The Foundations of US Air Doctrine* (Maxwell Air Force Base, Ala.: Air University Press, 1984), 7–11.

12. Williamson Murray, *Strategy for Defeat: The Luftwaffe 1933–1945* (Maxwell Air Force Base, Ala.: Air University Press, 1983), 19. Malcolm Smith, *British Air Strategy Between the Wars* (Oxford: Clarendon Press, 1984), 70. However, it should be noted that the British did consider the possibility of defense more seriously than the Americans or the Germans.

13. The term *Stuka* derives from the German *Sturzkampfflugzeug* (dive bomber) and actually applied to two aircraft. One was the single-engined JU-87. However, the twin-engined JU-88 also served in the role. See Jeffrey I. Ethell et al., *The Great Book of World War II Airplanes* (Tokyo: Zokeisha Publications, 1984), 523.

14. Murray, *Strategy for Defeat,* 45.

15. John Ellis, *Brute Force: Allied Strategy and Tactics in the Second World War* (New York: Viking, 1990), 171.

16. Ibid., 177.

17. Ibid., 186.

18. P. M. S. Blackett, *Studies of War* (London: Oliver & Boyd, 1962), 224–25.

19. Ellis, *Brute Force,* 172.

20. Kennett, *History of Strategic Bombardment,* 135.

21. Thomas H. Greer, *The Development of Air Doctrine in the Army Air Arm 1917–1941* (Washington, D.C.: Office of Air Force History, United States Air Force, 1985), 10–11.

22. Futrell, *Ideas, Concepts, Doctrine,* 68.

23. Greer, *Development of Air Doctrine,* 115.

24. Futrell, *Ideas, Concepts, Doctrine,* 80–81.

25. Specifications are for the B-17G, the last model produced in World War II. See David Mondey, *American Aircraft of World War II* (New York: Hamlyn/Aerospace, 1982), 20–27.

26. Ellis, *Brute Force,* table 43.

27. Ibid., 221.

28. Kennett, *History of Strategic Bombardment,* 153.

29. Ibid., 46.

30. Haywood S. Hansell Jr., *The Air Plan That Defeated Hitler* (New York: Arno Press, 1980), 252.

31. And its cousin, the Sperry bombsight.

32. C. V. Glines, "The Blue Ox," *Air Force Magazine,* August 1992, 74.

33. Ibid.

34. Ellis, *Brute Force,* 175.

35. Kennett, *History of Strategic Bombardment,* 156.

36. Hansell, *Air Plan That Defeated Hitler,* 289–94.

37. *U.S. Strategic Bombing Survey,* Washington, D.C., 1945, 2.

38. Ibid., 11.

39. Kennett, *History of Strategic Bombardment,* 160.

40. Ibid., 167.

41. David A. Anderton, *History of the U.S. Air Force* (New York: Military Press, 1981), 124.

42. Ibid., 124–25.

43. Kennett, *History of Strategic Bombardment,* 172.

44. On the Soviet inability to develop a bomber force, see John C. Baker, "The Long-Range Bomber in Soviet Military Planning," in Paul J. Murphy, *The Soviet Air Forces* (Jefferson, N.C.: McFarland, 1984), 182–86.

10   Rethinking Failure

1. On the failure of American strategists to think through the Vietnam War, see Harry Summers, *On Strategy: A Critical Analysis of the Vietnam War* (New York: Dell, 1982).

2. Mark Clodfelter, *The Limits of Air Power: The American Bombing of North Vietnam* (New York: Free Press, 1989), 50.

3. Ibid.

4. Richard H. Kohn and Joseph P. Harahan, eds., *Strategic Air Warfare: An Interview with Generals Curtis E. LeMay, Leon W. Hohnson, David A. Burchinal and Jack J. Catton* (Washington, D.C.: Office of Air Force History, United States Air Force, 1988), 125–26.

5. Bruce Palmer Jr., *The 25-Year War: America's Military Role in Vietnam* (New York: Simon & Schuster, 1984), 43. Palmer argues that this was the consistent view of the CIA from 1964 onward.

6. A. J. C. Lavalle, ed., *The Tale of Two Bridges and the Battle for the Skies Over North Vietnam,* USAF Southeast Asia Monograph Series, vol. 1 (Maxwell Air Force Base, Ala.: Air University Press, 1976), 31–42.

7. Ibid., 46.

8. In Summers, *On Strategy,* 159.

9. Palmer, *The 25-Year War,* 178.

10. Donald J. Mrozek, *Air Power and the Ground War in Vietnam* (Maxwell Air Force Base, Ala.: Air University Press, 1988), 103.

11. Lavalle, *Tale of Two Bridges,* 79.

12. On the historical development of these weapons, see David R. Mets, *The Quest for a Surgical Strike: The United States Air Force and Laser Guided Bombs* (Eglin Air Force Base, Fla.: Office of History, Air Force Systems Command, 1987).

13. Lavalle, *Tale of Two Bridges,* 80.

14. Ibid., 84.

15. Ibid., 85.

16. On this, see Mrozek, *Air Power and the Ground War in Vietnam.*

17. On airpower as tactical firepower in Vietnam, see Robert H. Scales Jr., *Firepower in Limited War* (Washington, D.C.: National Defense University Press, 1990), 63–154. On the problem of intelligence, see pp. 114–19.

18. Ronald H. Spector, *Eagle Against the Sun* (New York: Vintage, 1985), 46.

19. Lanchester models are only one type of model. In addition, there are simulation models, which attempt to reproduce what is actually happening on the battlefield and are used most often in analyzing single-weapon combat—plane vs. plane, for example. Their basic problem is that it is difficult to create a model that faithfully reproduces all the variables. There are also firepower-score models, which aggregate firepower on each side and perform relatively simple arithmetic functions to determine winners and losers—a highly simplistic approach. Finally, there are heuristic models, in which equations are derived from a set of expectations generated by specialists, rather than on any formally justified theory. The Lanchester equations are by far the most widely used, and widely criticized. See John A. Battilega and Judith K. Grange, *The Military Applications of Modeling* (Wright Patterson Air Force Base, Ohio: Air Force Institute of Technology Press, 1984), 75–143, for an overview of these differing approaches.

20. On air-ground interaction, see John Erickson, Lynn Hansen, and William Schneider, *Soviet Ground Forces: An Operational Assessment* (Boulder, Colo.: Westview Press, 1986), 181–206.

21. On operational testing and evaluation models see Battilega and Grange, *Military Applications of Modeling*, 220–26.

22. Norman R. Augustine, *Augustine's Laws* (New York: American Institute of Aeronautics and Astronautics, 1982).

23. Walter Kross, *Military Reform: The High-Tech Debate in Tactical Air Forces* (Washington, D.C.: National Defense University Press, 1985), 46–47.

24. Jim Bussert, "Can the USSR Build and Support High Technology Fighters?" *Defense Electronics*, April 1985, 122.

25. United States Naval Institute Database. For an opposed view on the MiG-29, see Richard D. Ward, an analyst with General Dynamics, who showed how this design philosophy has persisted, writing that "close examination of the MiG-29 Fulcrum at Farnborough resurrected an old controversy in the West and prompted questions about how the Soviets match Western performance with less sophisticated machines and whether the West should apply Soviet developmental criteria to aircraft design. Western engineers must live with weapons design constraints different from those imposed on Soviet engineers. Soviet weapons design keeps aircraft widely deployable, simply maintainable, and rapidly employable. Taking into account the Soviet approach to war-fighting clears up much of what is difficult to understand about Soviet design practice. Basically, they expect battles in the next war to be short and intense, needing a massive flow of replacements. Weapons must be reliable, but only for the short term with minimum support—and, very important, always available in great numbers. Acknowledging that aircraft can survive only a short time in a major war is realistic. There is little point in designing for an operational life of several thousand hours." ("Realistic Aircraft Design," *Aerospace America*, May 1989, 5.)

26. Pierre M. Sprey, "Land-Based Tactical Aviation," in *Reforming the Military*, ed. Jeffrey G. Barlow (Washington, D.C.: Heritage Foundation, 1981). For a more general argument, see James Fallows, *National Defense* (New York: Random House, 1981).

27. These examples are drawn from Kross, *Military Reform*. 138–39.

11   Dawn Breaks

1.   "Potential Casualties Put at 10,000," *Los Angeles Times,* September 5, 1990, part P, p. 2.

2.   Jon Sawyer, "Calculations Figures Hard to Pin Down," *St. Louis Post-Dispatch,* November 19, 1990, 1A.

3.   *U.S. News & World Report,* September 10, 1990. After the war, defense and national security officials admitted that they had real fears that casualties would run this high.

4.   *Newsday,* August 23, 1990.

5.   Timothy J. McNulty, "War and Peace," *Chicago Tribune,* December 2, 1990, 1.

6.   Hearing of the House Armed Services Committee, April 30, 1991.

7.   John A. Warden, *The Air Campaign: Planning for Combat* (Washington, D.C.: National Defense University Press, 1988).

8.   Ibid., 9.

9.   Ibid., 6: "Our focus will be on the employment of air forces at the operational level in a theater of war. Depending on the goals of the war, the theater may extend from the front to the enemy's heartland, as it did for the Western Allies after the Normandy invasion in World War II. Conversely, the theater may be a relatively isolated area, as in the desert war between Britain and the Axis in North Africa prior to November, 1942."

10.   Richard P. Hallion, *Storm Over Iraq* (Washington, D.C.: Smithsonian Press, 1992), 169.

11.   Department of Defense, *Conduct of the Persian Gulf War* (Washington, D.C.: GPO, April 1992), 11.

12.   Hallion, *Storm Over Iraq,* 195.

13.   Ibid., 188.

14.   Thomas A. Keaney and Eliot A. Cohen, *Gulf War Air Power Survey (GWAPS)* (Washington, D.C.: Office of Secretary of the Air Force, 1993), vol. 5, table 190.

15.   John F. Morton, "TACAIR—lesson from the Gulf: status of Air Force and Navy programs; tactical aircraft," *Defense Daily,* April 17, 1991, S1.

16.   Hallion, *Storm Over Iraq,* 166–67. See also U.S. Naval Institute (USNI) database entry on Hellfire missile.

17.   Kenneth P. Werrell, *Archie, Flak, AAA and SAM* (Maxwell Air Force Base, Ala.: Air University Press, 1988), 182–83.

18.   Ibid., 120–21.

19.   Don Logan and Jay Miller, *Rockwell International B-1A/B,* Aerofax Minigraph 24 (Arlington, Tex.: Aerofax, 1988), 2.

20.   On radar cross section, see P. S. Hall, "Principles of Radar Operation," in P. S. Hall et al., *Radar* (London: Brassey's, 1991), 24.

21.   USNI database, F-117 entry. The resin is said to be a boron-fiber and polymer material called Fibaloy, produced by Dow Chemical.

22. Bill Sweetman, "Stealth Techniques Detailed," *International Defense Review,* February 1992, 159–60.

23. Ibid., 161.

24. "Hill Exposes DOD's Counter-Stealth Unit," *Defense & Aerospace Electronics,* June 8, 1992, 1.

25. Ulf Ivarsson, "Multistatic Radars Promise Stealth Detection," *International Defense Review,* July 1993, 584. It should be noted that other claimed radar breakthroughs, such as the use of wide-band radar, have failed to prove themselves. See "Impulse radar fails to defeat stealth technology in tests," *Aviation Week and Space Technology,* October 19, 1992, 47.

26. DoD, *Conduct of the Persian Gulf War,* 116.

27. USNI database, F-117A entry.

28. Hallion, *Storm Over Iraq,* 174.

29. "Pentagon backs away from Desert Storm reports," *Aerospace America,* May 1992, 1.

30. Associated Press, "Commander: F-111s Achieved Nearly Same Success as Stealth," May 12, 1991.

31. Barton Gellman, "Gulf War Workhorses Suffer in Analysis," *Los Angeles Times,* April 10, 1992, 39.

32. David R. Mets, *The Quest for a Surgical Strike: The United States Air Force and Laser Guided Bombs* (Elgin Air Force Base, Fla.: Office of History, Air Force Systems Command, 1987), 9–10.

33. Ibid., 12–20.

34. Ibid., 23–31. See also Alfred Price, "Bridge Busting," *Air Force Magazine,* December 1993, 48.

35. USNI database, Paveway Laser Guided Bomb Series.

36. The bomb carries a spot seeker tuned to the 1.064-micron wavelength of the neodymium: yttrium-aluminum-garnet laser. See USNI entry on Paveway.

37. Mets, *Quest for a Surgical Strike,* 67.

38. Bill Sweetman, "Modern Bombs in the Gulf," *Jane's Defence Weekly,* February 9, 1991, 178. See also Roy Braybrook, "Airborne Night Vision Equipment," Armada, 7.

39. USNI database.

40. Sweetman, "Modern Bombs in the Gulf," 178. Mark Hawish, Bill Sweetman, and Anthony Robinson, "Precision Guided Munitions Come of Age," *International Defense Review,* May 1991, 460.

41. Discussion of the Maverick missile drawn from USNI database.

42. On the Matador/Mace, and the history of cruise missiles in general, see Kenneth P. Werrel, *The Evolution of the Cruise Missile* (Maxwell Air Force Base, Ala.: Air University Press, 1985).

43. *The Impact of Cruise Technology* (Toronto: Canadian Institute of Strategic Studies, 1987), 3–6.

44. Office of the Historian, *From Snark to Peacekeeper* (Offutt Air Force Base, Nebr.: Headquarters, Strategic Air Command, 1990), 69.

45. Thomas O'Toole, "Tomahawk Makes Its Wartime Debut," *Aerospace America,* March 1991, 13.

46. Data on Tomahawk from USNI database.

47. David A. Fulghum, "Secret Carbon-Fiber Warheads Blinded Iraqi Air Defenses," *Aviation Week and Space Technology,* April 27, 1992, 18.

48. *GWAPS,* vol. 2, 288–89.

49. Ibid.

50. Ibid., vol. 5, statistical summary, table 190.

51. Michael R. Rip, "How Navstar Became Indispensable," *Air Force Magazine,* November 1993.

12 End Game

1. On Air Force reluctance concerning cruise missiles, see Edward R. Harshberger, "Long Range Conventional Missiles: Issues for Near Term Development" (Ph.D. diss., Rand Graduate School, Santa Monica, Calif. 1992), 78–83.

2. Committee on Hypersonic Technology for Military Applications, Air Force Studies Board, Commission on Engineering and Technical Systems, National Research Council, *Hypersonic Technology for Military Application* (Washington, D.C.: National Academy Press, 1989), 56.

3. James Fallows, *National Defense* (New York: Random House, 1981), 45.

4. USNI database, SA-10 entry.

5. USNI database, SA-14 entry.

6. Clifford Beal, Bill Sweetman, and Gerard Turbe, "Smarter, Faster, Further: Air-to-Air Missile Developments," *International Defense Review,* April 1992, 343–44.

7. On SAM guidance systems, see P. S. Hall et al., *Radar* (London: Brassey's, 1991), 110–12.

8. USNI database, EA-6B entry.

9. USNI database, HARM entry.

10. Stephen M. Hardy and Zachery A. Zum, "No Clear Channels: The Current State of Communications Jamming," *Journal of Electronic Defense,* January 1993, 31.

11. Myron Struck, "CECOM's EW/RSTA: Developing a New Generation of Electronic Warfare Systems," *Defense Electronics,* October 1991, 29.

12. Neil Munro, *The Quick and the Dead* (New York: St. Martin's Press, 1991), 100.

13. USNI database, entry on ALQ-165 airborne self-protection jammer.

14. Sheldon B. Hershkowitz, "Electronic Warfare in the Year 2000 and Beyond," *Microwave Journal,* September 1991, 48.

15. Philip J. Klass, "Westinghouse P-MAWS to Spot True Threats," *Aviation Week and Space Technology,* September 20, 1993, 88.

16. David A. Fulghum, "New Missile Threats Drive EF-111 Program," *Aviation Week and Space Technology,* May 10, 1993, 24.

17. USNI database, B-2 entry.

18. USNI database, B-2 and B-52 entries.

19. The Air Force estimated, in 1990, that the total Soviet investment in air defenses was over $400 billion, and included ten thousand radars, eight thousand SAM launchers, and three thousand interceptor aircraft. See "USAF Paper Says Stealth Fleet Necessary to Maintaining Nuclear Triad," *Aviation Week and Space Technology,* June 25, 1990, 23.

20. Mark V. Anderson, *The B-2 Bomber: A Comparative Assessment.* Program in Science and Technology for International Security, Report #21 (Cambridge, Mass.: MIT, July 1989), 68.

21. Department of the Air Force, SAF/AQ, *B-2 Survivability Against Air Defense Systems* (Arlington, Va.: Pentagon, March 1990), 10.

22. "Official Says Russians Could Detect Stealth Bomber," Reuters, February 18, 1994.

23. "USAF Commander Admits Soviets Could See B-2," *Flight International,* October 23, 1991.

24. Neil Munro and Barbara Opall, "European Officials: Electro-Optics Will Counter Stealth," *Defense News,* June 24, 1991.

25. "CNA Says Stealth Aircraft Can Be Uncloaked," *Defense & Aerospace Electronics,* August 17, 1992, 4.

26. "B-2 Avionics May Get Help," *Defense & Aerospace Electronics,* January 25, 1993, 5.

27. See George L. Donohue, *The Role of the B-2 in the New U.S. Defense Strategy* (Santa Monica, Calif.: Rand Corporation, P-7744, 1991), for a defense of the B-2.

28. James F. O'Bryon, "Unlocking G-LOC," *Aerospace America,* September 1991, 61.

29. Robert E. Van Patten, "G-Lock and the Fighter Jock," *Air Force Magazine,* October 1991, 51.

30. "NADC Counters G-Forces," *Defense & Aerospace Electronics,* July 22, 1991, 4.

31. "Collision Course," *Flight International,* July 7, 1993.

32. Neil Munro and Barbara Opall, "Air Force Resists Flight Safety Computers," *Defense News,* July 1, 1991, 3.

33. Stacey Evers, "Unmanned Fighters: Flight Without Limits," *Jane's Defence Weekly,* April 10, 1996, 28.

34. "Periscope Daily Defense News Capsule," March 5, 1996.

Introduction: The Logic of Space Warfare

1. On satellite communications in the Gulf, see S. A. Masud, "DCA Rigged High-Tech Comm System Quickly in the Gulf," *Government Computer News,* June 24, 1991, 29; T. W. Bill Carr, "Victory in Space: Satellite Success in the Gulf War," *Defense & Foreign Affairs,* May 1992, 11; and Peter Anson and Dennis Cummings, "The First Space War," *RUSI Journal,* winter 1991.

13   A New Foundation

1. That was one of the reasons why Castro's victory in Cuba was a strategic triumph for the Soviets. For the first time, Soviet aircraft could leave the Kola Peninsula, pass

between Scotland and Iceland, travel down the East Coast examining electronic signals and monitoring communications, and land in Cuba for rest and refueling.

2. For a discussion of the role of strategic bombardment in Soviet military theory, see John C. Baker, "The Long-Range Bomber in Soviet Military Planning," in Paul J. Murphy, ed., *The Soviet Air Forces* (Jefferson, N.C.: McFarland, 1984). On strategic aircraft in World War II, see p. 178.

3. USNI database, U-2 entry.

4. William V. Kennedy, *Intelligence Warfare* (New York: Crescent Books, 1987), 133–34.

5. It is interesting to recall that, mindful of its aerial reconnaissance capability, the United States had been pressing for an open-skies program—the establishment of the principle that nations had the right to overfly each other for reconnaissance purposes. The Soviets, having no long-range intercontinental aircraft, nor bases in the Western Hemisphere during the 1950s, rejected the proposal. However, by launching *Sputnik,* the Soviets were obviously accepting and asserting the principle that national sovereignty did not extend into space. In space, at least, open skies applied. It is not clear whether the Soviets realized the implications of this position. Perhaps they felt their lead in space was so enormous that the United States would be hard-pressed to catch up. More likely, they were so eager to launch a satellite that they were not concerned about the international law they were writing. As it happened, the Soviets created the very open-sky policy in space that they had opposed in the atmosphere. In the end, they opened the door to a vast American effort based on substantially superior technology—a superiority that was not always apparent during the failures of the 1950s but that was nevertheless very real.

6. In Jeffrey T. Richelson, *America's Secret Eyes in Space: The U.S. Keyhole Spy Satellite Program* (New York: Harper & Row, 1990), 3–4.

7. Ibid., 87.

8. Ibid., 58–59.

9. Edward Horton, *Submarine* (London: Sidgwick and Jackson, 1974), 154.

10. Office of the Historian, *From Snark to Peacemaker* (Offat Air Force Base, Nebr., Headquarters, Strategic Air Command, 1990), 13–14.

11. Richelson, *America's Secret Eyes in Space,* 68.

12. Craig Covault, "Recon Satellites Lead Allied Intelligence Effort," *Aviation Week and Space Technology,* February 4, 1991, 25.

13. There were weaknesses in the system. Time was wasted by having to route information through Colorado Springs, which relayed it to CENTCOM. General Horner, Air Force chief of staff, testified to Congress: "The modified DSP functioned near the limits of ungraded design capability throughout the Gulf War and benefited from exceptionally unique and favorable geographic, weather, and operational conditions—conditions that are unlikely to be duplicated in any future conflict." What he meant was that DSP worked splendidly, but he still wanted a new system, called the Future Early Warning System, or FEWS.

14. William Burrows and John Free, "Space Spies," *Popular Science* 236, no. 3. (March 1990): 61.

15. On the CCD, see James R. Janesick and Morley M. Blouke, "Sky on a Chip: The Fabulous CCD," *Sky and Telescope,* September 1987, 238–42. Also see their article "Introduction to Charge Coupled Device Image Sensors," in Kosta Tsipis, ed., *Arms*

Control Verification: The Technologies That Make It Possible (New York: Pergamon-Brassey's, 1985).

16. James Bamford, "America's Supersecret Eyes in Space," *New York Times Magazine,* January 13, 1985, 39. Lyndon Johnson acknowledged the existence of satellite photography accidentally.

17. Lyn Dutton, *Military Space* (London: Brassey's, 1990), 96.

18. Jon Trux, "Desert Storm: A Space Age War," *New Scientist,* July 27, 1991, 30.

19. Covault, "Recon Satellites Lead Allied Intelligence Effort," 25.

20. Earl Lane, "Sentries in Orbit," *Newsday,* September 11, 1990, 7.

21. Covault, "Recon Satellites Lead Allied Intelligence Effort," 25.

22. Burrows and Free, "Space Spies."

23. Jeffrey T. Richelson, "The Future of Space Reconnaissance," *Scientific American,* January 1991, 39.

24. Trux, "Desert Storm," 32.

25. During the war, the USAF became the world's single largest user of commercial space-based images, even receiving Souzkarta five-meter images from the Soviets. See Dana J. Johnson, Max Nelson, and Robert J. Lempert, *U.S. Space-Based Remote Sensing: Challenges and Prospects* (Santa Monica, Calif.: Rand Corporation, Note N-3589-AF/A/OSD, 1993), 13–16.

26. "KH-11 Recons Modified," *Aviation Week and Space Technology,* October 9, 1995, 28.

27. Johnson et al., *U.S. Space-Based Remote Sensing,* 32.

28. William E. Burrows, *Deep Black: Space Espionage and National Security* (New York: Random House, 1987), reports the presence of photomultipliers on board the KH-11. Photomultiplication (PM) differs from infrared in that, where IR senses emissions in the infrared band, photomultiplication takes available visible light—in the infinitesimal quantities present even in pitch black—and multiples the reaction of highly sensitive CCDs thousands of times. IR and PM serve much the same purpose but in utterly different ways.

29. Dutton, *Military Space,* 110.

30. "Titan Deployed Advanced Photo Reconnaissance Satellite," *Aerospace Daily,* November 28, 1990, 342.

31. Richelson, *America's Secret Eyes in Space,* 218–20.

32. Ike Chang, *The Rise of Active Element Phased Array Radar* (Santa Monica, Calif.: Rand Corporation, P-7747-RGS, n.d.), 6.

33. Stafan T. Possony, "Synthetic Aperture Radar: A New Era," *Defense & Foreign Affairs,* January/February 1983, 29.

34. Hall et al., *Radar,* 49. Also see Burrows and Free, "Space Spies," 60.

35. R. Jeffrey Smith, "Sky High Spies: Keeping Track Just Keeps Getting Easier," *Chicago Tribune,* November 24, 1985, 17.

36. Bob Woodward and Patrick E. Tyler, "Eavesdropping System Betrayed," *Washington Post,* May 21, 1986, 1.

37. "Storm Support From Space," *Flight International,* April 3, 1991.

38. Bamford, "America's Supersecret Eyes in Space," 39.

39. George Lardner Jr., "National Security Agency: Turning On and Tuning In," *Washington Post*, March 18, 1990, 1.

40. "Spying on Saddam," *Flight International*, August 20, 1990.

41. Craig Couvalt, "Space Recon of Iraq Taxes CIA Operations," *Aviation Week and Space Technology*, September 3, 1990, 30.

42. Neil Munro, "Pentagon to Merge Spy Data Networks," *Defense News*, October 28, 1991, 3.

43. During the war, there was substantial tension between intelligence officials and Air Force officers required to fly missions with limited intelligence. See "Intel Collection Good but Analysis Wanting During Gulf War," *Aerospace Daily*, August 16, 1993, 265.

44. Testimony of Gen. Norman Schwarzkopf to the Senate Armed Services Committee, *Hearing on the Conduct of the Gulf War*, June 12, 1991.

45. "JDW Interview," *Jane's Defence Weekly*, February 9, 1991, 200.

46. Bruce D. Nordwall, "EW Goal Is Improved Situational Awareness," *Aviation Week and Space Technology*, July 5, 1993, 59.

14 Space and the Future of American Strategy

1. Global-position systems (GPS), which use satellites to provide precise information on location, simultaneously provide information in formats useful to humans and cruise missiles. In a hypersonic age, reconnaissance data will have to be provided at hyperspeeds, and in multiple formats—pictures that can be understood by humans and digital arrays useful to missiles.

2. William B. Scott, "Space Warfare Center Supports Warfighters," *Aviation Week and Space Technology*, March 28, 1994, 64.

3. Darryl Gehly, "Controlling the Battlefield: US Electronic Warfare and Command and Control Systems," *Journal of Electronic Defense*, June 1993, 42.

4. James Adams, "Spies in the Skies," *Sunday Times* (London), January 27, 1991.

5. For an overview of UAVs, see Louis C. Gerken, *UAV—Unmanned Aerial Vehicles* (Chula Vista: Calif.: American Scientific Corporation, 1991). For the Army's plans for utilizing UAVs, see Miles A. Libbey III and Patrick A. Putigano, "See Deep, Shoot Deep: UAVs on the Future Battlefield," *Military Review*, February 1991, 38–47.

6. On power for UAVs, see Clifford Beal, "Power on Demand," *International Defense Review*, February 1992, 188; "Microwaves Lift Plane Aloft," *Design News*, December 21, 1987, 35; Lloyd D. Resnick and Martin R. Stiglitz, "An Airplane Powered by Microwave Radiation," *Microwave Journal*, February 1988, 66–71; and Barbara Starr, "Solar RAPTOR to show its TALONS," *Jane's Defence Weekly*, September 12, 1992, 19.

7. According to the RFP (request for proposal):
   Hypersonic Aerodynamic Weapons (HAWs) show a strong payoff as a future weapon system in a precision ground strike role for both strategic and tactical applications. HAWs incorporate high hypersonic lift-to-drag ratios allowing long range glides or extensive terminal maneuvers. They also include guidance, navigation and control, sensors, warhead, fusing and other applicable subsystems appropriate for precision strike. A preliminary pre-

cision ground strike target set has been defined with two general CEP goals as follows: CEP Goal I—Chemical, Nuclear, and Biological facilities, Depots and Airfields, and Terrorist Camps. CEP Goal II—National Command and Control facilities, Power Plants, and Communication Centers.

U.S. Government Procurements, *Commerce Business Daily*, no. PSA-0547 (March 9, 1992).

8.   Barbara Opall, "AF Pursues Strike Weapon," *Defense News*, April 5, 1992, 3.

9.   "General Dynamics Creates Hypersonics Unit at Ft. Worth: Targets HAW," *Aerospace Daily*, June 17, 1992, 446.

10.   Office of Technology Assessment (OTA), U.S. Congress, *SDI: Technology, Survivability and Software* (Washington, D.C.: GPO, 1988), 12.

11.   Department of Defense (DoD), Strategic Defense Initiative Organization, *1989 Report to the Congress on the Strategic Defense Initiative* (Washington, D.C.: GPO, March 13, 1989), sec. 5.3–11.

12.   See, for example, Lowell Wood and Roderick Hyde, *Brilliant Pebbles and Ultravelocity Slings: A Robust, Treaty Compliant Accidental Launch Protection System* (Livermore, Calif.: Lawrence Livermore National Laboratory, May 20, 1988).

13.   See Louis R. Bertolini et al., *SHARP: A First Step Towards a Full Sized Jules Verne Launcher* (Livermore, Calif.: Lawrence Livermore National Laboratory).

14.   "Gunning for Space," *Flight International*, November 18, 1992.

15.   Stephen Ashley, "Shooting for the Moon; Super High Altitude Research Project Gun," *Mechanical Engineering–CIME* (November 1992): 116.

16.   Lawrence M. Fisher, "Gas Guns Could Launch Spacecraft, Testers Say," *New York Times*, February 15, 1994, sec. C, p. 8. See also, "Rockwell flies gun-launched scramjets to Mach 8," *Defense Daily*, December 14, 1993, 381.

17.   Bill Sweetman, "Hypersonic Aurora: A Secret Dawning," *Jane's Defence Weekly*, December 12, 1992, 14–16. It was also noted by *Jane's* that retiring the SR-71 Blackbird, our only other strategic reconnaissance aircraft, made no sense unless a follow-on aircraft like the Aurora existed.

18.   "Coincidence?" *Aerospace Daily*, November 8, 1993, 229.

19.   William B. Scott, "Recent Sightings of XB-70 like Aircraft Reinforces 1990 Reports from Edwards Area," *Aviation Week and Space Technology*, August 24, 1992, 23; and in the same issue, Michael A. Dornheim, "United 747 Crew Reports Near-Collision with Mysterious Supersonic Aircraft," 24.

20.   Committee on Hypersonic Technology for Military Applications, Air Force Studies Board, Commission on Engineering and Technical Systems, National Research Council, *Hypersonic Technology for Military Application* (Washington, D.C.: National Academy Press, 1989), 56.

21.   Of course, NASP is a very different idea from the hypersonic aerodynamic weapon. On one hand, it is much simpler since it passes out of the atmosphere and beyond the stresses and frictions of hypersonic movement through air. But it is also a much more complex system, since it is intended to carry humans and must therefore be large and limited in acceleration and maneuverability.

22.   T. A. Heppenheimer, *Hypersonic Technologies and the National Aerospace Plane* (Arlington, Va.: Pasha Publications, 1989), 142–45.

23. "No Propulsion Barriers, but Practical Hypersonics Still Distant," *Aerospace Propulsion,* June 11, 1992, 2. Six new technologies were identified as of critical importance: lightweight, high-temperature composite materials; high-temperature lubricants; high-performance fuel; injectors/flame holders, inlet and exhaust nozzle systems; cryogenic fuel storage systems. In other words, the entire system needs to be invented, or at least perfected.

24. This was a problem that stymied Japanese designers of an advanced jet engine called LACE—not a true scramjet. See Yoichi Iwamoto, "Cooling Dilemma Keeps Space Plane's Engine on Ice," *Nikkei Weekly,* September 19, 1992, 14.

25. James J. Robinson, "Getting the Space Plane off the Ground: A Materials Enterprise," *Journal of Metals,* July 1987, 8–9. On coatings for hypersonic surfaces, see Josephine Covino, Karl Klemm, and John Dykema, "Coatings for Hypersonic Applications," *Materials Performance,* May 1991, 75–79.

26. The apparent approach has been to cool the windows cryogenically, using small tubes in the window. See "Army Awards Aerojet $9.1 Million for Hypersonic Windows Research," *Aerospace Daily,* August 7, 1990.

27. On the beginnings of mass production of key materials, see "Bulk Manufacture of X-30 Materials in Offing as NASP Design Matures," *Aviation Week and Space Technology,* April 15, 1991, 69.

28. From *New World Vistas,* cited in *Intelligence Newsletter,* March 7, 1996.

29. William B. Scott, "USAF, NASA Programs Push Hypersonic Boundaries," *Aviation Week and Space Technology,* May 6, 1996, 22.

30. John M. Collins, *Military Space Forces: The Next 50 Years,* commissioned by the U.S. Congress (Washington, D.C.: Pergamon-Brassey, 1989), 14.

31. Ibid., 15. Recently, a third belt, much weaker, near low earth orbit, has been discovered. It does not appear that the new belt will have a significant impact on space operations.

32. P. R. K. Chetty, *Satellite Technology and Its Applications* (Blue Ridge Summit, Pa.: Tab Professional and Reference Books, 1991), 20.

33. Lynn Dutton, *Military Space* (London: Brassey's, 1990), 9–29.

34. On the problems inherent in the eccentricities of geosynchronous objects, see Jill L. Tabor and John D. Vedder, "Long-Term Evolution of Uncontrolled Geosynchronous Orbits: Orbital Debris Implications," *Journal of the Astronautical Sciences* 40, no. 3 (July-September 1992): 407–18.

35. John Overstreet, "Inclined Orbits: Extending Satellite Life," *Satellite Communications,* July 1991, 25–26.

36. John M. Hanson, Maria J. Evans, and Ronald E. Turner, "Designing Good Partial Coverage Satellite Constellations," *Journal of Astronautical Sciences* (April-June 1992): 224.

37. Mark A. Stein, "Elbowing for a Piece of Space," *Los Angeles Times,* September 20, 1993, 1. Theoretically, only 180 communications satellites can be operational in geostationary orbit at any one time. This has already led to major disputes between minor powers. For example, Indonesia and Tonga—which, with American partners, have boosted a satellite to permit transpacific communications—have been in a nasty battle over choice space. There has been an urgent search for technical solutions, but commercial pressure will inevitably make the allocation and control of

critical space positions more complex. See, for example, H. J. Weiss, "Maximizing Access to the Geostationary Satellite Orbit," *Telecommunications Journal,* August 1986.

38. Dutton, *Military Space,* 15–17. It should always be remembered that in terms of geography the Russians are in a much worse position than the United States. Much farther north, they cannot avoid firing over land—unless they are prepared to go even farther north, to the Pacific or even Arctic coasts. Thus, the Russians built their launch sites in Central Asia, as far south as possible. Their primary launch facility at Tyuratam is just off the Aral Sea, at 46.5° N. As bad as that is, there is another problem. The Chinese province of Sinkiang bulges up to east, so a 46.5° launch is impossible. To avoid China, all Russian launches must have an inclination of 64°! Of course they have to build massive rockets.

39. For a history of MAD as a doctrine and strategy, see Simon P. Worden, *SDI and the Alternatives* (Washington, D.C.: National Defense University Press), 17–56.

40. Quoted in Crockett L. Grabbe, *Space Weapons and the Strategic Defense Initiative* (Ames, Iowa: Iowa State University Press, 1991), 154.

41. Dr. Hans Bethe, a Nobel Prize–winning physicist summed up his opposition to Star Wars as follows:

> *These people want to eliminate the danger of nuclear weapons by technical means. I think this is futile. The only way to eliminate it is by having a wise policy. That means going back to the policy of the six Presidents preceding Reagan. We need to try to understand the other fellow and negotiate and try to come to some agreement about the common danger. That is what's been forgotten. The solution can only be political. It would be terribly comfortable for the President and the Secretary of Defense if there was a technical solution. But there isn't any.*

Quoted in William J. Broad, "Star Wars Is Coming, but Where Is It Going?" *New York Times Magazine,* December 6, 1987, 80.

42. OTA, *SDI,* 6.

43. DoD, Strategic Defense Initiative Organization, *1989 Report,* Sec. 5.6; U.S. Department of Defense, Office of Technology Assessment, Heritage Foundation, *Anti-Missile and Anti-Satellite Technologies and Programs* (Park Ridge, N.J.: Noyes Publications, 1986), 6–10.

44. *Science and Technology of Directed Energy Weapons: Report to the American Physical Society Study Group* (American Physical Society, April 1987), 2.

45. "Hype, Skepticism Surround Katyusha-Busting Laser," *Defense Week,* April 29, 1996, 4.

46. OTA, *SDI,* 108–9.

47. David L. Chandler, "The Patriot," *Boston Globe,* April 4, 1994, 27.

48. D. G. King-Hele et al., *The RAE Earth Satellites 1957–1986* (Farnborough, England: Royal Aircraft Establishment, 1987), 859.

49. Bruce D. Nordwall, "Air Force Uses Optics to Track Space Objects," *Aviation Week and Space Technology,* August 16, 1993, 66.

50. Description of announcement, *Aerospace Daily,* April 5, 1994, 24.

51. *Commerce Business Daily,* no. PSA-0862, SOL PRDA 93-08 (June 8, 1993).

52. This is the full announcement, as it appeared on Lexis-Nexis information service. Note the incredibly small amount of the contract.

April 1, 1992
SECTION: CONTRACT AWARDS; Supplies, Equipment and Material; Issue No. PSA-0564
LENGTH: 50 words
CATEGORY: 87—Agricultural Supplies
CONTRACTING OFFICE: Commander, U.S. Army Missile Command, Research, Development and Engineering Center (RDEC) Procurement Office, Redstone Arsenal, AL 35898-5275
SUBJECT: 87—PERFORMANCE ANALYSES AND CONCEPT DEFINITION FOR THE STARFIRE OPTICAL RANGE (SOR) 3.5 METER TELESCOPE ACTIVE/PASSIVE BEAM CONTROL SUBSYSTEM
BODY:
POC (RDPC) Zina Long, Contract Specialist, AMSMI-RD-PC-JB, (205) 842-7424, Earnest L. Taylor, Jr., Contracting Officer, (205) 876-7500 Synopsis # R087-92. CNT DAAHO1-89-D-0065-0011 AMT $ 99,907.00 DTD 032792 TO Science Applications, International Corp., 10260 Campus Pt Dr, San Diego, CA 92121

53. "BMDO Defends Brilliant Eyes Capability, but Says It's Not Mandatory," *Aerospace Daily,* July 11,1993, 435.

54. "Brilliant Eyes to Use Mix of Sensors for Dual Missions," *Aerospace Daily*, December 23, 1992, 455.

55. Eric Raiten and Kosta Tsipis, *Conventional Anti-Satellite Systems,* Program in Science and Technology for International Security, no. 10 (Cambridge, Mass.: MIT, March 1984), 21.

56. Ibid., 99. 33–34.

57. Philip Finegan, "DoD Antisatellite Weapon Program May Continue Despite Restrictions," *Defense News,* August 5, 1991, 37.

58. Hence an article by D. Nahrstedt, "The Case for a Submarine-Based Anti-Satellite System," *Submarine Review,* January 1991, 50–55.

59. G. Harry Stine, *Confrontation in Space* (Englewood Cliffs, N.J.: Prentice-Hall, 1981), 82.

60. For a definition of electric propulsion, see John R. Beattie and Jay Penn, "Electric Propulsion as a National Capability," *Aerospace America,* July 1990, 56.

61. Takayuki Matsunaga, "Satellite Propulsion Engine Claimed More Efficient," *Nikkei Weekly,* October 3, 1992, 15.

62. Peter J. Turchi et al., "Electric Propulsion: The Future Is Now," *Aerospace America,* July 1992, 38.

63. "NASA Ponders Large Lockheed Military Satellite Bus for Low-End Station," *Aerospace Daily,* April 6, 1993, 27.

64. William J. Broad, "Russian Scientists to Detail Plans for a Fast Nuclear Space Rocket," *New York Times,* January 13, 1992, 1.

65. John Lawrence, "Nuclear Power Source in Space: A Historical Perspective," *Nuclear News,* November 1991, 87.

15 The Return of the Poor, Bloody Infantry

1. Paul F. Gorman, *Supertroop via I-Port: Distributed Simulation Technology for Combat Development and Training Development,* IDA paper P-2374 (Arlington, Va.: August 1990) prepared for DARPA. The author had been CINC South in the 1980s.

2.　Robert A. Heinlein, *Starship Trooper* (New York: G. P. Putnam Sons, 1959).

3.　Neil Munro, "Army Tests High-Tech Gear for Infantryman," *Defense News,* July 6, 1992, 12.

4.　United States Infantry School, *Infantry 2000: The Force That Leads the Way* (Fort Benning, Ga.: Department of the Army, October 3, 1991), sec. 7.

5.　Mark Trapscott, "At Night: First Look, First Shot Wins," *Defense Electronics,* April 1992, 50.

6.　National Research Council, *Star 21: Strategic Technologies for the Army of the Twenty-first Century* (Washington, D.C.: National Academy Press, 1992), 47.

7.　A similar system is being developed for infantrymen under the SIPE program. It will have local area communication, enhanced acoustics, electro-optical and night capability, as well as ballistic protection. It will not, however, have a virtual-reality capability. See Barbara Starr, "Enhanced Soldier for the U.S. Army," *Jane's Defence Weekly,* January 30, 1993, 25.

8.　Roger Lesser, "Electronic Soldier Concept Moves Closer to Reality," *Defense Electronics,* October 1992, 39.

9.　Claude J. Bauer, "Future soldiers train on virtual battlefields," *Government Computer News,* September 20, 1993, 43.

10.　Business Wire, November 23, 1993.

11.　Gorman, *Supertroop via I-Port.* This was General Gorman's point in his paper on Supertroop. Gorman argued that the training technologies developed under the name SimNet (simulated networks) and its individual interface—I-Port—would evolve into an actual Supertroop.

12.　Lesser, "Electronic Soldier Concept," 39.

13.　Ramon Lopez, "Rifles on the Front Line," *Jane's Defence Weekly,* May 30,1992, 936.

14.　National Research Council, *Star 21,* 65.

15.　A report issued by the U.S. Army Infantry School at Fort Benning describes the planned-for capabilities in the next-generation personal weapon that will replace the rifle:

> *The Objective Individual Combat Weapon will enable the individual soldier to engage individual personnel, groups of personnel, unarmored vehicles, and personnel in hasty fortifications, as well as to provide self-defense against weapon systems mounted on lightly armored vehicles and slow aircraft.*

U.S. Infantry School, *Infantry 2000,* sec. 7.

16.　Gorman discusses the problem of a soldier's burden with eloquent examples. See *Supertroop via I-Port,* sec. 8, p. 5–6.

17.　Gorman, *Supertroop via I-Port,* sec. 7, p. 1. Gorman's idea is to use the emerging training devices—I-Port—as the basis for a new fighting system. Gorman's discussion was to a great extent based on a paper by Jeffrey A. Moore, *PITMAN: A Powered Exoskeletal Suit for Infantryman* (Los Alamos, N.M.: Los Alamos National Laboratory, LA-10761MS, June 1986 and June 1988). Moore's suggestion, the most radical, was that the suit's movement depend on magnetic sensing of brain waves—necessary for efficient mobility. This moved the suit out of the reach of current technology, or so his superiors seemed to feel. Thus, PITMAN was not pursued, in favor of more modest and less imaginative initiatives under the SIPE program at Nattick Labs. Nevertheless, PITMAN represents the basic framework for an armored, powered suit. It is also reminiscent of an earlier idea by Robert Heinlein, the great science-fiction writer, whose book *Starship Trooper* proposed a similar scheme. This is reminiscent of H. G. Wells's landships and the development of the modern tank.

18. Ann-Marie Cunningham, "Robo-Soldiers: PITMAN thinking armor," *Technology Review* (April 1988): 11.

19. The power source is here, as in so many evolving technologies, the real bottleneck. The same work on power storage that is being used in UAVs and in electromagnetic and electrothermal guns will be applied here. One irony—the battery will have to lift itself. On military battery technology, see "Military Batteries: Desert Storm, Glasnost and Military Spending," *Battery & EV Technology*, February 1991, including a discussion of the PITMAN technology.

20. "Foot Soldiers to Get Suits of Armor," *Machine Design*, October 9, 1986, 20.

21. The original PITMAN was intended to be two hundred pounds. The protective ensemble is under development by the U.S. Army in the SIPE program. The weapons are being developed under TEISS.

22. National Research Council, *Star 21*, 66.

23. On SIPE, see Mike Nugent, "The Soldier Integrated Protective Ensemble," *Army Research, Development and Acquisition Bulletin* (September-October 1990): 1–5. SIPE incorporates only a sketchy notion of exoskeletons, compared to PITMAN. A civilian undertaking in this area was attempted by a Minneapolis company, Ergon, Inc., which sought to build a "muscle multiplier" in the early 1990s. See Kathe Connair Palmquist, "A Start-up Puts on Muscle," *Business Dateline*, September 1990. With so many initiatives developing, one, private or public, will shortly succeed.

24. National Research Council, *Star 21*, 71–73.

25. Ray Bonds, ed., *Modern American Weapons* (New York: Prentice-Hall, 1986), 98.

26. Ibid., 96.

27. From George Friedman and Meredith LeBard, "A New Kind of War," *Command Magazine*, May-June 1993.

Conclusion: The Permanent Dilemma

1. Testimony of Richard Davis, director, National Security Analysis, United States General Accounting, April 26, 1994, in House Armed Services Committee, *Military Acquisition Fy95 Defense Authorization.*

2. *Cruise Missile,* General Accounting Office report, April 21, 1995, 8.

Index